Studies in Computational Intelligence

Volume 620

Series editor

Janusz Kacprzyk, Polish Academy of Sciences, Warsaw, Poland
e-mail: kacprzyk@ibspan.waw.pl

About this Series

The series "Studies in Computational Intelligence" (SCI) publishes new developments and advances in the various areas of computational intelligence—quickly and with a high quality. The intent is to cover the theory, applications, and design methods of computational intelligence, as embedded in the fields of engineering, computer science, physics and life sciences, as well as the methodologies behind them. The series contains monographs, lecture notes and edited volumes in computational intelligence spanning the areas of neural networks, connectionist systems, genetic algorithms, evolutionary computation, artificial intelligence, cellular automata, self-organizing systems, soft computing, fuzzy systems, and hybrid intelligent systems. Of particular value to both the contributors and the readership are the short publication timeframe and the worldwide distribution, which enable both wide and rapid dissemination of research output.

More information about this series at http://www.springer.com/series/7092

Juan Julian Merelo · Agostinho Rosa
José M. Cadenas · António Dourado
Kurosh Madani · Joaquim Filipe
Editors

Computational Intelligence

International Joint Conference, IJCCI 2014
Rome, Italy, October 22–24, 2014 Revised
Selected Papers

Springer

Editors
Juan Julian Merelo
Ingenería Informática
Escuela Técnica Superior de
Granada
Spain

Agostinho Rosa
aSEEB-ISR-IST
Technical University of Lisbon (IST)
Lisbon
Portugal

José M. Cadenas
Facultad de Informática
University of Murcia
Murcia
Spain

António Dourado
University of Coimbra
Coimbra
Portugal

Kurosh Madani
Images, Signals and Intelligence
University PARIS-EST Creteil (UPEC)
Créteil
France

Joaquim Filipe
Instituto Politécnico de Setúbal (IPS)
Setúbal
Portugal

ISSN 1860-949X   ISSN 1860-9503   (electronic)
Studies in Computational Intelligence
ISBN 978-3-319-26391-5   ISBN 978-3-319-26393-9   (eBook)
DOI 10.1007/978-3-319-26393-9

Library of Congress Control Number: 2015950009

Springer Cham Heidelberg New York Dordrecht London

Springer International Publishing AG Switzerland is part of Springer Science+Business Media
(www.springer.com)

Preface

The present book includes extended and revised versions of a set of selected papers from the Sixth International Joint Conference on Computational Intelligence (IJCCI 2014). IJCCI was sponsored by the Institute for Systems and Technologies of Information, Control and Communication (INSTICC). This conference was held in Rome, Italy, from October 22 to 24, 2014.

IJCCI was technically co-sponsored by IEEE Computational Intelligence Society, co-sponsored by International Federation of Automatic Control (IFAC), and held in cooperation with the ACM Special Interest Group on Artificial Intelligence (ACM SIGART), Association for the Advancement of Artificial Intelligence (AAAI), Asia Pacific Neural Network Assembly (APNNA), European Society for Fuzzy Logic and Technology (EUSFLAT), International Neural Network Society, and the International Fuzzy Systems Association.

Since its first edition in 2009, the purpose of the International Joint Conference on Computational Intelligence (IJCCI) has been to bring together researchers, engineers, and practitioners in computational technologies, especially those related to the areas of fuzzy computation, evolutionary computation, and neural computation. IJCCI is composed of three co-located conferences with each one specialized in one of the aforementioned areas. Namely:

- International Conference on Evolutionary Computation Theory and Applications (ECTA)
- International Conference on Fuzzy Computation Theory and Applications (FCTA)
- International Conference on Neural Computation Theory and Applications (NCTA)

Their aim is to provide major forums for scientists, engineers, and practitioners interested in the study, analysis, design, and application of these techniques to all fields of human activity.

In ECTA, modeling and implementation of bio-inspired systems, both theoretically and in a broad range of application fields, is the central scope. Considered as a subfield of computational intelligence focused on optimization problems, evolutionary computation is associated with systems that use computational models of evolutionary processes as the key elements in design and implementation, i.e., computational techniques which are inspired by the evolution of biological life in the natural world. A number of bio-inspired models have been proposed, including genetic programming, genetic algorithms, evolution strategies, evolutionary programming, swarm optimization, and ant colony optimization.

In FCTA, results and perspectives of modeling and implementation of fuzzy systems, in a broad range of fields, are presented and discussed. Fuzzy computation, based on the theory of fuzzy sets and fuzzy logic, is dedicated to the solution of information processing, system analysis, knowledge extraction from data, and decision problems. Fuzzy computation is taking advantages of the powerful available technologies to find useful solutions for problems in many fields, such as medical diagnosis, automated learning, image processing and understanding, and systems control.

NCTA is focused on modeling and implementation of neural-based computation and related issue as those dealing with artificial neural networks and brain's structure issued architectures. Neural computation and artificial neural networks have seen a continuous explosion of interest in recent decades, and are being successfully applied across an impressive range of problem domains, including areas as diverse as finance, medicine, engineering, geology, and physics, providing appealing solutions to problems as varied as prediction, classification, decision-making, or control. Numerous architectures, learning strategies and algorithms have been introduced in this highly dynamic field in the last couple of decades.

The joint conference IJCCI received 210 paper submissions from 51 countries, of which 15 % were presented as full papers. The high quality of the papers received imposed difficult choices in the review process. To evaluate each submission, a double-blind paper evaluation method was used: each paper was reviewed by at least two experts from the independent international Program Committee, in a double-blind review process, and most papers had three reviews or more. This book includes revised and extended versions of a strict selection of the best papers presented at the conference.

On behalf of the Conference Organizing Committee, we would like to thank all participants. First of all to the authors, whose quality work is the essence of the conference, and to the members of the Program Committee, who helped us with their expertise and diligence in reviewing the papers. As we all know, producing a post-conference book, within the high technical level exigency, requires efforts of many individuals. We wish to thank also all the members of our Organizing Committee, whose work and commitment were invaluable.

April 2015

Juan Julian Merelo
Agostinho Rosa
José M. Cadenas
António Dourado
Kurosh Madani
Joaquim Filipe

Organization

Conference Chair

Joaquim Filipe, Polytechnic Institute of Setúbal/INSTICC, Portugal

Program Co-chairs

ECTA
Juan Julian Merelo, University of Granada, Spain
Agostinho Rosa, IST, Portugal

FCTA
José M. Cadenas, University of Murcia, Spain
António Dourado, University of Coimbra, Portugal

NCTA
Kurosh Madani, University of Paris-EST Créteil (UPEC), France

Organizing Committee

Helder Coelhas, INSTICC, Portugal
Vera Coelho, INSTICC, Portugal
Lucia Gomes, INSTICC, Portugal
Ana Guerreiro, INSTICC, Portugal
André Lista, INSTICC, Portugal
Andreia Moita, INSTICC, Portugal
Vitor Pedrosa, INSTICC, Portugal
Andreia Pereira, INSTICC, Portugal

João Ribeiro, INSTICC, Portugal
Susana Ribeiro, INSTICC, Portugal
Sara Santiago, INSTICC, Portugal
Mara Silva, INSTICC, Portugal
José Varela, INSTICC, Portugal
Pedro Varela, INSTICC, Portugal

ECTA Program Committee

Arturo Hernandez Aguirre, Centre for Research in Mathematics, Mexico
Chang Wook Ahn, Sungkyunkwan University (SKKU), Korea, Republic of
Mehmet Emin Aydin, University of Bedfordshire, UK
Thomas Baeck, Leiden University, The Netherlands
Michal Bidlo, Brno University of Technology, Faculty of Information Technology,
Czech Republic
Tim Blackwell, University of London, UK
Christian Blum, IKERBASQUE and University of the Basque Country, Spain
William R. Buckley, California Evolution Institute, USA
Edmund Burke, University of Nottingham, UK
David Cairns, University of Stirling, UK
Pedro Castillo, University of Granada, Spain
Pei-Chann Chang, Yuan Ze University, Taiwan
Sung-Bae Cho, Yonsei University, Korea, Republic of
Chi-Yin Chow, City University of Hong Kong, Hong Kong
Antonio Della Cioppa, University of Salerno, Italy
Ernesto Costa, Universidade De Coimbra, Portugal
Kyriakos Deliparaschos, National Technical University of Athens (ntua), Cyprus
Peter Duerr, Sony Corporation, Japan
Marc Ebner, Ernst-Moritz-Arndt-Universität Greifswald, Germany
Andries Engelbrecht, University of Pretoria, South Africa
Fabio Fassetti, DIMES, University of Calabria, Italy
Carlos M. Fernandes, University of Granada, Spain
Stefka Fidanova, Bulgarian Academy of Sciences, Bulgaria
Bogdan Filipic, Jozef Stefan Institute, Slovenia
Dalila B.M.M. Fontes, Faculdade de Economia and LIAAD-INESC TEC,
Universidade do Porto, Portugal
Marcus Gallagher, The University of Queensland, Australia
David Geoffrey Green, Monash University, Australia
Steven Guan, Xian Jiaotong-Liverpool University, China
Jörg Hähner, Universität Augsburg, Germany
Lutz Hamel, University of Rhode Island, USA
Thomas Hanne, University of Applied Arts and Sciences Northwestern Switzer-
land, Switzerland

Mohsin Bilal Hashmi, Umm Al-Qura University, Saudi Arabia
Wei-Chiang Hong, Hangzhou Dianzi University, Taiwan
Jeffrey Horn, Northern Michigan University, USA
Seiya Imoto, University of Tokyo, Japan
Liu Jing, Xidian University, China
Colin Johnson, University of Kent, UK
Iwona Karcz-Duleba, Wroclaw University of Technology, Poland
Ed Keedwell, University of Exeter, UK
Joanna Kolodziej, Cracow University of Technology, Poland
Mario Köppen, Kyushu Institute of Technology, Japan
Ondrej Krejcar, University of Hradec Kralove, Czech Republic
Pavel Krömer, VSB Ostrava, Czech Republic
Jiri Kubalik, Czech Technical University, Czech Republic
Halina Kwasnicka, Wroclaw University of Technology, Poland
Dario Landa-Silva, University of Nottingham, UK
Piotr Lipinski, University of Wroclaw, Poland
Shih-Hsi Liu, California State University, Fresno, USA
Wenjian Luo, University of Science and Technology of China, China
Rainer Malaka, Bremen University, Germany
Jörn Mehnen, Cranfield University, UK
Marjan Mernik, University of Maribor, Slovenia
Konstantinos Michail, Cyprus University of Technology, Cyprus
Chilukuri Mohan, Syracuse University, USA
Antonio Mora, University of Granada, Spain
Sanaz Mostaghim, Otto-von-Guericke-Universität Magdeburg, Germany
Luiza de Macedo Mourelle, State University of Rio de Janeiro, Brazil
Christine L. Mumford, Cardiff University, UK
Pawel B. Myszkowski, Wroclaw University of Technology, Poland
Kei Ohnishi, Kyushu Institute of Technology, Japan
Schütze Oliver, CINVESTAV-IPN, Mexico
Ender Özcan, University of Nottingham, UK
Gary B. Parker, Connecticut College, USA
Kalin Penev, Southampton Solent University, UK
Francisco Pereira, Instituto Politécnico de Coimbra, Portugal
Petrica Pop, North University of Baia Mare, Romania
Aurora Pozo, Federal University of Parana, Brazil
Roy Rada, University of Maryland Baltimore County, USA
José Risco-Martín, Universidad Complutense de Madrid, Spain
Mateen Rizki, Wright State University, USA
Katya Rodriguez, Instituto de Investigaciones en Matemáticas Aplicadas y en
Sistemas (IIMAS), Mexico
Olympia Roeva, Institute of Biophysics and Biomedical Engineering, Bulgarian
Academy of Sciences, Bulgaria
Agostinho Rosa, ISR—Instituto de Sistemas e Robótica, Portugal
Suman Roychoudhury, Tata Consultancy Services, India

Filipe Azinhais Santos, ISCTE-IUL, Portugal
Miguel A. Sanz-Bobi, Pontificia Comillas University, Spain
Emmanuel Sapin, University of Exeter, UK
Ruhul Amin Sarker, University of New South Wales, Australia
Robert Schaefer, AGH University of Science and Technology, Poland
Giovanni Stracquadanio, Johns Hopkins University, USA
Jonathan Thompson, Cardiff University, UK
Krzysztof Trojanowski, Institute of Computer Science Polish Academy of Sciences, Poland
George Tsihrintzis, University of Piraeus, Greece
Elio Tuci, Aberystwyth University, UK
Lucia Vacariu, Technical University of Cluj Napoca, Romania
Neal Wagner, Massachusetts Institute of Technology Lincoln Laboratory, USA
Junzo Watada, Waseda University, Japan
Peter Whigham, University of Otago, New Zealand
Gary Yen, Oklahoma State University, USA

ECTA Auxiliary Reviewers

Xuelei Hu, University of Queensland, Australia
Juan L.J. Laredo, University of Luxembourg, Luxembourg
Antonio Mora, University of Granada, Spain
Pablo García Sánchez, University of Granada, Spain
Victor Rivas Santos, University of Jaen, Spain
Jihua Wang, FedEx Corporate Services, USA

FCTA Program Committee

Rafael Alcala, University of Granada, Spain
Sansanee Auephanwiriyakul, Chiang Mai University, Thailand
Ahmed Bufardi, Ecole Polytechnique Federale de Lausanne, Switzerland
Jinhai Cai, University of South Australia, Australia
Daniel Antonio Callegari, PUC-RS Pontificia Universidade Catolica do Rio Grande do Sul, Brazil
Heloisa Camargo, UFSCar, Brazil
João Paulo Carvalho, INESC-ID/Instituto Superior Técnico, Portugal
Giovanna Castellano, University of Bari, Italy
France Cheong, RMIT University, Australia
Mikael Collan, Lappeenranta University of Technology, Finland
Shuang Cong, University of Science and Technology of China, China
Keeley Crockett, Manchester Metropolitan University, UK

Martina Dankova, University of Ostrava, Czech Republic
Scott Dick, University of Alberta, Canada
Pietro Ducange, University of Pisa, Italy
Mario Fedrizzi, Univerisity of Trento, Italy
János Fodor, Obuda University, Hungary
Yoshikazu Fukuyama, Meiji University, Japan
Alexander Gegov, University of Portsmouth, UK
Brunella Gerla, University of Insubria, Italy
Maria Angeles Gil, University of Oviedo, Spain
Sarah Greenfield, De Montfort University, UK
Masafumi Hagiwara, Keio University, Japan
Susana Muñoz Hernández, Universidad Politécnica de Madrid (UPM), Spain
Katsuhiro Honda, Osaka Prefecture University, Japan
Chih-Cheng Hung, Southern Polytechnic State University, USA
Cengiz Kahraman, Istanbul Technical University, Turkey
Uzay Kaymak, Eindhoven University of Technology, The Netherlands
Frank Klawonn, Ostfalia University of Applied Sciences, Germany
Li-Wei (Leo) Ko, National Chiao Tung University, Taiwan
Laszlo T. Koczy, Szechenyi Istvan University, Hungary
Anne Laurent, Lirmm, University Montpellier 2, France
Chin-Teng Lin, National Chiao Tung University, Taiwan
Ahmad Lotfi, Nottingham Trent University, UK
Edwin Lughofer, Johannes Kepler University, Austria
Luis Magdalena, European Centre for Soft Computing, Spain
Francesco Marcelloni, University of Pisa, Italy
Corrado Mencar, University of Bari, Italy
Ludmil Mikhailov, University of Manchester, UK
Javier Montero, Complutense University of Madrid, Spain
Alejandro Carrasco Muñoz, University of Seville, Spain
Hiroshi Nakajima, Omron Corporation, Japan
Vesa Niskanen, Univ. of Helsinki, Finland
Yusuke Nojima, Osaka Prefecture University, Japan
Vilém Novák, University of Ostrava, Czech Republic
Parag Pendharkar, Pennsylvania State University, USA
Irina Perfilieva, University of Ostrava, Czech Republic
David Picado, European Centre for Soft Computing, Spain
Valentina Plekhanova, University of Sunderland, UK
Daowen Qiu, Sun Yat-sen University, China
Jordi Recasens, Universitat Politècnica de Catalunya, Spain
Antonello Rizzi, University of Rome "La Sapienza", Italy
Alessandra Russo, Imperial College London, UK
Jurek Sasiadek, Carleton University, Canada
Steven Schockaert, Cardiff University, UK
Hirosato Seki, Kwansei Gakuin University, Japan
Dipti Srinivasan, NUS, Singapore

Petr Štemberk, Czech Technical University, Czech Republic
Hooman Tahayori, Ryerson University, Canada
Dat Tran, University of Canberra, Australia
Hamid Vakilzadian, University of Nebraska-Lincoln, USA
Maria Luiza Velloso, Rio de Janeiro State University, Brazil
Wen-June Wang, National Central University, Taiwan
Dongrui Wu, GE Global Research, USA
Chung-Hsing Yeh, Monash University, Australia
Jianqiang Yi, Institute of Automation, Chinese Academy of Sciences, China
Hans-Jürgen Zimmermann, European Laboratory for Intelligent Techniques
Engineering (ELITE), Germany

FCTA Auxiliary Reviewer

Robert Fuller, Obuda University, Hungary

NCTA Program Committee

Shigeo Abe, Kobe University, Japan
Veronique Amarger, University PARIS-EST Creteil (UPEC), France
Davide Anguita, University of Genoa, Italy
Vijayan Asari, University of Dayton, USA
Gilles Bernard, PARIS 8 University, France
Daniel Berrar, Tokyo Institute of Technology, Japan
Yevgeniy Bodyanskiy, Kharkiv National University of Radio Electronics, Ukraine
Ivo Bukovsky, Czech Technical University in Prague, Faculty of Mechanical
Engineering, Czech Republic
Jinde Cao, Southeast University/King Abdulaziz University, China
Mu-Song Chen, Da-Yeh University, Taiwan
Ning Chen, Instituto Superior de Engenharia do Porto, Portugal
Amine Chohra, Paris-East University (UPEC), France
Catalina Cocianu, The Bucharest University of Economic Studies, Faculty of
Cybernetics, Statistics and Informatics in Economy, Romania
Netta Cohen, University of Leeds, UK
Shuang Cong, University of Science and Technology of China, China
Sorin Dan Cotofana, Delft University of Technology, The Netherlands
Marc Ebner, Ernst-Moritz-Arndt-Universität Greifswald, Germany
Leonardo Franco, Universidad de Málaga, Spain
Josep Freixas, Escola Politècnica Superior d'Enginyeria de Manresa, Spain
John Qiang Gan, University of Essex, UK
Marcos Gestal, University of A Coruña, Spain

Michel Verleysen, Université Catholique de Louvain, Belgium
Brijesh Verma, Central Queensland University, Australia
Ricardo Vigário, Aalto University School of Science, Finland
Michael N. Vrahatis, University of Patras, Greece
Shuai Wan, Northwestern Polytechnical University, China
Ling Wang, Tsinghua University, China
Shengrui Wang, University of Sherbrooke, Canada
Junzo Watada, Waseda University, Japan
Weiwei Yu, Northwestern Polytechnical University, China
Wenwu Yu, Southeast University, China
Cleber Zanchettin, Federal University of Pernambuco, Brazil
Huiyu Zhou, Queen's University Belfast, UK

NCTA Auxiliary Reviewers

Peter Benes, Czech Technical University in Prague, Czech Republic
Changlin Chen, Delft Univerisity of Technology, The Netherlands
Chris Christodoulou, University of Cyprus, Cyprus
Costas Neocleous, Cyprus University of Technology, Cyprus

Invited Speakers

Marie-Jeanne Lesot, Université Pierre et Marie Curie—LIP6, France
Danil Prokhorov, Toyota Tech Center, USA
Bernard De Baets, Ghent University, Belgium
Cesare Alippi, Politecnico di Milano, Italy

Contents

Part I
Evolutionary Computation Theory
and Applications

Evolutionary Tuning of Optimal PID Controllers for Second Order Systems Plus Time Delay

Jesús-Antonio Hernández-Riveros, Jorge-Humberto Urrea-Quintero and Cindy-Vanessa Carmona-Cadavid

Abstract PID stands for "proportional, integral, derivative" The PID controller is the most widely used industrial device for monitoring and controlling processes. Those three elements are the basics of a PID Controller. Each one performs a different task and has a different effect on the functioning of a system. The expected behavior of a system depends of the setting of those parameters. There are alternatives to the traditional rules of PID tuning, but there is not yet a study showing that the use of heuristic algorithms it is indeed better than using classic methods of optimal tuning. This is developed in this paper. An evolutionary algorithm MAGO (*Multidynamics Algorithm for Global Optimization*) is used to optimize the controller parameters minimizing the ITAE performance index. The procedure is applied to a set of benchmark problems modeled as *Second Order Systems Plus Time Delay* (SOSPD) plants. The evolutionary approach gets a better overall performance comparing to traditional methods (Bohl and McAvoy, ITAE-Hassan, ITAE-Sung), regardless the plant used and its operating mode (servo or regulator), covering all restrictions of the traditional methods and extending the maximum and minimum boundaries between them.

Keywords PID tuning · Evolutionary computation · Heuristic algorithm · Integral performance criterion · Multidynamics optimization · MAGO

J.-A. Hernández-Riveros (✉) · J.-H. Urrea-Quintero · C.-V. Carmona-Cadavid
Facultad De Minas, Universidad Nacional de Colombia – Sede Medellín, Medellín, Colombia
e-mail: jahernan@unal.edu.co

J.-H. Urrea-Quintero
e-mail: jhurreaq@unal.edu.co

C.-V. Carmona-Cadavid
e-mail: cvcarmonac@unal.edu.co

© Springer International Publishing Switzerland 2016
J.J. Merelo et al. (eds.), *Computational Intelligence*,
Studies in Computational Intelligence 620, DOI 10.1007/978-3-319-26393-9_1

1 Introduction

The order of a system specifies the number of integrations between the input and the output of the system being controlled (i.e., the plant). There is only a proportional relation between the input and the output in a zero-order system (this is zero integrations). A first order system takes into account that proportional relationship and also the rate (velocity) of the output. Besides these two relations, in a second-order system there is also a proportional relation between the input and the acceleration of the output. The difficulty of learning and using a control system increases with the order of the plant. The dynamics of a second order system is typical of many real-world control tasks and is more representative of problems that have inertia. The inertial dynamic properties are described as lags of the plant and alter the shape of the output. Additionally, in real systems could be the presence of time delays. Time delays are due to the time of transmission information through some medium. Time delays affect only the temporal relation between input and output and have no effect on the shape of the actual response, relative to the immediate input. Most real plants are modeled as second order systems with both lags and time delays. This behavior is a challenge for control systems. The problem of controlling Second Order System Plus time Delay (SOSPD) is recognized as difficult [1].

In [2], a comparative study of performance of different tuning classical methods for PID (proportional-integral-derivative) controllers is achieved. This study concludes that tuning methods that require a SOSPD model perform better than those that require a First Order Lag Plus time Delay model (FOLPD). O'Dwyer [3] reports that 90 % of the tuning rules developed are based on a model of first and second order plus time delay. Current trends in controller parameter estimation minimize an integral performance criterion. The most frequently tuning rules used are not based on an integral performance criterion. The optimal tuning rules based on second-order models are just 14 of the 84 reported until 2009. In general, those rules are based on several conditions of the parameters defining the process model. The SOSPD model was selected in this paper as representing the plants in order to compare the performance of a heuristic algorithm with the "best" techniques developed for PID controllers optimal tuning. For SOSPD general models 147 tuning rules have been defined based on the ideal PID structure [3]. In [4, 5] the performance and robustness of some tuning rules are evaluated, and a complete analysis of the methods of tuning controllers based on SOSPD is made. Each of the developed tuning rules for PID controllers has only been applied to a certain group of processes. Usual tuning methodologies, such as design based on the root locus, pole-zero cancellation, location of the closed-loop poles, among others, require cumbersome procedures and specialized knowledge. Additionally, most methods for optimal tuning of SOSPD require extra system information from experiments carried out directly on the plant; activities that are not always possible to perform because the presence of extreme stresses and oscillations which may create instability and damage to the system.

The studies mentioned suggest the lack of a general rule for tuning PID controllers. Due to the large number of existing tuning rules it is necessary to find a tuning method that best satisfies the requirements of each problem and also ensures optimal values for the controller parameters according to the selected performance criterion.

There is a trend to develop new methods for tuning PI and PID controllers [5–7], posed as a nonlinear optimization problem. In reviewing the literature is found that evolutionary algorithms (EA) are applied to the tuning of controllers on particular cases and not in the general cases as in this paper. Nor are compared with traditional methods that minimize some tuning performance index [8–11]. This implies that although there are alternatives to the traditional rules of tuning, there is not yet a study showing that the use of heuristic algorithms it is indeed better than using the traditional rules of optimal tuning. Hence, this matter is addressed. Other applications of the EA in control systems, among them, are system identification [12] and optimal configuration of sensors [13]. The use of an EA for tuning PID controllers in processes represented by SOSPD models is proposed in this paper.

This paper is concerned with PID controllers for processes modeled as SOSPD, optimizing the ITAE (Integral of Time Multiplied by Absolute Error) and not requiring additional system information.

EA are a proven tool for solving nonlinear systems and optimization problems. The weaknesses of these algorithms are in the large number of control parameters of the EA to be determined by the analyst and the lack of a solid mathematical foundation [14]. Looking address these weaknesses arise recently the Estimation of Distribution Algorithms, EDA [15]. These algorithms do not use genetic operators, but are based on statistics calculated on samples of the population, which is constantly evolving. This variant when introduce statistics operators provides a strong way to demonstrate the evolution. Nevertheless, they are difficult to manage and do not eliminate the large number of control parameters of classical EA. Set a classic EA is itself a difficult optimization problem; the analyst must try with probabilities of crossover, mutation, replication, operator forms, legal individuals, loss of diversity, etc. Whereas, the EDA require expert skills as the formulation of simultaneous complex distributions or the Bayesian networks structure. For its part, Multi dynamics Algorithm for Global Optimization (MAGO) also works with statistics from the evolution of the population [16]. MAGO is a heuristic algorithm resulting from the combination of Lagrangian Evolution, Statistical Control and Estimation of Distribution. MAGO has shown to be an efficient and effective tool to solve problems whose search space is complex [17] and works with a real-valued representation. MAGO only requires two parameters provided by the analyst: the number of generations and the population size. The traditional EA, additionally to the number of generations and the population size requires from the user the definition of the selection strategy, the individuals' representation, probabilities of mutation, crossover, replication, as well as, the crossover type, the locus of crossing, among others. Depending of its design, some EA also have extra parameters of tuning as control variables, number of branch and nodes, global step size, time constant, etc. [18]. Because of that, MAGO becomes a good choice as a tool for solving controller tuning as an optimization problem.

The results obtained by MAGO are compared with traditional tuning methods not requiring additional system information. The ITAE is optimized to penalize the error. As it is further shown, the system model used makes no difference for the MAGO, because only input and output signals from the closed loop system are required to calculate the controller parameters. Regardless of the relationship between the parameters of the system (time delay, constant time, etc.) the results obtained by MAGO overcome those from the traditional methods of optimal tuning.

This paper begins with an introduction of controller parameters estimation and performance index calculation. The tuning of PID controllers on SOSPD using both the traditional methods and the evolutionary algorithm MAGO follow. A results analysis and some conclusions come after.

2   PID Controller Tuning

The control policy of an ideal PID controller is shown in Eq. (1), where $E(s) = (R(s)-Y(s))$. The current value $Y(s)$ of the controlled variable is compared to its desired value R(s), to obtain an error signal $E(s)$ (feedback). This error is processed to calculate the necessary change in the manipulated variable $U(s)$ (control action).

$$U(s) = Kc\left[1 + \frac{1}{T_i s} + T_d s\right] E(s) \tag{1}$$

Some rules of tuning controllers are based on critical system information, on reaction curves and on closed loop tests [19].

This paper is concerned to PID controllers for processes modeled as SOSPD, optimizing the ITAE and not requiring additional system information.

In [3], it is indicated that 20.7 % of the rules of tuning PID controllers have been developed from SOSPD models (with or without a zero in the numerator). This implies 84 rules, 66 of them do not include the zero in the numerator. Of these, only 14 optimize an integral performance criterion, from which 4 rules propose selecting controller parameters by means of tables and other 6 require additional system information (ultimate gain, Ku; ultimate frequency, Tu). There are only 4 tuning rules that optimize an integral performance criterion and are only function of the SOSPD parameters. For regulators these rules are: Bohl and McAvoy, Minimum ITAE—Hassan, Minimum ITAE—Sung; for servomechanisms: Minimum ITAE—Sung. Table 1 shows the summary of the study, the chosen rules are shadowed. The equations for the calculation of proportional gain, Kc; integral time, Ti and derivative time, Td can be consulted in [20–23]. These tuning rules define restrictions on the behavior of the plant, expressed in the range of validity.

Table 1 Tuning methods for PID controllers optimizing an integral criterion on SOSPD system

Method	Type of operation	Type of plant	Range of pertinence	Observation
Minimum IAE —Wills	Regulator	(6)	Tm2 = τ = 0.1 Tml	Requires critical system information (Ku, Tu)
Minimum IAE —López	Regulator	(5)	$0.5 < \xi < 4$; $0.1 < \tau/$ Tml < 10	Tuningrule base on tables
Minimum IAE —Shinskey	Regulator	(6)	Tm2/(Tm2 + τ) = 0.25, 0.5, 0.75	Requires critical system information (Ku. Tu)
Minimum IAE —Kang	Regulator	(6)	τ/Tinl, Tm2/Tml	Tuningrule base on tables
Minimum ITAE—López	Regulator	(5)	$0.5 < \xi < 4$; $0.1 < \tau/$ Tml < 1	Tuningrule base on tables
Bohl and Mc Avoy	Regulator	(6)	$0.12 <$ Tml/Tm2 < 0.9; $0.1 < \tau/$Tml < 0.5	Tuningrule requiring only SOSPD model parameters
Minimum ITAE—Hassan	Regulator	(5)	$0.5 < \xi < 2$; $0.1 < \tau/$ Tml < 4	Tuningrule requiring only SOSPD model parameters
Minimum ITAE—Sung	Regulator	(5)	$0.05 < \tau/$Tml < 2	Tuningrule requiring only SOSPD model parameters
Nearly minimum IAE, ISE, ITAE— Hwang	Regulator	(5)	$0.6 < \xi < 4.2$; $0.2 < \tau/$ Tml < 2	Requires critical system information (Ku, Tu)
Minimum IAE —Wills	Servomechanism	(6)	Tm2 = τ = 0.1 Tml	Requires critical system information (Ku, Tu)
Minimum IAE —Gallier and Otto	Servomechanism	(5) & (6)	$0.05 < \tau/2$Tml < 4	Tuningrule base on tables
Minimum ITAE—Wills	Servomechanism	(6)	Tml = Tm2; τ = 0.1 Tml	Requires critical system information (Ku, Tu)
Minimum ITAE—Sung	Servomechanism	(5)	$0.05 < \tau/$Tml < 2	Tuningrule requiring only SOSPD model parameters
Nearly minimum IAE, ISE, ITAE— Hwang	Servomechanism	(5)	$0.6 < \xi < 4.2$; $0.2 < \tau/$ Tml < 2	Requires critical system information (Ku, Tu)

2.1 Performance Criteria of PID Controllers

The criterion used for tuning a controller is directly related to the expected performance of the control loop. It can be based on desired characteristics of the response, in time or frequency. Searching for a way to quantify the behavior of control loops led to the establishment of performance indexes based on the error signal, $e(t)$ (feedback). The objective is to determine the controller setting that minimizes the chosen cost function. The parameters are optimal under fixed performance criteria. Of these, the best known are the so-called *integral criteria* [19], defined in Eqs. (2) and (3).

$$\text{Integral of Absolute Error } IAE = \int_0^\infty |e(t)|dt \qquad (2)$$

$$\text{Integral of Time Multiplied by Absolute Error } ITAE = \int_0^\infty t|e(t)|dt \qquad (3)$$

$$\text{Where the error is given by: } e(t) = r(t) - y(t) \qquad (4)$$

$r(t)$ is the reference value, and $y(t)$ is the current value of the controlled variable, both expressed in time.

2.2 Plant Parameters and Performance Indexes

To compare the performance of the studied controllers it is necessary to tune them with the same plants. The plant models used are given in Eqs. (5) and (6), [24].

$$G(s) = \frac{K_p e^{-\tau_m s}}{T_{m1}^2 s^2 + 2\xi_m T_{m1} s + 1} \qquad (5)$$

$$G(s) = \frac{K_p e^{-\tau_m s}}{(1 + T_{m1}s)(1 + T_{m2}s)} \qquad (6)$$

The following considerations are taken for Eq. (5): $Kp = 1$, $\tau m = 1$, $\xi = 1$ and Tm1 ranging from 1, 10 and 20. For Eq. (6), the following considerations are taken: $Kp = 1$, $\tau m = 1$, $Tm1 = 1$ and $Tm2 = a*Tm1$, where $a \leq 1$. Tables 2 and 3 presents

Table 2 Transfer functions of plants 1, Eq. (5), for the tuning

Servomechanism	Regulator
$G_{p1\_servo1}(s) = \frac{e^{-s}}{s^2 + 2s + 1}$	$G_{p1\_reg1}(s) = G_{p1\_servo1}(s)$
$G_{p1\_servo2}(s) = \frac{e^{-s}}{100s^2 + 20s + 1}$	$G_{p1\_reg2}(s) = G_{p1\_servo2}(s)$
$G_{p1\_servo3}(s) = \frac{e^{-s}}{400s^2 + 40s + 1}$	$G_{p1\_reg3}(s) = G_{p1\_servo3}(s)$

Table 3 Transfer functions of plants 2, Eq. (6), for the tuning

Servomechanism	Regulator
$G_{p2\_servo1}(s) = \frac{e^{-s}}{(1+s)(1+0.1s)}$	$G_{p2\_reg1}(s) = G_{p2\_servo1}(s)$
$G_{p2\_servo2}(s) = \frac{e^{-s}}{(1+s)(1+0.5s)}$	$G_{p2\_reg2}(s) = G_{p2\_servo2}(s)$
$G_{p2\_servo3}(s) = \frac{e^{-s}}{(1+s)(1+s)}$	$G_{p2\_reg3}(s) = G_{p2\_servo3}(s)$

a set of transfer functions according to the parameter values of each plant given by Eqs. (5) and (6).

The values of the PID controller parameters for each selected tuning rules are presented, further on, on Table 4. The parameters are calculated according to the formulas proposed for each kind of plant. The selected methods for tuning controllers minimize the integral performance criterion, ITAE. Therefore, in Table 4, besides the values of controller parameters, the ITAE is also reported. The ITAE is calculated in all cases using the commercial software MATLAB® function "trapz".

Table 4 PID controller parameters

Plant (2)	PID operating as regulator						ITAE	
	Kc		Ti		Td			
	B&M	MAGO	B&M	MAGO	B&M	MAGO	B&M	MAGO
$G_{P2\text{-}reg1}(s)$	1.7183	1.4296	1.8978	1.5433	1.8988	0.3341	7.7760	3.1052
$G_{P2\text{-}reg2}(s)$	1.0300	1.4656	1.4164	1.5552	1.6702	0.5597	6.8722	3.6071
$G_{P2\text{-}reg3}(s)$	0.3092	1.8527	0.5854	1.7791	0.7286	0.7575	3.8073	3.6738
Plant (2)	PID operating as servomechanism						ITAE	
	Kc		Ti		Td			
	Hassan	MAGO	Hassan	MAGO	Hassan	MAGO	Hassan	MAGO
$G_{P2\text{-}servo1}(s)$	NC*	0.5658	NC*	1.6705	NC*	1.0318	NC*	72.6860
$G_{P2\text{-}servo2}(s)$	NC*	0.2731	NC*	1.0966	NC*	0.4871	NC*	69.4943
$G_{P2\text{-}servo3}(s)$	NC*	0.9074	NC*	2.0666	NC*	0.5258	NC*	63.2413
Plant (1)	PID operating as servomechanism						ITAE	
	Kc		Ti		Td			
	SUNG	MAGO	SUNG	MAGO	SUNG	MAGO	SUNG	MAGO
$G_{P1\text{-}servo1}(s)$	1.2420	1.2318	2.0550	2.1167	0.6555	0.6050	2.0986	2.0486
$G_{P1\text{-}servo2}(s)$	9.0500	10.3237	18.009	16.8942	4.9386	5.5162	3.7911	2.8532
$G_{P1\text{-}servo3}(s)$	16.4953	19.7929	35.689	29.7905	9.5595	10.7718	3.7937	2.7827
Plant (1)	PID operating as regulator						ITAE	
	Kc		Ti		Td			
	SUNG	MAGO	SUNG	MAGO	SUNG	MAGO	SUNG	MAGO
$G_{P1\text{-}reg1}(s)$	1.8160	1.8557	1.9120	1.7563	0.7073	0.7518	3.8100	3.6623
$G_{P1\text{-}reg2}(s)$	12.8460	17.3252	16.7995	7.4691	−1.99e − 6	2.3730	894.5522	3.6427
$G_{P1\text{-}reg3}(s)$	21.8276	31.8262	37.7393	11.0993	−1.17e − 4	3.7005	314.5554	4.4240

NC* = Not converged
B&M* = Bohl and McAvoy

For the Hassan method, the controller parameter values are not reported because there was no convergence in the closed loop system response for the selected plants given by Eq. (5), operating as regulator.

3 Tuning Optimal PID Controllers Using an Evolutionary Algorithms

Different solutions there may exist in optimization problems, therefore a criterion for discriminating between them, and finding the best, is required. The tuning of controllers that minimize an integral performance criterion can be seen as an optimization problem, inasmuch as the ultimate goal is to find the combination of parameters Kc, Ti and Td, such that the value of the integration of a variable of interest is minimal (error between the actual output of the plant and the desired value).

EA are widely studied as a heuristic tool for solving optimization problems. They have shown to be effective in problems that exhibit noise, random variation and multimodality. Genetic algorithms, for example, have proven to be valuable in both obtaining the optimal values of the PID controller parameters, and in computational cost [23]. One of the recent trends in EA is Estimation of Distribution Algorithms [15]. These do not use genetic operators but are based on statistics from the same evolving population. The Multidynamics Algorithm for Global Optimization (MAGO) [16] also works with statistics from the evolving population. MAGO is autonomous in the sense that it regulates its own behavior and does not need human intervention.

3.1 Optimization and Evolutionary Algorithms

There are techniques used to obtain better results (general or specific) for a problem. The results can greatly improve the performance of a process, which is why this kind of tools is known as optimization. When speaking of an optimization problem is to minimize or maximize depending of the design requirements.

These mean representative criteria of the system efficiency. The chosen criterion is called objective function. The design of an optimization problem is subject to specific restraints of the system, decision variables and design objectives, which leads to an expression such that the optimizer can interpret. Given its nature of global optimizer, an evolutionary algorithm (EA) is used in this work. EA have been used in engineering problems [25] and the tuning of PID controllers [8, 26]. The late is the case tries in this work, where successful results have been obtained. The tuning of controllers that minimize an objective function can be formulated as an optimization problem; it is a case of optimal control [27]. The optimal control consists in selecting a control structure (including a PID controller) and adjusts its parameters such that a criterion of overall performance is minimized. In the case of a PID controller Eq. (1), the ultimate goal is to find the combination of the Kc, Ti and Td parameters, given some restrictions, such that the value of the integral of a variable of interest (error between the plant's actual output and the desired value or control effort) is minimal. The problem consists of minimizing an objective function, where its minimum is the result of obtaining a suitable combination of the three parameters of PID controller.

3.2 Multidynamics Algorithm for Global Optimization—MAGO

MAGO inspires by statistical quality control for a self-adapting management of the population. In control charts it is assumed that if the mean of the process is out of some limits, the process is suspicious of being out of control. Then, some actions should be taken to drive the process inside the control limits [28]. MAGO takes advantage of the concept of control limits to produce individuals on each generation simultaneously from three distinct subgroups, each one with different dynamics. MAGO starts with a population of possible solutions randomly distributed throughout the search space. The size of the whole population is fixed, but the cardinality of each sub-group changes in each generation according to the first, second and third deviation of the actual population. The exploration is performed by creating new individuals from these three sub-populations. For the exploitation MAGO uses a greedy criterion in one subset looking for the goal.

In every generation, the average location and the first, second and third deviations of the whole population are calculated to form the groups. The first subgroup of the population is composed of improved elite which seeks solutions in a neighborhood near the best of all the current individuals. $N1$ individuals within one standard deviation of the average location of the current population of individuals are displaced in a straight line toward the best of all, suffering a mutation that incorporates information from the best one. The mutation is a simplex search as the Nelder–Mead method [18] but only two individuals are used, the best one and the trial one. A movement in a straight line of a fit individual toward the best one occurs. If this movement generates a better individual, the new one passes to the next generation; otherwise its predecessor passes on with no changes. This method does not require gradient information. For each trial individual $X_i^{(j)}$ at generation j a shifted one is created according to the rule in Eq. (7).

$$X_T^{(j)} = X_i^{(j)} + F^{(j)}(X_B^{(j)} - X_m^{(j)})$$
$$F^{(j)} = S^{(j)} / \|S^{(j)}\|$$

(7)

where $X_B^{(j)}$ is the best individual, $X_m^{(j)}$ is an individual randomly selected. To incorporate information of the current relations among the variables, the factor $F^{(j)}$ depending on the covariance matrix is chosen in each generation. $S^{(j)}$ is the population covariance matrix at generation j. This procedure compiles the differences among the best individuals and the very best one. The covariance matrix of the current population takes into account the effect of the evolution. This information is propagated on new individuals. Each mutant is compared to his father and the one with better performance is maintained for the next generation. This subgroup, called *Emergent Dynamics*, has the function of making faster convergence of the algorithm.

The second group, called *Crowd Dynamics*, is formed by creating *N2* individuals from a uniform distribution determined by the upper and lower limits of the second deviation of the current population of individuals. This subgroup seeks possible solutions in a neighborhood close to the population mean. At first, the neighborhood around the mean can be large, but as evolution proceeds it reduced, so that across the search space the population mean is getting closer to the optimal. The third group, or *Accidental Dynamics*, is the smaller one in relation to its operation on the population. *N3* individuals are created from a uniform distribution throughout the search space, as in the initial population. This dynamic has two functions: maintaining the diversity of the population, and ensuring numerical stability of the algorithm.

The Island Model Genetic Algorithm also works with subpopulations [29]. But in the Island model, more parameters are added to the genetic algorithm: number of islands, migration size, migration interval, which island migrate, how migrants are selected and how to replace individuals. Instead, in MAGO only two parameters are needed: number of generations and population size. On another hand, the use of a covariance matrix to set an exploring distribution can also be found in [30], where, in only one dynamics to explore the promising region, new individuals are created sampling from a Gaussian distribution with an intricate adapted covariance matrix. In MAGO a simpler distribution is used.

To get the cardinality of each dynamics, consider the covariance matrix of the population, $S^{(j)}$, at generation j, and its diagonal, $diag(S^{(j)})$. If $Pob^{(j)}$ is the set of potential solutions being considered at generation j, the three groups can be defined as in Eq. (8), where: XM(j) = mean of the actual population. If *N1*, *N2* and *N3* are the cardinalities of the sets *G1*, *G2* and *G3*, the cardinalities of the Emergent Dynamics, the Crowd Dynamics and the Accidental Dynamics are set, respectively, and *Pob(j) = G1 U G2 U G3*.

$$G_1 = \left\{ x \in Pob(j) / XM(j) - \sqrt{diag(S(j))} \le x \le XM(j) + \sqrt{diag(S(j))} \right\}$$

$$G_2 = \left\{ x \in Pob(j) / or \begin{array}{c} XM(j) - 2\sqrt{diag(S(j))} < x \le XM(j) - \sqrt{diag(S(j))} \\ \\ XM(j) + \sqrt{diag(S(j))} \le x < XM(j) + 2\sqrt{diag(S(j))} \end{array} \right\} \qquad (8)$$

$$G_3 = \left\{ x \in Pob(j) / or \begin{array}{c} x(j) - 2\sqrt{diag(S(j))} \\ \\ x \ge XM(j) + 2\sqrt{diag(S(j))} \end{array} \right\}$$

This way of defining the elements of each group is dynamical by nature. The cardinalities depend on the whole population dispersion in the generation j. The Emergent Dynamics tends to concentrate $N1$ individuals around the best one. The Crowd Dynamics concentrates $N2$ individuals around the mean of the actual population. These actions are reflected in lower values of the standard deviation in each of the problem variables. The Accidental Dynamics, with $N3$ individuals, keeps the population dispersion at an adequate level. The locus of the best individual is different from the population's mean. As the evolution advances, the location of both the best individual and of the population's mean could be closer between themselves. This is used to self-control the population diversity. Figure 1 shows the flow diagrams of a traditional evolutionary algorithm and the MAGO. Following is MAGO's pseudo code.

```
1: j = 0, Random initial population generation uniformly
   distributed over the search space.
2: Repeat.
3: Evaluate each individual with the objective function.
4: Calculate the population covariance matrix and the
   first, second and third dispersion.
5: Calculate the cardinalities N1, N2 and N3 of the groups
   G1, G2 and G3.
6: Select N1 best individuals, modify them according to
   Eq (7), make them compete and translate the winners
   towards the best one. Pass the fittest to the
   generation j + 1.
7: Sample from a uniform distribution in hyper-rectangle
   [LB(j), UB(j)] N2 individuals, pass to generation j+ 1.
8: Sample N3 individuals from a uniform distribution over
   the whole search space and pass to generation j+1.
9: j = j + 1
10: Until an ending criterion is satisfied.
```

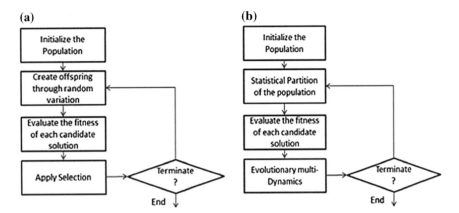

Fig. 1 Scheme of: **a** Evolutionary algorithm, **b** MAGO

3.3 Statement of the Problem

An EA represents a reliable approach when adjusting controllers is proposed as an optimization problem [25]. Given their nature of global optimizers, EA could face non-convex, nonlinear and highly restrictive optimization problems [7, 31, 32]. The MAGO has been shown as a very efficient instrument to solve problems in a continuous domain [17]. Thus, the MAGO is applied as a tool for estimating the parameters of a PID controller that minimizes an integral performance index. In the case where the system is operating as servomechanism, the control problem consists of minimizing the integral of the error multiplied by the time (ITAE). This involves finding the values for the parameters Kc, Ti y Td, such that the system gets the desired $r(t)$ value as fast as possible and with few oscillations. In the case where the system operates as a regulator, the reference is a constant R, but the control problem is also to minimize the ITAE index. This implies, again, finding the values of the parameters Kc, Ti and Td, but the goal in this mode is that at the appearance of a disturbance the system returns as quickly as possible to the point of operation. The optimization problem is defined in Eq. (9).

$$J\left(\underbrace{K_c,\quad T_i,\quad T_d}_{x}\right) = \min_{x}\left\{J_{ITAE} = \int_0^\infty t|e(t)|dt\right\} \qquad (9)$$

3.4 Evolutionary Design of PID Controller

The controller design is made for the modes servo and regulator. For the servo, a change in a unit step reference is applied. For the regulator, the same change is

Table 5 Structure of the evolutionary individual	$Kc \in R^+$	$Ti \in R^+$	$Td \in R^+$

applied but as a unit step disturbance to the second-order plant. The controllers are tuned for the six plants defined in Tables 2 and 3. The two parameters of MAGO: number of generations (ng) and number of individuals (n), are very low and fixed for all cases ($ng = 150$, $n = 100$). MAGO is a real-valued evolutionary algorithm, so that the representation of the individual is a vector containing the controller parameters. The parameters are positive values in a continuous domain (Table 5).

The fitness function is in Eq. (9). The error is calculated as the difference between the system output and the reference signal. The error is calculated for each point of time throughout the measurement horizon. MAGO does not use genetic operators as crossover or mutation. The adaptation of the population is based on moving $N1$ individuals to the best one with a Simplex Search, creating $N2$ individuals over the average location of the actual population and creating $N3$ individuals through a uniform distribution over the whole search space, as previously discussed.

3.5 Controller Parameters and Performance Indexes

The comparison between the PID controller parameters obtained with the traditional tuning rules and the MAGO algorithm are shown in Table 4. These values minimize the ITAE. Figure 2 illustrates the time response, in closed loop, for the plants given in Tables 2 and 3. Figure 3 illustrates the time response of the plants defined by Eq. (6), given in Table 3. For this mode of operation, in the literature review, no tuning rule has been found that could compute the PID controller parameters requiring only the parameters of the plant. However, with MAGO is possible to find controller parameters that minimize the ITAE, without additional information and regardless of the operating mode. The closed-loop system simulations from which the controller was tuned using the MAGO are presented.

4 Analysis of Results

The study of traditional tuning methods shows that despite the large amount of available tuning rules, there is no one that is effective for the solution of all control problems based on SISO systems. It is evident that a single tuning rule applies only to a small number of problems. A tendency to develop new methods for tuning PID controllers [5–7, 32] has been noticed. The most recent are focused on controller's parameter calculation achieving a desired performance, where this index is one of

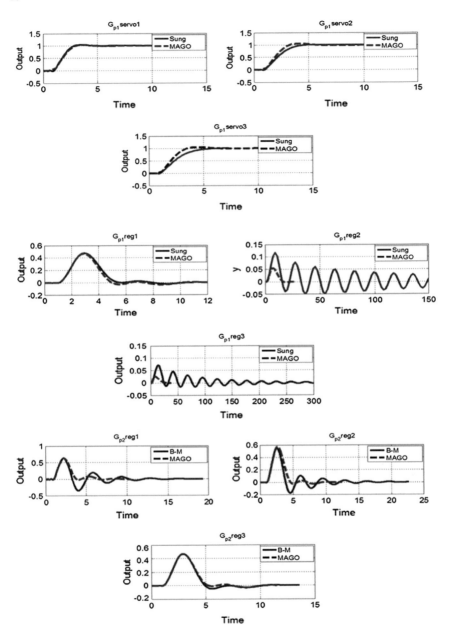

Fig. 2 Time response of the plants given by Eqs. (5) and (6), operating as servomechanism and regulator

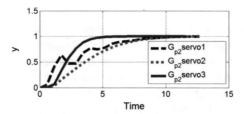

Fig. 3 Response to step change in the input of the plant 2, Eq. (6), as servomechanism (MAGO only)

those mentioned before (IAE, ITAE). Table 4 shows the results when tuning PID controllers for different plant models based on Eqs. (5) and (6). The parameters obtained minimize the ITAE criterion. In the case of plants based on the model of Eq. (5), when the system operates as servomechanism, the tuning rules used are those proposed by Sung. Obtaining an ITAE close to 3, the response behavior of the system is a smooth one, free of oscillations (Fig. 2). For the system operating as a regulator the rules by Sung are employed. In this case the ITAE value is considerably higher for plants Gp1_servo3 and Gp1_servo2, and the system presents oscillations. From this result, it has to be concluded that the rules proposed by Sung are a good choice for the system operating as a servomechanism; while for the case where the system operates as a regulator the use of these rules should be reconsidered.

On another hand, in the case of plants operating as regulators, whose model is given by the Eq. (6), the rules proposed by Bohl and McAvoy were used to calculate the controller parameters. The results for this experiment are reported in Table 4.

The response of the closed loop system is smooth using the parameters found by this method. The value for the ITAE performance index, in all cases, is below 10. Due to the features that the control problem has, where the objective is to minimize a function by a suitable combination of controller parameters which can be expressed as a function of cost, the solution is presented as an optimization problem. The algorithm MAGO is used to calculate the controller parameters seeking to minimize the ITAE.

The results, reported in Table 4, are compared with those obtained by the traditional tuning rules.

The results obtained by MAGO were very satisfactory for all cases. The ITAE performance index is low when the controller parameters are calculated by the MAGO, whatever the plant is represented by Eqs. (5) or (6), and for the two modes of operation, servo and regulator. Additional to the above, the responses of closed loop systems where the controller parameters are obtained using the MAGO could be observed in Fig. 2. These responses are softer and exhibit less oscillation with respect to the response where controllers are calculated with traditional methods. It can be appreciate in the Sung case as regulator, that the addressed problem has a big variability. Table 4 also reports the results obtained for the plant based on Eq. (6).

For this case no comparative data are available, because the only traditional tuning rule found that minimizes the performance index ITAE and requires no additional system information is proposed by Hassan (Table 1). However, in the experiments with this tuning rule it was not possible to obtain convergence to a real value of the parameters of the controller and thus it was not possible to calculate the ITAE. Whereas with MAGO, requiring only the minimum information of the model, it was possible to find the controller parameters reaching an acceptable answer, because in a finite time less than the open-loop system settling time the reference value is achieved, Fig. 3.

5 Conclusion

A method for optimal tuning of PID controllers through the evolutionary algorithm MAGO has been successfully developed and implemented. The process resolves the controller tuning as an optimization problem. MAGO calculates the parameters of PID controllers minimizing the ITAE performance index, and penalizing the error between the reference value and the output of the plant. It must be noted that the PID controller tuning was made for SOSPD, without additional knowledge of the plant.

The results showed that MAGO, operating on servo and regulator modes, gets a better overall performance comparing to traditional methods [20–22]. Each of these traditional methods is restricted to certain values on the behavior of the plant and is limited to an only one type of operation. The solution obtained with the evolutionary approach cover all these restrictions and extends the maximum and minimum between them. Finally, it should be noted that the MAGO successful results are obtained regardless of, both, the plant or controller models used.

References

1. Niculescu, S.I., Verriest, E.I., Dugard, L., Dion, J.D.: Stability and robust stability of time-delay systems: a guided tour. Lecture notes in control and information sciences. In: Stability and Control of Time-Delay Systems, vol. 228, pp. 1–71. Springer, London (1998)
2. Desanti, J.: Robustness of tuning methods of based on models of first-order plus dead time PI and PID controllers (in Spanish). Universidad de Costa Rica, Escuela de Ingeniería Eléctrica (2004)
3. O'Dwyer, A.: Handbook of PI and PID controller tuning rules, vol. 2. Imperial College Press, London (2009)
4. Mora, J.: Performance and robustness of the methods based on second order models plus dead time tuning PID controllers. (in Spanish). Escuela de Ingeniería Eléctrica, Universidad de Costa Rica. (2004)
5. Solera Saborío, E.: PI/PID controller tuning with IAE and ITAE criteria for double pole plants. (in Spanish). Escuela de Ingeniería Eléctrica, Universidad de Costa Rica. (2005)

6. Liu, G.P., Daley, S.: Optimal-tuning PID control for industrial systems. Control Eng. Pract. **9**(11), 1185–1194 (2001)
7. Tavakoli, S., Griffin, I., Fleming, P.J.: Multi-objective optimization approach to the PI tuning problem. IEEE C. Evol. Computat. 3165–3171 (2007)
8. Chang, W.D., Yan, J.J.: Optimum setting of PID controllers based on using evolutionary programming algorithm. J. Chin. Inst. Eng. **27**(3), 439–442 (2004)
9. Junli, L., Jianlin, M., Guanghui, Z.: Evolutionary algorithms based parameters tuning of PID controller. Chinese Control and Decision Conference (CCDC), pp. 416–420 (2011)
10. Saad, M.S., Jamaluddin, H., Darus, I.Z.M.: PID controller tuning using evolutionary algorithms. WSEAS Trans. Syst. Contr. **4** (2012)
11. Saad, M.S., Jamaluddin, H., Darus, I.Z.M.: Implementation of PID controller tuning using differential evolution and genetic algorithms. Int. J. Innov. Comput. I. Control. **8**(11), 7761–7779 (2012)
12. Hernández-Riveros, J., Arboleda-Gómez, A.: Multi-criteria decision and multi-objective optimization for constructing and selecting models for systems identification. WIT Trans. Model Sim. **55** (2013)
13. Michail, K., Zolotas, A.C., Goodall, R.M., Whidborne, J.F.: Optimised configuration of sensors for fault tolerant control of an electro-magnetic suspension system. Int. J. Syst. Sci. **43**(10), 1785–1804 (2012)
14. Whitley, D.: An overview of evolutionary algorithms: practical issues and common pitfalls. Inform. Softw. Tech. **43**(14) (2001)
15. Lozano, J.A., Larrañaga, P., Inza, I., Bengoetxea, E. (eds.): Towards a new evolutionary computation: advances on estimation of distribution algorithms. In: Studies in Fuzziness and Soft Computing, vol. 192. Springer, Heidelberg (2006)
16. Hernández, J.A., Ospina, J.D.: A multi dynamics algorithm for global optimization. Math. Comput. Model. **52**(7), 1271–1278 (2010)
17. Hernández-Riveros, J.-A., Villada-Cano, D.: Sensitivity analysis of an autonomous evolutionary algorithm. Lecture Notes in Computer Science. In: Advances in Artificial Intelligence, vol. 763, pp. 271–280. Springer, Heidelberg (2012)
18. Yu, X., Gen, M.: Introduction to evolutionary algorithms. Springer, London (2010)
19. Åström, K.J., Hägglund, T.: PID controllers: theory, design and tuning. Instrument Society of America (1995)
20. Bohl, A.H., McAvoy, T.J.: Linear feedback vs. time optima control. II. The regulator problem. Ind. Eng. Chem. Proc. Dd. **15**(1), 30–33 (1976)
21. Hassan, G.A.: Computer-aided tuning of analog and digital controllers. Control Comput. **21** (1), 1–6 (1993)
22. Sung, S.W.: Lee, O.J., Lee, I.B., Yi, S.H.: Automatic tuning of PID controller using second-order plus time delay model. J. Chem. Eng. Jpn. **29**(6), 990–999 (1996)
23. Lagunas, J.: Tuning of PID controllers using a multi-objective genetic algorithm, (NSGA-II). (in Spanish) (Doctoral dissertation) Departamento de Control Automático. Centro de Investigación y de Estudios Avanzados, México (2004)
24. Åström, K.J., Hägglund, T.: Benchmark systems for PID control. IFAC Workshop on Digital Control. Spain. 181–182 (2000)
25. Fleming, P.J., Purshouse, R.C.: Evolutionary algorithms in control systems engineering: a survey. Control Eng. Pract. **10**(11), 1223–1241 (2002)
26. Li, Y., Ang, K.H., Chong, G.C.: PID control system analysis and design. IEEE Contr. Syst. Mag. **26**(1), 32–41 (2006)
27. Vinter, R.: Optimal Control. Systems & Control: Foundations & Applications. Springer, London (2000)
28. Montgomery, D.C.: Introduction to statistical quality control. Wiley, New York (2008)
29. Skolicki, Z., De Jong, K.: The influence of migration sizes and intervals on island models. In: Proceedings of the conference on Genetic and evolutionary computation ACM, pp. 1295–1302 (2005)

30. Hansen, N.: Towards a new evolutionary computation: the CMA evolution strategy: a comparing review. In: Studies in Fuzziness and Soft Computing, vol. 192. Springer, Heidelberg (2006)
31. Herreros, A., Baeyens, E., Perán, J.R.: Design of PID-type controllers using multiobjective genetic algorithms. ISA T. **41**(4), 457–472 (2002)
32. Iruthayarajan, M.W., Baskar, S.: Evolutionary algorithms based design of multivariable PID controller. Expert Syst. Appl. **36**(5) (2009)

Evolution of Graphs for Unsupervised Learning

Christina Chrysouli and Anastasios Tefas

Abstract In this paper, we propose a novel method which adopts evolutionary techniques so as to optimize a graph structure. The method that was developed has been applied in clustering problems, where spectral graph clustering technique has been used. In order to use evolutionary algorithms initial population has been created consisting of nearest neighbor graphs and variations of these graphs, which have been properly altered in order to form chromosomes. Since it was observed that initial population is crucial for the performance of the algorithm, several techniques have been considered for the creation. A fitness function was used in order to decide about the appropriateness of the chromosomes. The major advantage of our approach is that the algorithm is generic and can be used to all problems that are, or can be, modeled as graphs, such as dimensionality reduction and classification. Experiments have been conducted on a traditional dance dataset and other multidimensional datasets, providing encouraging results.

Keywords Spectral clustering · Similarity graphs · Evolutionary algorithms

1 Introduction

The aim of clustering is to discover the natural grouping of a dataset, such that similar samples are placed in the same group, while dissimilar samples are placed into different ones. Clustering has been used in order to solve a diversity of problems, including bioinformatics, data mining, image analysis, information retrieval etc. A detailed survey on clustering applications can be found in [1] and a more recent study in [2].

C. Chrysouli · A. Tefas (✉)
Department of Informatics, Aristotle University of Thessaloniki,
University Campus, 54124 Thessaloniki, Greece
e-mail: tefas@aiia.csd.auth.gr

C. Chrysouli
e-mail: chrysouli@aiia.csd.auth.gr

© Springer International Publishing Switzerland 2016
J.J. Merelo et al. (eds.), *Computational Intelligence*,
Studies in Computational Intelligence 620, DOI 10.1007/978-3-319-26393-9_2

Spectral graph clustering is widely used and have received a lot of attention nowadays, as it can be applied to a wide variety of practical problems, such as computer vision and speech analysis. Spectral graph clustering [3] refers to a class of graph techniques, that rely on eigenanalysis of the Laplacian matrix of a similarity graph, aiming to divide graph nodes in disjoint clusters. In all clustering techniques, and thus also in spectral clustering, nodes that originate from the same cluster should have high similarity values, whereas nodes from different clusters should have low similarity values. In [4] the authors summarize some of the applications of spectral graph clustering.

So far, some evolutionary-based approaches to the problem of clustering have been proposed throughout the years. In [5] the authors proposed a genetic algorithm in order to search for the cluster centers by minimizing a clustering metric, while in [6] authors aim to find the optimal partition of the data, using a genetic algorithm, without searching all possible partitions. A more detailed survey of evolutionary algorithms for clustering is presented in [7].

In the proposed approach, similarity graphs are evolved, which have been transformed properly in order to play the role of the chromosomes in the employed genetic algorithm [8]. In order to use evolutionary algorithms we construct the initial population with the aid of k-nearest neighbor graphs which, then, are transformed to one-dimensional binary strings and undergo genetic operators.

The remainder of this paper is organized as follows. In Sect. 2, the problem that we attempt to solve is stated and some general aspects that concern the algorithm are discussed, including similarity graph construction, and spectral clustering issues. Section 3, presents the proposed evolutionary algorithm in detail. In Sect. 4, experimental results of the algorithm are described. Finally, in Sect. 5, conclusions are drawn and future work is discussed.

2 Problem Statement

Clustering is the process of partitioning a usually large dataset into groups, according to a similarity (or dissimilarity) measure. The goal is to place samples that have a small distance from each another, to the same cluster, whereas samples that are at a large distance from each another are placed to different clusters. Clustering is usually not a trivial task, as the only information we have about the data, is the data itself. In order to obtain some information about the structure of the data, we usually construct similarity matrices.

2.1 Similarity Functions and Similarity Graphs

Similarities of data samples can be represented as a similarity graph $G = (V, E)$, where V, E represent vertices (or nodes) and edges of the graph, respectively. If we assume that each vertex v_i represents a data sample, then two nodes v_i, v_j are connected if the similarity $s_{i,j}$ between them is positive or larger than a threshold, and the edge is weighted by $s_{i,j}$. The problem of clustering may now be reformulated as finding a partition of the graph such that the weights within a cluster have high values, whereas weights between different clusters have low values.

Before constructing a similarity graph, we need to define a similarity function on the data. The most common similarity function $\mathbf{S}$ is the Gaussian similarity function (heat kernel). Heat kernel between two graph nodes is defined as:

$$\mathbf{S} = \mathbf{h}_{i,j} = \mathbf{exp}\left(-\frac{\left\|\mathbf{v_i} - \mathbf{v_j}\right\|^2}{\sigma^2}\right),\tag{1}$$

where σ is a parameter that defines the width of the neighborhood.

Generally, the most common choice of similarity graphs are k-nearest neighbor graphs (to be called k-nn graphs) because of their simplicity as well as their sparsity. The aim of a k-nn graph $\mathbf{A}$ is to connect node v_i with node v_j if v_j is among the k nearest neighbors of v_i. This results in a directed graph which is easily transformed to an undirected by simply ignoring the directions of the edges. In the proposed method, an undirected graph was used, in order to construct the similarity graph.

However, it is well known that spectral clustering is very sensitive to the choice of the similarity graph that is used for constructing the Laplacian [9]. Indeed, selecting a fixed k parameter for the k-nn graph is very difficult and different values lead to dramatically different clusterings. Optimizing the clustering over the graph structure is not a trivial task, since the clustering criteria are not differentiable with respect to the graph structure. Thus, we propose in this paper to use evolutionary algorithms in order to optimize specific clustering criteria, that are considered as fitness functions, with respect to the underlying graph, which is transformed to a chromosome solution.

2.2 Spectral Graph Clustering

Spectral graph clustering [3], refers to a class of graph techniques, which rely on the eigenanalysis of a matrix, in order to partition graph nodes in disjoint clusters and is commonly used in many clustering applications [4].

Let $\mathbf{D}$ be a diagonal $N \times N$ matrix having the sum $d_{ii} = \sum_j W_{i,j}$ on its main diagonal. Then, the generalized eigenvalue problem is defined as:

$$(\mathbf{D} - \mathbf{W})\mathbf{v} = \lambda\mathbf{Dv},\tag{2}$$

where $\mathbf{W}$ is the adjacency matrix, and $\mathbf{v}$, λ are the eigenvectors and eigenvalues respectively.

$$
\begin{array}{c}
\mathbf{S} \\
\begin{bmatrix}
1 & 0.1 & 0.4 & 0.6 & 0.8 & 0.7 \\
0.1 & 1 & 0.5 & 0.8 & 0.1 & 0.4 \\
0.4 & 0.5 & 1 & 0.6 & 0.9 & 0.5 \\
0.6 & 0.8 & 0.6 & 1 & 0.6 & 0.9 \\
0.8 & 0.1 & 0.9 & 0.6 & 1 & 0.2 \\
0.7 & 0.4 & 0.5 & 0.9 & 0.2 & 1
\end{bmatrix}
\end{array}
\odot
\begin{array}{c}
\mathbf{A} \\
\begin{bmatrix}
1 & 0 & 0 & 0 & 1 & 0 \\
0 & 1 & 0 & 1 & 0 & 0 \\
0 & 0 & 1 & 1 & 1 & 0 \\
0 & 1 & 1 & 1 & 0 & 1 \\
1 & 0 & 1 & 0 & 1 & 0 \\
0 & 0 & 0 & 1 & 0 & 1
\end{bmatrix}
\end{array}
=
\begin{array}{c}
\mathbf{W} \\
\begin{bmatrix}
1 & 0 & 0 & 0 & 0.8 & 0 \\
0 & 1 & 0 & 0.8 & 0 & 0 \\
0 & 0 & 1 & 0.6 & 0.9 & 0 \\
0 & 0.8 & 0.6 & 1 & 0 & 0.9 \\
0.8 & 0 & 0.9 & 0 & 1 & 0 \\
0 & 0 & 0 & 0.9 & 0 & 1
\end{bmatrix}
\end{array}
$$

Fig. 1 The **S** matrix represents the full similarity matrix constructed using (1). The **A** matrix represents a k-nn graph, which has undergone genetic operators. The $\odot$ operator performs element-wise multiplication, resulting in a sparse matrix **W**, which only contains elements in places where **A** matrix contains elements

Although many variations of graph Laplacians exist [9], we focus on the normalized graph Laplacian **L** [10] defined as:

$$\mathbf{L} = \mathbf{I} - \mathbf{D}^{-1/2}\mathbf{W}\mathbf{D}^{-1/2} \tag{3}$$

where **W** is the adjacency matrix, with $w_{i,j} = w_{j,i} \geq 0$, **D** is the degree matrix and **I** is the identity matrix. The smallest eigenvalue of **L** is 0, which corresponds to the eigenvector $\mathbf{D}^{-1/2}\mathbf{1}$. The **L** matrix is always positive semi-definite and has n non-negative real-valued eigenvalues $\lambda_1 \leq \cdots \leq \lambda_n$. The computational cost of spectral clustering algorithms is quite low when matrices are sparse. Luckily, we make use of k-nn graphs which are in fact sparse.

In the proposed method, we perform eigenanalysis on **L** matrix, where **W** is defined as:

$$\mathbf{W} = \mathbf{S} \odot \mathbf{A}, \tag{4}$$

S represents the full similarity matrix obtained using (1) and **A** represents an undirected k-nn matrix, which is a sparse matrix. The $\odot$ operator performs element-wise multiplication. This process results in a sparse matrix **W**, only containing elements in places where **A** matrix contains elements. An example of the $\odot$ operator is illustrated in Fig. 1. Eigenvalues are always ordered increasingly, respecting multiplicities, and the first k eigenvectors correspond to the k smallest eigenvalues. Once the eigenanalysis has been performed and the new representation of the data has been obtained, the k-means algorithm is used in order to attach a cluster to every data sample.

3 The Proposed Algorithm

In order to partition a dataset into clusters, spectral graph clustering has been applied on evolving k-nn similarity graphs. In more detail, we evolve a number of k-nn similarity graphs with the aid of a genetic algorithm, in order to optimize the structure of the graph, by optimizing a clustering criterion. In this paper, clustering criteria were employed as fitness functions. Moreover, k-nn similarity graphs are transformed properly into chromosome solutions, in order to be used in the genetic algorithm.

Let J be a clustering criterion that depends on the similarity graph $\mathbf{W}$. However, the optimization problem is not convex and moreover the fitness function is not differentiable with respect to $\mathbf{W}$. Since $\mathbf{S}$ is considered constant after selecting a specific similarity function and through the definition of $\mathbf{W}$ in (4), the optimization problem is defined as:

$$\underset{\mathbf{A}}{optimize}\ J(\mathbf{A}), \tag{5}$$

where $\mathbf{A}_{i,j} \in 0, 1$ is a k-nn graph.

3.1 Construction of Initial Population

In our algorithm, we use the sparse matrices that originate from k-nn graphs, resulting in an initial population that consists of matrices with binary elements. In this method, a Gaussian function has been employed as a similarity measure, in order to obtain the similarity matrix $\mathbf{S}$, which is calculated pairwise for all the data in a database of our choice, using (1). Our experiments showed that the value of σ has a decisive role to the performance of the algorithm, thus, several, arbitrary rules exist; in the proposed method, we have used multiples of the data diameter.

First, we calculate k-nearest neighbor matrices $\mathbf{A}$, with $k = 3, \ldots, 8$, which constitute the backbone of the initial population. Next step is to enrich the population with nearly k-nearest neighbor matrices. In order to achieve that, we alter the k-nearest neighbor matrices that have already been calculated, by converting a small proportion of 0's, from $\mathbf{A}$ matrices, to 1's and vice versa. This process guarantees that the proportion of 1's and 0's will remain the same in the new matrix. It is important not to alter the k-nn graphs completely, so as to keep all the good properties. Finally, a small proportion of completely random matrices are added, in order to increase the population diversity, in which the number of 1's are equal to the number of 1's that a 5-nn graph would have.

From the various experiments conducted, we have concluded that the selection of the parameter k of the nearest neighbor graphs is crucial to the clustering results, as illustrated in Fig. 2. Figure 2a presents a dataset that consists of two classes with each one having a different color. Figure 2b, c represent the clustering results when a 3 and a 5-nearest neighbor graph were used, respectively.

Before proceeding to the algorithm, we must define the way that the k-nn matrices, and variants of these matrices, in the initial population are transformed into chromosomes, thus, we need to define how a square matrix becomes a one-dimensional vector. As the k-nn graphs $\mathbf{A}$ are constructed in such a way to be symmetrical, we may only keep the elements of the upper triangular matrix, with no loss of information. Then, the remaining elements are accessed in rows, forming the one-dimensional vector (Fig. 3).

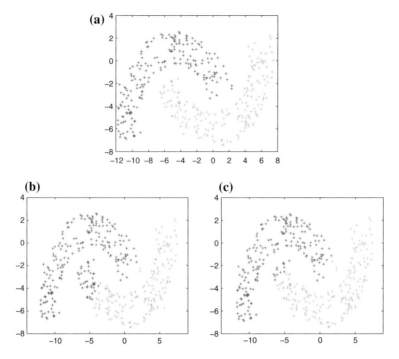

Fig. 2 The effect of k-nearest neighbor graphs in clustering. In Fig. 2a the two classes of the dataset are presented. Figure 2b, c represent the clustering results when a 3 and a 5-nearest neighbor graph were used, respectively. Notice the difference in clustering results especially when the data are close to both classes

$$
\begin{bmatrix}
1 & 0 & 0 & 0 & 1 & 0 \\
0 & 1 & 0 & 1 & 0 & 0 \\
0 & 0 & 1 & 1 & 1 & 0 \\
0 & 1 & 1 & 1 & 0 & 1 \\
1 & 0 & 1 & 0 & 1 & 0 \\
0 & 0 & 0 & 1 & 0 & 1
\end{bmatrix}
\rightarrow
\begin{bmatrix}
 & 0 & 0 & 0 & 1 & 0 \\
 & & 0 & 1 & 0 & 0 \\
 & & & 1 & 1 & 0 \\
 & & & & 0 & 1 \\
 & & & & & 0 \\
 & & & & &
\end{bmatrix}
\rightarrow
\begin{bmatrix} 000100100110010 \end{bmatrix}
$$

Fig. 3 The way a k-nn graph **A** is transformed into a, one-dimensional vector, chromosome. We only keep the elements of the *upper* diagonal, as the matrix is constructed to be symmetric, resulting in a matrix like the one in the *middle*. Then, this matrix is accessed horizontally, in order to obtain the desirable result, the chromosome

3.2 Optimization of the Solutions

The novelty of the proposed algorithm is based on the way that we select to optimize the solutions of the problem, by optimizing a clustering criterion J, as previously defined in (5). Clustering criteria are divided into two main categories, internal and external criteria. The calculation of internal criteria implies that we have no prior knowledge about the data and we can only depend on quantities and features inherent to the dataset, whereas calculation of external criteria implies that we have some knowledge about the dataset in advance (i.e. ground truth).

In the recent literature, many different clustering criteria [11] have been proposed. Some of the most common internal criteria are Calinski-Harabasz index [12], Davies-Bouldin index [13] and Dunn's index [14], whereas some external criteria are purity [15], F-measure [16], a measure based on Hungarian algorithm [17] and normalized mutual information [18]. Some of the aforementioned internal criteria have been used both for optimization and evaluating the performance of the algorithm, whereas the external criteria only for evaluation.

As the value of such criteria cannot be optimized, without the use of derivatives, we have employed evolutionary techniques in order to solve this problem. The optimization is performed by altering the chromosomes or, else, by altering the k-nn similarity matrices $\mathbf{A}$ as in (2).

3.3 The Genetic Cycle

Evolutionary algorithms solve problems based on operators inspired from biology. The first step of the genetic algorithm is to select the chromosomes which will undergo the crossover operator. For this purpose, a roulette wheel method has been employed [19], where a probability is associated with each chromosome, based on the value of the fitness function: the higher the value, the higher the probability to be selected. The probability p_i of the ith chromosome to be selected, if f_i is its fitness value is defined as:

$$p_i = \frac{f_i}{\Sigma_{j=1}^{N} f_j}.$$ (6)

Next, we combine the selected chromosomes, based on the crossover rate which was set to 0.7, in order to produce new ones. In the proposed algorithm, a single crossover point is randomly selected for every set of chromosomes and the subsequences that are formed are exchanged respectively. Then, we randomly choose a small proportion of the chromosomes, based on the mutation rate which was set to 0.4, to undergo mutation, that is the random change of some elements of a chromosome. In order to guarantee that the newly produced chromosomes will not have been altered too much we perform mutation by converting 1 % of 0's to 1's and vice versa.

After the application of genetic operations to the chromosomes, the new generation has been formed. In order to perform spectral clustering (Sect. 2.2), we need to reconstruct the k-nearest neighbor matrix $\mathbf{A}$, which will consist of binary digits, from the one-dimensional vector chromosome. Then we apply the similarity matrix $\mathbf{S}$ on $\mathbf{A}$ using the $\odot$ operator, in order to obtain the $\mathbf{W}$ as illustrated in Fig. 1. Spectral clustering [10] may now be performed on $\mathbf{L}$ as in (3).

The next step is to calculate the fitness values of all the newly produced chromosomes, and place them along with the parent-chromosomes. Then, elitism is performed: we sort all chromosomes, with the fittest being on the top, and we keep only those chromosomes with the highest fitness value, so as the number of the chromosomes kept to remain unchanged after every generation.

The proposed algorithm terminates when a maximum of 50 generations has been reached, or when the optimized criterion has not been altered for 5 consecutive generations.

4 Experiments

In order to evaluate the proposed algorithm, we have conducted several experiments using 3 different datasets and exploiting several input parameters. The characteristics of the datasets used, are described in Table 1.

Datasets "Movie 1" and "Movie 2" consist mainly of facial images originate from movies, detected using a face detector. In the experiments the images were scaled, in order to have the same size, considering all the detected facial images of the movie clip and using a mean bounding box, from all bounding boxes that the face detector provided. A problem that might arise is that of anisotropic scaling: the images returned by the detector might have different height and width, which is problematic when scaling towards a mean bounding box, thus we calculate the bigger dimension of the bounding box and then we take the square box that equals this dimension centered to the original bounding box center. Lastly, the initial "Folk dances" dataset consists of videos of 5 different traditional dances: Lotzia, Capetan Loukas, Ramna, Stankena and Zablitsena with 180, 220, 220, 201 and 192 videos respectively, from which histograms were extracted according to [20]. An example of the dataset is illustrated in Fig. 4.

The size of the populations remained unchanged for all the experiments conducted and was set to 200 chromosomes. Every experiment was executed 3 times, so the results presented here are the average of these runs. We should highlight here that, in every experiment, only one clustering criterion c is being optimized. The values of the rest of the criteria are also calculated during every experiment only for evaluation reasons. In other words, the values of the rest of the criteria are not their best

Table 1 Datasets used

Dataset	Duration	Classes	Size of dataset	# of features
Movie 1	02 : 06 : 21	21	1,222	152×152
Movie 2	01 : 11 : 31	41	1,435	150×150
Folk dances	–	5	1012	1000

Fig. 4 An example of Ramna dance, from the "Folk dances" dataset

values as if they were being optimized themselves. Instead, their values depend on the clustering obtained by optimizing the criterion c. Moreover, the optimization of a single criterion does not necessarily mean that the rest of the criteria will also be improved, especially when the way in which the criteria are calculated differs a lot.

In tables presented here, we have attempted to summarize some of the results of the datasets. The results of the proposed method are represented under the label "best", while "5nn" represent the results of the clustering if the 5-nn graph would have been employed to the data. For Tables 2, 3 and 4 the criterion being optimized is highlighted. The σ parameter is the heat kernel parameter as in (1) and C is the

Table 2 Folk dances dataset. Optimizing Calinski-Harabasz criterion

σ	C	Calinski-Harabasz		Davies-Bouldin		NMI		Purity	
		Best	5nn	Best	5nn	Best	5nn	Best	5nn
0.45	5	**77.803**	40.665	**2.116**	3.317	**0.32**	0.255	**0.468**	0.434
0.9	5	**71.026**	38.309	**2.745**	3.252	**0.281**	0.271	**0.441**	0.434
1.8	5	**74.923**	43.649	**2.292**	3.013	**0.312**	0.291	**0.469**	0.463

Table 3 Movie 1. Optimizing Calinski-Harabasz criterion

σ	C	Calinski-Harabasz		Davies-Bouldin		Hungarian		Purity	
		Best	5nn	Best	5nn	Best	5nn	Best	5nn
5000	21	**161.239**	121.659	1.165	**1.162**	**20.922**	20.758	0.468	**0.475**
15000	21	**161.011**	123.922	1.208	**1.103**	**21.495**	21.167	0.462	**0.477**
20000	21	**149.195**	121.413	1.169	**1.072**	**21.113**	20.404	0.459	**0.475**

Table 4 Movie 2. Optimizing Calinski-Harabasz criterion

σ	C	Calinski-Harabasz		Davies-Bouldin		Hungarian		Purity	
		Best	5nn	Best	5nn	Best	5nn	Best	5nn
25	40	**81.917**	70.737	1.240	**1.204**	**15.889**	15.447	**0.400**	0.398
50	41	**76.269**	69.302	**1.144**	1.257	**16.353**	15.819	**0.410**	0.408
75	41	**78.449**	66.245	1.226	**1.200**	**16.121**	15.981	0.401	**0.402**
150	40	**82.090**	66.393	**1.183**	1.248	**16.167**	15.772	**0.403**	0.391

number of clusters. We observe that in almost all cases the external criteria agrees with the internal optimized criterion that the clustering that was performed did actually grouped the data better than if the 5-nn graph would have been employed. In most cases, the other internal criterion also agrees to this conclusion.

5 Conclusion

We have presented a novel algorithm that makes use of evolutionary algorithms in order to achieve good clustering results, with the aid of nearest neighbor graphs. It is important to remark that the algorithm is general and can be used to manipulate a wide variety of different problems, such as clustering and dimensionality reduction. The technique of using nearest neighbor graphs as initial population appears to yield satisfactory results, in terms of both internal and external criteria.

In the future, we aim to improve the proposed evolutionary algorithm, by optimizing even different criteria, or even use multiple of them in order to decide which chromosome is best. We shall also focus our efforts on creating an even better initial population, for example by including more than only random variations of the nearest neighbor graphs.

Acknowledgments This research has been co–financed by the European Union (European Social Fund—ESF) and Greek national funds through the Operation Program "Education and Lifelong Learning" of the National Strategic Reference Framework (NSRF)—Research Funding Program: THALIS–UOA–ERASITECHNIS MIS 375435.

References

1. Jain, A.K., Murty, M.N., Flynn, P.J.: Data clustering: a review. ACM Comput. Surv. (CSUR) 31 **3**, 264–323 (1999)
2. Jain, A.K.: Data clustering: 50 years beyond K means. Pattern Recognition Letters 31(8), 651–666 (2010)
3. Jordan, M.I., Bach, F.R.: Learning spectral clustering. In: Advances in Neural Information Processing Systems 16 (2003)

4. Satu S.E.: Graph clustering. Comput. Sci. Rev. 1 **1**, 27–64 (2007)
5. Maulik, U., Bandyopadhyay, S.: Genetic algorithm-based clustering technique. Pattern Recognit. **33**(9), 1455–1465 (2000)
6. Murthy, C.A., Chowdhury, N.: In search of optimal clusters using genetic algorithms. Pattern Recognit. Lett. **17**(8), 825–832 (1996)
7. Eduardo H.R., Campello, R.G.B., Freitas, A.A., De Carvalho, A.C.P.L.F.: A survey of evolutionary algorithms for clustering. IEEE Trans. Syst. Man Cybern. Part C Appl. Rev. **39**(2), 133–155 (2009)
8. John H.H.: Adaptation in natural and artificial systems: an introductory analysis with applications to biology, control, and artific. 211 (1992)
9. Von, U.L.: A tutorial on spectral clustering. Stat. Comput. **17**(4), 395–416 (2007)
10. Andrew Y.Ng., Jordan, M.I., Yair, W.: On spectral clustering: analysis and an algorithm. Adv. Neural Inf. Process. Syst. **2**, 849–856 (2002)
11. Vendramin, L., Campello, R.J.G.B., Hruschka, E.R.: On the Comparison of Relative Clustering Validity Criteria, In: SDM, pp. 733–744 (2009)
12. Caliński, T., Harabasz, J.: A dendrite method for cluster analysis. Commun. Stat. Theory Methods **3**(1), 1–27 (1974)
13. Davies, D.L., Bouldin, D.W.: A cluster separation measure. IEEE Trans. Pattern Anal. Mach. Intell. **2**, 224–227 (1979)
14. Dunn, J.C.: Well-separated clusters and optimal fuzzy partitions. J. Cybern. **4**(1), 95–104 (1974)
15. Ying, Z., Karypis, G.: Criterion functions for document clustering: experiments and analysis. Technical Report (2001)
16. Eissen, Z., Meyer, B.S.S., Wißbrock, F.: On cluster validity and the information need of users, pp. 216–221. ACTA Press (2003)
17. Munkres, J.: Algorithms for the assignment and transportation problems. J. Soc. Ind. Appl. Math. **5**(1), 32–38 (1957)
18. He, Z., Xu, X., Deng, S.: K-ANMI: a mutual information based clustering algorithm for categorical data (2005)
19. De Jong, K.A.: An analysis of the behavior of a class of genetic adaptive systems. PhD diss., PhD thesis, University of Michigan, Dissertation Abstracts International. **36**(10) (1975)
20. Iosifidis, A., Tefas, A., Ioannis, P.: Minimum class variance extreme learning machine for human action recognition. IEEE Trans. Circ. Syst. Video Technol. **23**(11), 1968–1979 (2013)

Sequence Analysis with Motif-Preserving Genetic Algorithm for Iterated Parrondo Games

Degang Wu and Kwok Yip Szeto

Abstract Comparison of simple genetic algorithm with motif preserving genetic algorithm is made for the sequence analysis of Parrondo games, which is an analogue to the flashing Brownian ratchet in non-equilibrium statistical physics. The original Parrondo game consists of two individual games: game A and game B. Here game A is a coin-tossing game with winning probability p_A slightly less than half, so that its persistent usage will be losing in the long run. Game B has two coins, and an integer parameter M. If the current cumulative capital (in discrete unit) is a multiple of M, an unfavorable coin with winning probability $p_b < 0.5$ is used, otherwise a favorable coin with winning probability $p_g > 0.5$ is used. Game B is also a losing game if played alone. Paradoxically, combination of game A and game B could lead to a winning game, either through random mixture, or deterministic switching. The resolution of this paradox can be made using Markov Chain analysis [1]. In this paper, we are interested in the analysis of finite deterministic switching sequences of N Parrondo game, so the number of possible sequences is 2^N. For small N, exhaustive search and backward induction have been applied successfully to short sequences. However, for long but finite deterministic sequence, the optimal ordering of games requires the use of combinatorial optimization techniques. Here we employ genetic algorithm to find the optimal game sequence. The structure found in short sequences using a problem-independent genetic algorithm, such as ABABB, is a motif. Using these motifs we invent motif-preserving point mutation operator and one-point crossover operator to exploit the structure of the optimal game sequences for large N. We show by numerical results the adapted motif-preserving genetic algorithm has great improvement in solution quality over simple genetic algorithm. The technique of motif preserving genetic algorithm can be applied to similar problem in sequence analysis once the condition of optimality is defined.

Keywords Genetic algorithm · Parrondo game · Optimization · Game theory

D. Wu (✉) · K.Y. Szeto
Department of Physics, The Hong Kong University of Science and Technology,
Clear Water Bay, Kowloon, Hong Kong
e-mail: dwuab@ust.hk

K.Y. Szeto
e-mail: phszeto@ust.hk

© Springer International Publishing Switzerland 2016
J.J. Merelo et al. (eds.), *Computational Intelligence*,
Studies in Computational Intelligence 620, DOI 10.1007/978-3-319-26393-9_3

1 Introduction

The Parrondo paradox of combining two losing games when played alone into a winning one is the invention inspired by the flashing ratchet [2]. In the original setting, the two games are game A and B, so that when the sequence of games contains only A or B will lead to loss, but a skillful combination, be it a random mixture, or a periodic one, can lead to positive gain. This paradox invented by Parrondo is resolved using Markov Chain analysis, and can also be described by discrete Fokker-Planck equation, thus a more rigorous relation between Parrondo game and Brownian ratchet was established [3, 4]. The simplest version of Parrondo game can be defined by four parameters, (p_A, p_g, p_b, M). Here one can imagine these two games are two slot machines installed in a casino. In both slot machines, the gambler will win one or lose one dollar in each game played. The gambler observes that if he continues playing on one slot machine, he will lose. On the other hand, he may end up winning if he once in a while changes his choice of slot machine. Here the parameter p_A is $0.5 - \epsilon$ is the winning probability of slot machine A and ϵ is a small number which can be interpreted as the commission charged by the casino. The other slot machine is rather tricky, as the casino will give a bad coin with winning probability $p_b < 0.5$ to the gambler if his cumulative capital in discrete unit is divisible by M. Otherwise the casino will give a good coin with winning probability $p_g > 0.5$ to the gambler though very likely he will still lose if he keeps playing on the same slot machine. The question one likes to resolve is the way the gambler will win most if he decides to play N games on these two slot machines, which he now called C and D. Of course, if he knows which slot machine gives him the game A, and the other giving him the game B and also the parameter M, then he can get the optimal winning sequence. For example, if the C slot machine gives him the slightly unfair coin p_A, and D slot machine gives him the tricky slot machine with $p_b = 0.1 - \epsilon$ and $p_g = 0.75 - \epsilon$, which are the parameters used most commonly in Parrondo paradox, then the gambler can win optimally by the following strategy: when his cumulative capital is divisible by M, then he plays on the C slot machine since his winning probability is $p_A = 0.5 - \epsilon > p_b = 0.1 - \epsilon$. When his cumulative capital is not divisible by M, then he plays on the D slot machine, since his winning probability is $p_g = 0.75 - \epsilon > p_A = 0.5 - \epsilon$. The problem facing the gambler is that he does not know which slot machine C stands for even if he knows the mathematical analysis of the Parrondo paradox. In this situation, what sequence of C and D should he plays? This is the problem of sequence analysis with the objective of maximizing the expected gain for the gambler after N games, when he does not know the nature of slot machine, albeit he knows the mathematics of Parrondo paradox.

This interesting problem in analyzing long but finite sequence of Parrondo games is a combinatorial optimization problem. For N games of A and B, the solution space is 2^N. For small N, exhaustive search and backward induction suffice. For example, with the above standard setting of parameters in Parrondo paradox,

($p_A = 0.5$, $p_g = 0.75$, $p_b = 0.1$, $M = 3$), and $N = 5$, the short sequence ABABB is optimal in terms of expected gain, though in actual realizations both CDCDD and DCDCC may be losing. The gambler can use CDCDD or DCDCC to try and with some luck he can also identify correctly the slot machines with the game A and B. However, for longer sequences of games, this may not be optimal and we may use genetic algorithm for the search of optimal gain sequence. There has been other techniques used, such as dynamical programming by Dinis [5]. It is worth noting that greedy algorithms do not work in finding the optimal game sequence for a long finite length. Multi-player versions of the games have been proposed, and they exhibit counterintuitive phenomena resembling those observed in game, control, and optimization theories or economics. Many researchers have found that greedy algorithms or strategies may lead to suboptimal or even losing solutions in these models [6–8]. Optimization problems of this type warrant the use of meta-heuristic algorithm such as genetic algorithms [9, 10].

The successful application of genetic algorithm has been demonstrated in fields such as biology [11–13], clusters [14–16] and glass transition [17] in condensed matter physics. In engineering, genetic algorithm has also been used with good results in problems such as cyclic-steam oil production optimization problem [18], speed control of brushless DC motor [19], airport scheduling [20], mobile robot motion control [21], modeling adaptive agents in stock markets [22, 23], and traveling salesman problem [24].

In this paper, we aim at developing a good version of genetic algorithm for sequence analysis for long sequence of games, making use of motifs found from short sequences. We start with Dinis algorithm for sequence of moderate length so as to gain insight of the structure of the optimal sequences, thereby collecting a database of good motifs. We then propose motif-preserving evolution operators (mutation and one-point crossover operators), coupled with an appropriate motif-preserving population initialization procedure to find the optimal sequence for longer game sequence. By motif-preserving condition, we mean that we impose certain constraints on the patterns that are observed in the optimal sequences of moderate length in the search process for longer sequence. In order to assess the effect of the constraints on the quality of the solution, we first use simple genetic algorithm as a benchmark sampling method in order to justify the adaptive approach we use later. Our paper is organized as follows. We first define the Parrondo game and the Dinis algorithm that reveals motifs found in the optimal sequence of moderate length in Sect. 2. We then propose the motif-preserving genetic operators: mutation and one-point crossover, as well as the motif-preserving population initialization procedure in Sect. 3. We summarize the design of the numerical experiments and their results in Sects. 4 and 5. In Sect. 6, we conclude with some comments on possible application of our motif preserving genetic algorithm for analysis of other sequences.

2 Parrondo Games

The original Parrondo game consists of two individual coin tossing games, namely
game A and game B. Game A has only one coin, whose winning probability is
$p_A = 1/2 - \epsilon$, where ϵ is a small and positive number. Let $X(t)$ be the cumulative
capital of the player at time t, an integer. If the player keeps playing game A, the
average capital satisfies

$$\langle X(t+1) \rangle = \langle X(t) \rangle + 2p_A + 1 \tag{1}$$

where $\langle \cdot \rangle$ is understood as ensemble average. Game B has two coins, one "good"
coin and one "bad" coin. Game B has an integer parameter M. If $X(t)$ is a multiple
of M, then $X(t+1)$ is determined by the "bad" coin with winning probability $p_b =
1/10 - \epsilon$, otherwise the "good" coin with winning probability $p_g = 3/4 - \epsilon$ is used.
Similar to game A, if the player keeps playing game B only, the average capital
satisfies

$$\langle X(t+1) \rangle = \langle X(t) \rangle + 2[\pi_0(t)p_b + (1-\pi_0(t))p_g] - 1 \tag{2}$$

which explicitly depends on $\pi_0(t)$, the probability that the capital $X(t) = 0 \mod M$.
Harmer and Abbott [25] showed that game B is a losing game when $p_b = 1/10 - \epsilon$,
$p_g = 3/4 - \epsilon$ and $M = 3$, with positive ϵ. In this paper, we only discuss the case
when $M = 3$. There is a recent article discussing the phase diagram of the more
complex situation where the games are two B games with different M [1]. Such
extended Parrondo game with multiple M exhibits interesting phenomena such as
weak and strong Parrondo effects when the sequence is infinite.

If we model the Parrondo game as a discrete-time Markov chain as in [26], we
can define the probability vector $\boldsymbol{\pi}(t) \equiv (\pi_0(t), \pi_1(t), \pi_2(t))^T$ (for $M = 3$). Accordingly, the transition matrix for game A is

$$\Pi_A = \begin{bmatrix} 0 & 1-p_A & p_A \\ p_A & 0 & 1-p_A \\ 1-p_A & p_A & 0 \end{bmatrix} \tag{3}$$

such that the time evolution equation is $\boldsymbol{\pi}(t+1) = \Pi_A \boldsymbol{\pi}(t)$. Similarly, the transition
matrix for game B is

$$\Pi_B = \begin{bmatrix} 0 & 1-p_g & p_g \\ p_b & 0 & 1-p_g \\ 1-p_b & p_g & 0 \end{bmatrix} \tag{4}$$

Parrondo game can be played according to a deterministic finite game sequence such
as ABABB, so at the end of the sequence, the probability vector $\boldsymbol{\pi}(5)$ is

$$\boldsymbol{\pi}(5) = \Pi_B \Pi_B \Pi_A \Pi_B \Pi_A \boldsymbol{\pi}(0). \tag{5}$$

Algorithm 1: Calculate the expected return for game sequence $\{S_i\}$.

---

function EXPECTED RETURN($\{S_i\}$)

   $g \leftarrow 0$

   $\pi = [1, 0, 0]^T$

   for $t \leftarrow 1, N$ **do**

      if $S_t = A$ **then**

         $g \leftarrow g + \pi^T [2p_A - 1, 2p_A - 1, 2p_A - 1]^T$

         $\pi \leftarrow \Pi_A \pi$

      else

         $g \leftarrow g + \pi^T [2p_b - 1, 2p_g - 1, 2p_g - 1]^T$

         $\pi \leftarrow \Pi_B \pi$

      end if

   end for

   return g

end function

Parrondo game has a seemingly paradoxical property that while game A and B are losing when they are played individually, the stochastic mixture of game A and B, or playing according to a deterministic sequence may lead to a winning combined game for small positive value of ϵ. For the detailed analysis of this paradox, please refer to [26]. For a finite game sequence with length N, the expected return at the end of the game sequence can be computed by Algorithm 1. Our task is to find the finite game sequence that has maximum cumulative gain. The expected return per game in the stationary state for a periodic sequence with length N can be computed using Algorithm 1 with two minor modifications: the initial value of π should be the equilibrium distribution of the transition matrix $\Pi_{\alpha(1)} \Pi_{\alpha(2)} \cdots \Pi_{\alpha(N)}$ and the final value of g should be divided by N.

Sequences up to period 12 have been studied using symbolic manipulators and exhaustive search [27], and the periodic sequence ABABB, or any of its permutations, has come up as the best in the sense that it provides the highest returns in the stationary state. However, exhaustive search for optimal sequences of finite length N is not feasible for large N. Using dynamic programming, Dinis discovered that optimal sequences with finite N "consist of several repetitions of the ABABB motif flanked by brief pieces of other sequences. Dinis [5] For example, the optimal sequence for $N = 20$ with initial condition $X(t) = 0$ and $\epsilon = 0$ is AB ABABBABABBABABB ABB. In fact, the structure of the optimal sequences is more specific: they are strings of AB and ABB. From these results, we have the following definition:

Definition 1 A game sequences is said to have a special structural-property if it is made of AB and/or ABB substrings exclusively.

Straightforward implementation of Dinis algorithm requires storage space that scales as N, in order to store all the numerical results from all intermediate steps. Moreover, the algorithm approximates two-dimensional plane with a discrete square grid (In [5], a 2000×2000 grid was used for each step). The error in the expected payoff due to grid approximation will accumulate across time and the total error in the expected payoff at the end of the sequence is at least a linear function of N. The effect

of this error to the correctness of the optimal sequence has not been investigated.
The approximation therefore constraints the computation of optimal sequences for
large N. This suggests that heuristic algorithms such as GA that demands much
less space and does not require approximation in the expected payoff are useful
for optimal sequences with long length. Moreover, by definition, Dinis' algorithm
does not provide any information of the suboptimal sequence. In contrast, while
searching for the optimal solution, GA also efficiently samples many suboptimal
solutions, which provide insights in the relation between the structure of a sequence
and its performance.

3 Motif-Preserving Genetic Algorithm

3.1 Representation

If we map game A to 0 and map game B to 1, a game sequence with length N can
be mapped to a binary string with the same length, and the order of the binary values corresponds to the order the games are played. The objective function is given
by Algorithm 1. For example, the game sequence ABABB is encoded as 01011. For
game sequences that are strings of AB and ABB, an *auxiliary representation* is useful:
0 for AB and 1 for ABB. For example, game sequence ABABB is encoded as 01 in
this auxiliary representation. However, game sequences under this auxiliary representation vary in length from sequence to sequence. To distinguish these two kinds
of representations, we call the one that maps game A to 0 the original representation.

3.2 Simple Genetic Algorithm

A simple genetic algorithm is used as the reference.

In the initialization, each locus of each chromosome is set to 1 or 0 with equal
probability. The new population P' is generated from the last one using deterministic
tournament: from previous generation, randomly draw k chromosomes with equal
probability, and select the one with the highest fitness value. This process is repeated
until the new population has the same number of chromosomes as the previous
population. The one-point crossover operator is denoted by XOVER(), and its rate is
denoted by p_c. The mutation rate per bit is denoted by p_m.

Algorithm 2: "Simple" Genetic Algorithm.

---

procedure SGA
   INIT(P)   ▷ P for the population
   repeat
      $P \leftarrow$ DET TOURNAMENT(P, k)
      for $i = 0$, SIZE(P)/2-1 **do**
         if RAND()$< p_c$ **then**
            XOVER($P(2i), P(2i + 1)$)
         end if
      end for
      for $i = 0$, SIZE(P)-1, $j = 0$, $N - 1$ **do**
         if RAND()$< p_m$ **then**
            FLIP($P(i, j)$)
         end if
      end for
      EVALUATE(P)   ▷ update the fitness values
   until Stopping criteria satisfied
end procedure

3.3 Motif-Preserving Mutation Operator

If a chromosome has special structural property, we can apply a mutation operator that preserves this property. We propose a motif-preserving mutation operator in Algorithm 3.

Algorithm 3: Motif-Preserving Mutation Operator.

---

function SPMUTATE(c)   ▷ c for a chromosome
   $c' \leftarrow$ ORIGINTOAUX(c)
   for $j = 0$, SIZE(c')-1 **do**
      if RAND()$< p_m$ **then**
         FLIP($c'(j)$)   ▷ $c'(j)$ is the j-th locus of c'
         $c \leftarrow$ AUXTOORIGINAL(c')
      end if
   end for
   return c
end function

However, when we use Algorithm 3, we must check if the sequence after mutation has length N. We therefore need an algorithm for chromosome repair, using first a conversion function AUXTOORIGINAL(c'). This function converts c' in the auxiliary representation back to a chromosome in the original representation, which is not always possible because the converted chromosome in original form may have a length greater or less than N. If SIZE(c) $> N$, we then pop 01 or 011 from the back of c until SIZE(c) $< N$. We append 01 or 011 to the back of c such that SIZE(c) $= N$.

3.4 Motif-Preserving Crossover Operator

If a chromosome has special structural property, we can apply an adapted one-point crossover operator that preserves this propose. The motif-preserving one-point crossover operator is shown in Algorithm 4.

Algorithm 4: Motif-preserving one-point crossover operator.

---

procedure SPXOVER(c_1, c_2)
    $c_1' \leftarrow$ AUXREPRE(c_1)
    $c_2' \leftarrow$ AUXREPRE(c_2)
    $xPt \leftarrow$ CHOOSEXPT(c_1', c_2')
    XOVER(c_1', c_2', xPt)
    $c_1' \leftarrow$ FIXCHROMOSOME(c_1')
    $c_2' \leftarrow$ FIXCHROMOSOME(c_2')
    $c_1 \leftarrow$ AUXTOORIGINAL(c_1')
    $c_2 \leftarrow$ AUXTOORIGINAL(c_2')
end procedure

Here the function CHOOSEXPT decides the crossover point in the auxiliary representation such that the two children chromosomes in the original representation are either one bit longer or shorter than N. The function FIXCHROMOSOME lengthens or shortens the chromosome by mutating 01 to 011, 011 to 01, converting 011 to 0101, or converting 0101 to 011.

3.5 Motif-Preserving Genetic Algorithm

In order to replace the mutation operator and one-point crossover operator with the motif-preserving versions, the chromosomes must have special structural property at all time. For this purpose, a motif-preserving initialization routine is propose in Algorithm 5.

As we will show later, this initialization routine per se greatly improves the quality of the initial chromosomes. In this paper, the motif-preserving genetic algorithm (SPGA) refers to ones that run in the framework of Algorithm 2, with population initialized by Algorithm 5, the mutation operator replaced by Algorithm 3 and the one-point crossover operator replaced by Algorithm 4. This combination of algorithms in SPGA reduces the search space substantially. One concern about this reduction is that it might fragment the solution space into disjoint subspace. We will discuss this issue in Sect. 5.

Algorithm 5: Motif-Preserving Initialization.

---

procedure SPINIT(P)
    for each c_i in P **do**
        empty c_i
        append 01 or 011 randomly to c until SIZE(c)==N
    end for
end procedure

---

4 Experimental Design

For comparison of the relative efficiency of the various components of our genetic algorithm, we consider the following versions of GAs.

- SGANoX: A GA that uses only the mutation operator, with uniform initialization and no crossover operators
- SGANoXwSPInit: SGANoX with motif-preserving initialization
- SGA: Simple Genetic Algorithm, with uniform initialization, standard mutation operator and one-point crossover operator
- SGAwSPInit: Simple Genetic Algorithm with motif-preserving initialization, standard mutation operator and one-point crossover operator
- SPGANoX: motif-preserving mutation only genetic algorithm with motif-preserving initialization and motif-preserving mutation operator, but without crossover
- SPGA: motif-preserving genetic algorithm with motif-preserving initialization, motif-preserving mutation operator and motif-preserving crossover operator

For simplicity, when we use the term SGAs, we mean the group of SGA, including SGANoX, SGA, SGANoXwSPInit, SGAwSPInit. Similarly, when we use the term SPGAs, we mean the group of SPGA, including SPGANoX, SPGA.

The size of the population, N_P, will be set to 100 for all versions of GAs. The size of the tournament, k, defined in Algorithm 2 in the deterministic tournament selection process will be set to 10. The crossover rate, p_c and the mutation rate per bit, p_m, both defined in Algorithm 2, will be chosen individually for each GA such that each GA has good performance.

5 Results and Analysis

First we investigate the effect of imposing motif-preserving evolution operators and population initialization on search space. Exhaustive search is not an option for sequences as long as $N = 80$. Instead we use SGA to perform a biased sampling on the search space by just running SGA with $p_m = 0.02$ and $p_c = 1.0$. The selection mechanism in SGA biases towards chromosomes having higher fitness values.

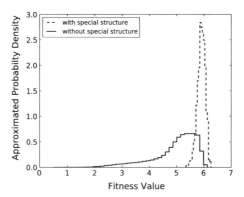

Fig. 1 Fitness probability density, approximated by histograms, for game sequences that satisfy special structural property and those who do not. Only distinct chromosomes are counted. Game sequences satisfying special structural property are strongly grouped together in terms of fitness values and they are among the chromosomes having the highest fitness values

Every chromosome that appears during 200 generations of evolution from 50 independent experiments are collected as samples to construct the fitness histogram and normalized to become the fitness probability density as shown in Fig. 1. We see two distinct probability density in the figure. The fitness distribution for game sequences satisfying special structural property are strongly grouped together at high fitness values, and they are among the chromosomes having fitness values mostly higher than the highest fitness of the probability density for sequences without special structure property. From this observation, we conclude that imposing motif-preserving evolution operators and population initialization on search space will not fragment the search space. Furthermore, the usage of motif-preserving evolution operators and population initialization can be justified due to the overall higher fitness value obtained.

Next we compare the performances of the various GAs mentioned in the last section on the search for optimal sequence of with $N = 80$. We first calculate the true expected return of the optimal sequence using Dinis' algorithm. We then measure the performance of various GAs by the optimality gap, the difference between the expected return of the optimal sequence and the expected return of the chromosome with the highest fitness value. We show the optimality gap as a function of number of generations in Fig. 2. The values of p_m and p_c are chosen on the basis of performance.

The motif-preserving initialization alone contributes significantly to the performance, as can be seen from the relative performances of SGANoX and SGA with or without motif-preserving population initialization in Fig. 2. In the case of problem-independent GAs, one-point crossover operator contributes to the early edge in performance. The problem-dependent motif-preserving GAs generally perform better. SPGANoX outperforms SGAwSPInit. SPGA performs the best.

Fig. 2 Optimality gap as a function of number of generations for $N = 80$. Every *curve* is an average from 100 independent experiments. Y-axis is in log-scale. The gap is smallest for SPGA, indicating it is the best performing GA

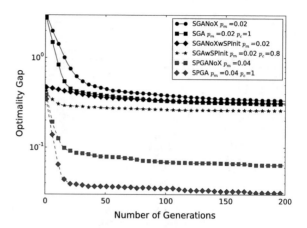

Since there is no guarantee that optimal game sequence will be obtained after 200 generations, we therefore need to assess the average quality of the best solutions after 200 generations. For N = 80, the optimal game sequence is

$$ABXXXXXXXXXXXXXXXXXABB, \qquad (6)$$

where X stands for ABABB. We call those AB (or ABB) substring that are not able to form ABABB with adjacent substring ungrouped AB (or ABB).

A typical suboptimal but still good solution resulted from SPGAs has the following structure:

$$XXXXXXABBXXXXABBXXXABXAB, \qquad (7)$$

which possesses the right number of AB and ABB substrings, and the SPGAs manage to figure out the overall structure. From this observation, we expect that a problem-specific local search can obtain the real optimal solution from the best solution obtained from SPGAs. In this context, we may introduce some alternative quality measures by the following enquiries:

1. Has the right number of AB and ABB been reached? (Here the right number refers to the number in the optimal sequence.) If not, calculate the deviation by

$$d_1(s) = |N_{AB}(s) - N_{AB}(s^*)| + |N_{ABB}(s) - N_{ABB}(s^*)|, \qquad (8)$$

where $N_{AB}(s)$ ($N_{ABB}(s)$) is the number of AB (ABB respectively) and s^* is the optimal sequence.

2. If the right number of AB and ABB has been reached, has the right number of ungrouped AB and ABB been reached? (Here, the right number again refers to the number in the optimal sequence.) If $N_{AB}^*(s)$ ($N_{ABB}^*(s)$) is the number of ungrouped AB (ABB respectively) and s^* is the optimal sequence, then this deviation can be written as

Table 1 Statistics of the best chromosomes obtained by various GAs at the end of 200-th generation for $N = 80$

GA variant	Prs	$d_1(s)$	$d_2(s)$
SGANoX	97/100	3.60825	7.87629
SGA	96/100	3.48958	6.84375
SGANoXwSPInit	97/100	4.89691	8.17526
SGAwSPInit	100/100	4.75	6.47
SPGANoX	100/100	0.25	0.89
SPGA	100/100	0.00	0.54

Data are collected from 100 independent experiments for each GA variant. The best chromosomes obtained from SGAs do not necessarily satisfy the special structural property, and therefore the numbers of instances of best chromosomes that satisfy the property are indicated. The quantity Prs stands for the Probability that Structural property is observed over 100 independent experiments. In the calculation of the quality measure, $d_1(s)$ and $d_2(s)$ for the case of SGA, we only count those chromosomes satisfying the special structural property

$$d_2(s) = |N^*_{AB}(s^*) - N^*_{AB}(s)| + |N^*_{ABB}(s^*) - N^*_{ABB}(s)|. \qquad (9)$$

For both measures, the smaller the values of $d_1(s)$ and $d_2(s)$ the higher is the quality of the sequence. We show the statistics of the best chromosomes obtained by various GAs at the end of 200-th generation in Table 1. Data are collected from 100 independent experiments for each GA variant. As can be seen from the table, SPGAs are better at figuring out the structure of the optimal sequence than other GAs.

We see that for medium size sequence ($N = 80$), our SPGA is good approximation to Dinis analysis, as shown in Fig. 2. Thus for medium size sequence, we do not see the advantage of our SPGA algorithm.

We now apply the same set of GAs on searching for the optimal sequences with longer length N. The values of p_m and p_c are the same as those used before. Note that the exact optimal sequence with length $N = 200$ is not easily available as straightforward implementation of Dinis algorithm needs huge amount of storage space and exhaustive search will require even more. Thus, we use an alternative measure of performance: we record the highest fitness value in the population as a function of the number of generations, averaged over 50 independent experiments and see which GA yields the best result. The performances of various GAs as a function of number of generation when searching for optimal sequence with length $N = 200$ are shown in Fig. 3. We see that SPGAs outperform problem-independent SGAs. In fact, for SGAs with motif-preserving initialization, the average best fitness value drops as the number of generation increases. Let us compare this curious feature of decreasing average fitness of SGAs for large $N(= 200)$ with the results for small $N(= 80)$ shown in Fig. 2. The optimality gaps are all decreasing with generation number in Fig. 2, implying that the average fitness of all GAs are increasing. We also see that SGAs are able to take advantage of the good quality initial chromosomes resulted from the structural-preserving initialization. However, when N is large, as shown in Fig. 3 for $N = 200$, SGAs no longer have the ability in exploiting the good quality

Fig. 3 Average best fitness as a function of number of generations for $N = 200$. Every *curve* is an average from 50 independent experiments

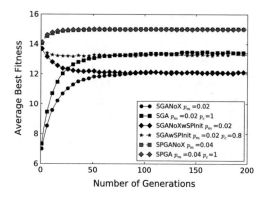

of the initial chromosomes. In Fig. 3, we observe the merging of several curves at large generation number:

1. SPGA($\diamond$) merges with SPGANoX($\square$)
2. SGAwSPInit (o) merges with SGA($\blacksquare$)
3. SGANoX($\bullet$) merges with SGANoXwSPInit($\blacklozenge$);

We see from 1 that SPGA with crossover operator, does not have noticeable advantage over SPGANoX in terms of best fitness value in the population, suggesting motif-preserving crossover operator does not provide additional advantage for large-size problem. From 2 we see that for SGA, the special initialization with motif preserving features is irrelevant for large N. From 3, we see similar effect as 2 without crossover for SGA, when N is large.

In Table 2, we see the average of the best fitness values at the end of 200-th generation and the standard deviation. Not only do SPGAs (SPGANoX and SPGA) achieve higher best fitness value, they also have smaller standard deviation in the best fitness value, which means that SPGAs are more reliable. Moreover, SPGANoX and SPGA have similar performance statistically, suggesting motif-preserving crossover operator does not provide noticeable additional advantage for large-size problem.

Results for optimal sequence search with length $N = 300$ exhibit similar qualitative features (Table 3). SPGAs outperform SGAs in terms of average best fitness

Table 2 Best fitness values statistics for $N = 200$ at the end of the experiment

GA variant	Avg. best fitness	Stand. deviat.
SGANoX	12.1	0.231
SGA	13.4	0.250
SGANoXwSPInit	12.1	0.228
SGAwSPInit	13.3	0.273
SPGANoX	15.01	0.0805
SPGA	15.00	0.0978

Data are collected from 50 experiments

Table 3 Best fitness values statistics for $N = 300$ at the end of the experiment

GA variant	Avg. best fitness	Stand. deviat.
SGANoX	15.9	0.364
SGA	18.0	0.360
SGANoXwSPInit	15.9	0.348
SGAwSPInit	17.7	0.334
SPGANoX	21.90	0.0907
SPGA	21.97	0.0893

Data are collected from 50 experiments

value. SGAs are not able to exploit the good quality chromosomes resulted from the motif-preserving initialization. Similar to the case of $N = 200$, the motif-preserving crossover operator does not provide additional benefit in performance.

6 Conclusions and Future Work

In this work on sequence analysis, we have developed a systematic way to solve the combinatorial optimization problem using genetic algorithm. We first explore the solution space using a problem-independent simple genetic algorithm to sample good but suboptimal solutions. Then we design a way to identify structures shared among the good candidate solutions. We call these structures the motifs, borrowing the terminology in bioinformatics. Next we incorporate these motifs by modifying the one-point mutation operator and the one-point crossover operator, resulting in motif-preserving genetic algorithm. This method of sequence analysis can be summarized in a meta-algorithm: first a problem-independent simple genetic algorithm is employed to sample the good solutions, and then techniques such as text-mining are used to automatically extract the structures shared among the good solutions, and finally an adaptive motif-preserving genetic algorithm is crafted to exploit the knowledge extracted and find a better solution more efficiently.

In this paper, the above meta-algorithm is tested on the analysis of long but finite Parrondo game sequences. Several numerical experiments to evaluate the relative efficiency of several genetic algorithms incorporating various features deemed important from observation of small N optimal sequences have been designed and we discover that the proposed motif-preserving initialization routine offers high-quality candidate solutions. Compared to the various simple genetic algorithms, we have also observed that both motif-preserving mutation operator and one-point crossover operator improve the performance substantially. Indeed, SPGAs are consistently better at figuring out the structure of optimal sequence for medium-size problem, e.g. $N = 80$ than simple genetic algorithm and motif-preserving crossover operator provides noticeable additional advantage, but then this turns out to yield little advantage for large-size problems, e.g. $N = 200$. This observation could be useful in using genetic algorithm to discover the structures of sequences for different lengths. In general, our analysis using genetic algorithm for Parrondo game sequences can be extended to the analysis of sequences from other fields. For the future works, we could develop a permutation-based evolution operator to further improve the performance of SPGAs.

From our numerous numerical experiments we may conclude that our motif-preserving genetic algorithm is useful in searching for the optimal sequence for Parrondo games. This is important for applications in other sequence analysis for which a good solution, not the mathematically optimal one, is needed. In the context of Parrondo game, for very long sequence, when Dinis algorithm is not easily implemented, our SPGA does provide a good solution efficiently. For application in other sequences, we must clearly define the condition of optimality and then use the meta-algorithm proposed in this work to obtain a good solution: first analyze the motif in short sequences where exhaustive search provide the optimal solution, followed by incorporating these motifs to obtain a good solution for long sequence with motif-preserving genetic algorithm.

Acknowledgments K.Y. Szeto acknowledges the support of grant FS GRF13SC25 and FS GRF14SC28. Degang Wu would like to thank Zhangyu Chang and Claus Aranha for helpful discussions.

References

1. Wu, D., Szeto, K.Y.: Extended parrondo's game and brownian ratchets: strong and weak parrondo effect. Phys. Rev. E **89**, 022142 (2014)
2. Parrondo, J.M.R.: How to Cheat a Bad Mathematician. EEC HC&M Network on Complexity and Chaos (1996)
3. Allison, A., Abbott, D.: The physical basis for parrondo's games. Fluctuation Noise Lett. **2**, L327–L341 (2002)
4. Toral, R., Amengual, P., Mangioni, S.: Parrondo's games as a discrete ratchet. Phys. A Stat. Mech. Appl. **327**, 105–110 (2003)
5. Dinis, L.: Optimal sequence for parrondo games. Phys. Rev. E **77**, 021124 (2008)
6. Dinis, L., Parrondo, J.M.R.: Optimal strategies in collective parrondo games. Europhys. Lett. (EPL) **63**, 319–325 (2003)
7. Dins, L., Parrondo, J.M.: Inefficiency of voting in parrondo games. Phys. A Stat. Mech. Appl. **343**, 701–711 (2004)
8. Parrondo, J.M.R., Dinis, L., Garca-Torao, E., Sotillo, B.: Collective decision making and paradoxical games. Eur. Phys. J. Special Topics **143**, 39–46 (2007)
9. Goldberg, D.E., Holland, J.H.: Genetic algorithms and machine learning. Mach. Learn. **3**, 9599 (1988)
10. Holland, J.H.: Adaptation in Natural and Artificial Systems: An Introductory Analysis with Applications to Biology, Control, and Artificial Intelligence. U Michigan Press (1975)
11. Clote, P.: An efficient algorithm to compute the landscape of locally optimal RNA secondary structures with respect to the nussinov-jacobson energy model. J. Comput. Biol. **12**, 83101 (2005)
12. Ding, Y.S., Zhang, T.L.: Using chous pseudo amino acid composition to predict subcellular localization of apoptosis proteins: An approach with immune genetic algorithm-based ensemble classifier. Pattern Recogn. Lett. **29**, 1887–1892 (2008)

13. Pond, S.L.K., Posada, D., Gravenor, M.B., Woelk, C.H., Frost, S.D.W.: GARD: a genetic algorithm for recombination detection. Bioinformatics **22**, 3096–3098 (2006)
14. Doye, J.P.: Network topology of a potential energy landscape: a static scale-free network. Phys. Rev. Lett. **88**, 238701 (2002)
15. Wales, D.J., Doye, J.P., Miller, M.A., Mortenson, P.N., Walsh, T.R.: Energy landscapes: from clusters to biomolecules. Adv. Chem. Phys. **115**, 1112 (2000)
16. Wales, D.J., Miller, M.A., Walsh, T.R.: Archetypal energy landscapes. Nature **394**, 758760 (1998)
17. Debenedetti, P.G., Stillinger, F.H.: Supercooled liquids and the glass transition. Nature **410**, 259267 (2001)
18. Patel, A.N., Davis, D., Guthrie, C.F., Tuk, D., Nguyen, T.T., Williams, J., et al.: Optimizing cyclic steam oil production with genetic algorithms. In: SPE Western Regional Meeting, Society of Petroleum Engineers (2005)
19. Xia, C., Guo, P., Shi, T., Wang, M.: Speed control of brushless DC motor using genetic algorithm based fuzzy controller. In: Proceeding of the 2004 International Conference on Intelligent Mechatronics and Automation, Chengdu, China, 3rd edn. A Treatise on Electricity and Magnetism, vol 2, pp. 68–73 (2004)
20. Shiu, K.L., Szeto, K.Y.: Self-adaptive mutation only genetic algorithm: an application on the optimization of airport capacity utilization. In: Intelligent Data Engineering and Automated Learning IDEAL 2008, pp. 428–435. Springer (2008)
21. Messom, C.: Genetic Algorithms for Auto-tuning Mobile Robot Motion Control. (2002)
22. Fong, L.Y., Szeto, K.Y.: Rules extraction in short memory time series using genetic algorithms. Eur. Phys. J. B **20**, 569572 (2001)
23. Szeto, K.Y., Fong, L.Y.: How adaptive agents in stock market perform in the presence of random news: a genetic algorithm approach. In: Intelligent Data Engineering and Automated LearningIDEAL 2000. Data Mining, Financial Engineering, and Intelligent Agents, pp. 505–510. Springer (2000)
24. Jiang, R., Szeto, K.Y., Luo, Y.P., Hu, D.C.: Distributed parallel genetic algorithm with path splitting scheme for the large traveling salesman problems. In: Proceedings of Conference on Intelligent Information Processing, 16th World Computer Congress, pp. 21–25 (2000)
25. Harmer, G.P., Abbott, D.: Game theory: losing strategies can win by parrondo's paradox. Nature **402**, 864–864 (1999)
26. Harmer, G.P., Abbott, D.: A review of parrondos paradox. Fluctuation and Noise Lett. **2**, R71–R107 (2002)
27. Wagon, S., Velleman, D.: Parrondo's paradox. Math. Educ. Res. **9**, 8590 (2001)

Evolutionary Learning of Linear Composite Dispatching Rules for Scheduling

Helga Ingimundardottir and Thomas Philip Runarsson

Abstract A prevalent approach to solving job shop scheduling problems is to combine several relatively simple dispatching rules such that they may benefit each other for a given problem space. Generally, this is done in an ad-hoc fashion, requiring expert knowledge from heuristics designers, or extensive exploration of suitable combinations of heuristics. The approach here is to automate that selection by translating dispatching rules into measurable features and optimising what their contribution should be via evolutionary search. The framework is straight forward and easy to implement and shows promising results. Various data distributions are investigated for both job shop and flow shop problems, as is scalability for higher dimensions. Moreover, the study shows that the choice of objective function for evolutionary search is worth investigating. Since the optimisation is based on minimising the expected mean of the fitness function over a large set of problem instances which can vary within the set, then normalising the objective function can stabilise the optimisation process away from local minima.

Keywords Job shop scheduling · Composite dispatching rules · Evolutionary search

1 Job Shop Scheduling

The job-shop scheduling problem (JSP) deals with the allocation of tasks of competing resources where the goal is to optimise a single or multiple objectives—in particular minimising a schedule's maximum completion time, i.e., the makespan, denoted C_{max}. Due to difficulty in solving this problem, heuristics are generally applied. Perhaps the simplest approach to generating good feasible solutions for JSP is by

H. Ingimundardottir · T.P. Runarsson (✉)
Industrial Engineering, Mechanical Engineering and Computer Science,
University of Iceland, Reykjavik, Iceland
e-mail: tpr@hi.is

H. Ingimundardottir
e-mail: hei2@hi.is

© Springer International Publishing Switzerland 2016
J.J. Merelo et al. (eds.), *Computational Intelligence*,
Studies in Computational Intelligence 620, DOI 10.1007/978-3-319-26393-9_4

applying dispatching rules (DR), e.g., choosing a task corresponding to longest or shortest processing time, most or least successors, or ranked positional weight, i.e., sum of processing times of its predecessors. Ties are broken in an arbitrary fashion or by another heuristic rule. Combining dispatching rules for JSP is promising, however, there is a large number of rules to choose from, thus its combinations rely on expert knowledge or extensive trial-and-error process to choose a suitable DR [21]. Hence given the diversity within the JSP paradigm, there is no "one-rule-fits-all" for all problem instances (or shop constraints), however single priority dispatching rules (SDR) based on job processing attributes have proven to be effective [8]. The classical dispatching rules are continually used in research; a summary of over 100 classical DRs for JSP can be found in [16]. However, careful combinations of such simple rules, i.e., composite dispatching rules (CDRs) can perform significantly better [12]. As a consequence, a linear composite of dispatching rules for JSP was presented in [10]. There the goal was to learn a set of weights, $\mathbf{w}$ via ordinal regression such that

$$h(\mathbf{x}_j) = \langle \mathbf{w} \cdot \boldsymbol{\phi}(\mathbf{x}_j) \rangle, \tag{1}$$

yields the preference estimate for dispatching job j that corresponds to post-decision state $\mathbf{x}_j$, where $\boldsymbol{\phi}(\mathbf{x}_j)$ denotes the feature mapping (cf. Sect. 4). In short, Eq. 1 is a simple linear combination of features found using a classifier which is trained by giving more weight to instances that are preferred w.r.t. optimality in a supervised learning fashion. As a result, the job dispatched is the following,

$$j^* = \arg \max_j \left\{ h(\mathbf{x}_j) \right\}. \tag{2}$$

A more popular approach in recent JSP literature is applying genetic algorithms (GAs) [17]. However, in that case an extensive number of schedules need to be evaluated, and even for low dimensional JSP, it can quickly become computationally infeasible. GAs can be used directly on schedules [1, 3, 4, 13, 22], however, then there are many concerns that need to be dealt with. To begin with there are nine encoding schemes for representing the schedules [3], in addition, special care must be taken when applying cross-over and mutation operators in order for schedules to still remain feasible. Moreover, in case of JSP, GAs are not adapt for fine-tuning around optima. Luckily a subsequent local search can mediate the optimisation [4].

The most predominant approach in hyper-heuristics, a framework of creating *new* heuristics from a set of predefined heuristics, is genetic programming [2]. Dispatching rules based genetic algorithms (DRGA) [5, 15, 23] are a special case of genetic programming [14], where GAs are applied indirectly to JSP via dispatching rules, i.e., where a solution is no longer a *proper* schedule but a *representation* of a schedule via applying certain DRs consecutively.

There are two main viewpoints on how to approach scheduling problems, (*a*) local level by building schedules for one problem instance at a time; and (*b*) global level by building schedules for all problem instances at once. For local level construction a simple construction heuristic is applied. The schedule's features are collected at each dispatch iteration from which a learning model will inspect the feature set to discriminate which operations are preferred to others via ordinal regression. The focus is essentially on creating a meaningful preference set composed of features and their ranks as the learning algorithm is only run once to find suitable operators for the value function. This is the approach taken in [10]. Expanding on that work, this study will explore a global level construction viewpoint where there is no feature set collected beforehand since the learning model is optimised directly via evolutionary search. This involves numerous costly value function evaluations. In fact it involves an indirect method of evaluation whether one learning model is preferable to another, w.r.t. which one yields a better expected mean.

2 Outline

In order to formulate the relationship between problem structure and heuristic efficiency, one can utilise Rice's framework for algorithm selection [18]. The framework consists of four fundamental components, namely,

Problem Space or Instance Space $\mathscr{P}$,
  set of problem instances;
Feature Space $\mathscr{F}$,
  measurable properties of the instances in $\mathscr{P}$;
Algorithm Space $\mathscr{A}$,
  set of all algorithms under inspection;
Performance Space $\mathscr{Y}$,
  the outcome for $\mathscr{P}$ using an algorithm from $\mathscr{A}$.

For a given problem instance $\mathbf{x} \in \mathscr{P}$ with k features $\boldsymbol{\phi}(\mathbf{x}) = \{\phi_1(\mathbf{x}), \ldots, \phi_k(\mathbf{x})\} \in \mathscr{F}$ and using algorithm $a \in \mathscr{A}$ the performance is $y = Y(a, \boldsymbol{\phi}(\mathbf{x})) \in \mathscr{Y}$, where $Y : \mathscr{A} \times \mathscr{F} \mapsto \mathscr{Y}$ is the mapping for algorithm and feature space onto the performance space. [11, 19, 20] formulate JSP in the following manner: (*a*) problem space $\mathscr{P}$ is defined as the union of N problem instances consisting of processing time and ordering matrices given in Sect. 3; (*b*) feature space $\mathscr{F}$, which is outlined in Sect. 4. Note, these are not the only possible set of features, however, they are built on the work by [10, 19] and deemed successful in capturing the essence of a JSP data structure; (*c*) algorithm space $\mathscr{A}$ is simply the scheduling policies under consideration and discussed in Sect. 5; (*d*) performance space is based on the resulting $C_{\max}$. Different fitness measures are investigated in Sect. 5.1; and (*e*) mapping Y is the step-by-step scheduling process.

   In the context of Rice's framework, and returning to the aforementioned approaches to scheduling problems, then the objective is to maximise its expected heuristic performance, i.e., (*a*) Local level

$$\max_{\mathscr{P}' \subset \mathscr{P}} \mathbb{E}\left[Y\left(a, \boldsymbol{\phi}(\mathbf{x})\right)\right] \tag{3}$$

where $\mathbf{x} \in \mathscr{P}'$ and algorithm a is obtained via ordinal regression based on the feature space $\mathscr{F}$, i.e., $\mathscr{F}|_{\mathscr{P}'} \mapsto \mathscr{A}$, such as the approach taken in [10], and will be used as a benchmark for the following,

(*b*) Global level

$$\max_{a \in \mathscr{A}} \mathbb{E}\left[Y\left(a, \boldsymbol{\phi}(\mathbf{x})\right)\right] \tag{4}$$

where training data $\mathbf{x} \in \mathscr{P}$ is guided by its algorithm a, i.e., $\mathscr{A} \mapsto \mathscr{P}$. This will be the focus of this study.

Note that the mappings $\boldsymbol{\phi} : \mathscr{P} \mapsto \mathscr{F}$ and $Y : \mathscr{A} \mapsto \mathscr{Y}$ are the same for both paradigms.

The paper concludes in Sect. 6 with discussion and conclusions.

3 Problem Space

For this study synthetic JSP and its subclass, permutation flow shop problem (PFSP), the scheduling task considered here is where n jobs are scheduled on a set of m machines, i.e., problem size $n \times m$, subject to the constraint that each job must follow a predefined machine order and that a machine can handle at most one job at a time. The pair (j, a) refers to the operation of dispatching job j on machine a. As a result, a total of $\ell = n \cdot m$ sequential operations need to be made for a complete schedule.

The objective is to schedule the jobs so as to minimize the maximum completion times, $C_{\max}$, also known as the makespan. For a mathematical formulation of JSP the reader is recommended [10].

There are two fundamental types of problem classes: non-structured versus structured. Firstly there are the "conventional" structured problem classes, where problem instances are generated stochastically by fixing the number of jobs and machines, as well as processing times are i.i.d. and sampled from a discrete uniform distribution from the interval $I = [u_1, u_2]$, i.e., $p \sim \mathscr{U}(u_1, u_2)$. Two different processing time distributions are explored, namely $\mathscr{P}_{j.rnd}$ where $I = [1, 99]$ and $\mathscr{P}_{j.rndn}$ where $I = [45, 55]$, referred to as random and random-narrow, respectively. The machine order is a random permutation of all of the machines in the job-shop.

Analogous to $\mathscr{P}_{j.rnd}$ and $\mathscr{P}_{j.rndn}$ the problem classes $\mathscr{P}_{f.rnd}$ and $\mathscr{P}_{f.rndn}$, respectively, correspond to the structured PFSP problem classes, however with a homogeneous machine order permutation. Secondly, there are structured problem classes of PFSP which are modelled after real-world flow-shop manufacturing namely job-correlated $\mathscr{P}_{f.jc}$ where job processing times are dependent on job index and

Evolutionary Learning of Linear Composite ...   53

Table 1 Problem space distributions used in Sect. 5

Name	Size	N_{train}	N_{test}	Note
Permutation flow shop problem (PFSP)				
$\mathcal{P}^{6\times5}_{f.rnd}$	6×5	500	–	Random
$\mathcal{P}^{6\times5}_{f.rndn}$	6×5	500	–	Random-narrow
$\mathcal{P}^{6\times5}_{f.jc}$	6×5	500	–	Job-correlated
$\mathcal{P}^{10\times10}_{f.rnd}$	10×10	–	500	Random
$\mathcal{P}^{10\times10}_{f.rndn}$	10×10	–	500	Random-narrow
$\mathcal{P}^{10\times10}_{f.jc}$	10×10	–	500	Job-correlated
Job shop problem (JSP)				
$\mathcal{P}^{6\times5}_{j.rnd}$	6×5	500	–	Random
$\mathcal{P}^{6\times5}_{j.rndn}$	6×5	500	–	Random-narrow
$\mathcal{P}^{10\times10}_{j.rnd}$	10×10	–	500	Random
$\mathcal{P}^{10\times10}_{j.rndn}$	10×10	–	500	Random-narrow

Note Problem instances are synthetic and each problem space is i.i.d. and '–' denotes not available

independent of machine index. Problem instances for PFSP are generated using [24] problem generator.[1]

For each JSP and PFSP class N_{train} and N_{test} instances were generated for training and testing, respectively. Values for N are given in Table 1. Note, difficult problem instances are not filtered out beforehand, such as the approach in [24].

4 Feature Space

When building a complete JSP schedule, a job is placed at the earliest available time slot for its next machine while still fulfilling constraints that each machine can handle at most one job at a time, and jobs need to have finished their previous machines according to its machine order. Unfinished jobs are dispatched one at a time according to some heuristic. After each dispatch the schedule's current features are updated. Features are used to grasp the essence of the current state of the schedule. As seen in Table 2, temporal scheduling features applied in this study are given for each possible post-decision state. An example of a schedule being built is given in Fig. 1, where there are a total of five possible jobs that could be chosen to be dispatched by some dispatching rule. These features would serve as the input for Eq. 1.

It's noted that some of the features directly correspond to a SDR commonly used in practice. For example, if the weights **w** in Eq. 1 were all zero, save for $w_6 = 1$, then Eq. 2 yields the job with the highest ϕ_6 value, i.e., equivalent to dispatching rule most work remaining (MWR).

---

[1]Both code, written in C++, and problem instances used in their experiments can be found at: http://www.cs.colostate.edu/sched/generator/.

Table 2 Feature space $\mathscr{F}$ for $\mathscr{P}$ given the resulting temporal schedule after dispatching an operation (j, a)

ϕ	Feature description
ϕ_1	Job j processing time
ϕ_2	Job j start-time
ϕ_3	Job j end-time
ϕ_4	When machine a is next free
ϕ_5	Current makespan
ϕ_6	Total work remaining for job j
ϕ_7	Most work remaining for all jobs
ϕ_8	Total idle time for machine a
ϕ_9	Total idle time for all machines
ϕ_{10}	ϕ_9 weighted w.r.t. number of assigned tasks
ϕ_{11}	Time job j had to wait
ϕ_{12}	Idle time created
ϕ_{13}	Total processing time for job j

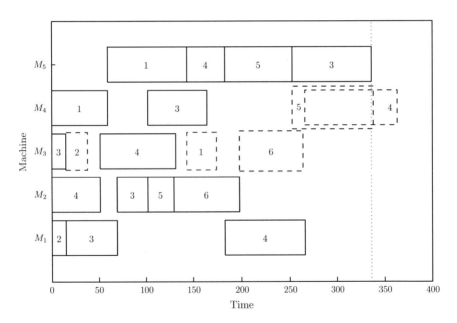

Fig. 1 Gantt chart of a partial JSP schedule after 15 operations: *Solid boxes* represent previously dispatched jobs, and *dashed boxes* represent the jobs that could be scheduled next. Current $C_{\max}$ denoted as *dotted line*

5 Experimental Study

The optimum makespan[2] is denoted $C_{\max}^{\text{opt}}$, and the makespan obtained from the heuristic model by $C_{\max}^{\text{model}}$. Since the optimal makespan varies between problem instances the performance measure is the following,

$$\rho := \frac{C_{\max}^{\text{model}} - C_{\max}^{\text{opt}}}{C_{\max}^{\text{opt}}} \cdot 100\,\% \tag{5}$$

which indicates the percentage relative deviation from optimality. Throughout a Kolmogorov-Smirnov test with $\alpha = 0.05$ is applied to determine statistical significance between methodologies.

Inspired by DRGA, the approach taken in this study is to optimise the weights $\mathbf{w}$ in Eq. 1 directly via evolutionary search such as covariance matrix adaptation evolution strategy (CMA-ES) [7]. This has been proven to be a very efficient numerical optimisation technique.

Using standard set-up of parameters of the CMA-ES optimisation, the runtime was limited to 288 h on a cluster for each training set given in Sect. 3 and in every case the optimisation reached its maximum walltime.

5.1 Performance Measures

Generally, evolutionary search only needs to minimise the expected fitness value. However, the approach in [10] was to use the known optimum to correctly label which operations' features were optimal when compared to other possible operations (Fig. 2). Therefore, it would be of interest to inspect if there is any performance edge gained by incorporating optimal labelling in evolutionary search. Therefore, two objective functions will be considered, namely,

$$ES_{C_{\max}} := \min \mathbb{E}[C_{\max}] \tag{6}$$

$$ES_\rho := \min \mathbb{E}[\rho] \tag{7}$$

Main statistics of the experimental run are given in Table 3 and depicted in Fig. 3 for both approaches. In addition, evolving decision variables, here weights $\mathbf{w}$ for Eq. 1, are depicted in Fig. 4.

In order to compare the two objective functions, the best weights reported were used for Eq. 1 on the corresponding training data. Its box-plot of percentage relative deviation from optimality, defined by Eq. 5, is depicted in Fig. 2 and Table 4 present its main statistics; mean, median, standard deviation, minimum and maximum values.

---

[2]Optimum values are obtained by using a commercial software package [6].

Fig. 2 Box-plot of training
data for percentage relative
deviation from optimality,
defined by Eq. (5), when
implementing the final
weights obtained from
CMA-ES optimisation, using
both objective functions
from Eqs. (6) and (7), *left*
and *right*, respectively

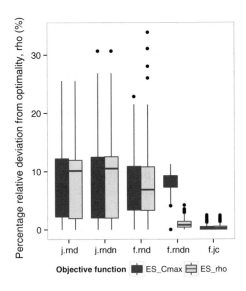

Table 3 Final results for CMA-ES optimisation; total number of generations and function evaluations and its resulting fitness value for both performance measures considered

(a) w.r.t. Eq. 6

$\mathscr{P}$	#gen	#eval	$ES_{C_{max}}$
j.rnd	4707	51788	448.612
j.rndn	4802	52833	449.942
f.rnd	5088	55979	571.394
f.rndn	5557	61138	544.764
f.jc	5984	65835	567.688

(b) w.r.t. Eq. 7

$\mathscr{P}$	#gen	#eval	ES_{ρ}
j.rnd	1944	21395	8.258
j.rndn	1974	21725	8.691
f.rnd	4546	50006	7.479
f.rndn	2701	29722	0.938
f.jc	1625	17886	0.361

In the case of $\mathscr{P}_{f.rndn}$, Eq. 6 gave a considerably worse results, since the optimisation got trapped in a local minima, as the erratic evolution of the weights in Fig. 4a suggest. For other problem spaces, Eq. 6 gave slightly better results than Eq. 7. However, there was no statistical difference between adopting either objective function. Therefore, minimisation of expectation of ρ, is preferred over simply using the unscaled resulting makespan.

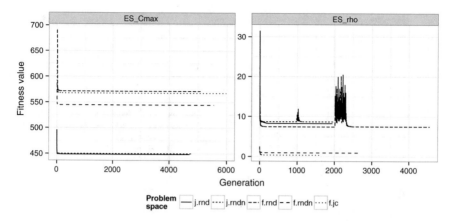

Fig. 3 Fitness for optimising (w.r.t. Eqs. (6) and (7) *above* and *below*, receptively), per generation of the CMA-ES optimisation

5.2 Problem Difficulty

The evolution of fitness per generation from the CMA-ES optimisation of Eq. 7 is depicted in Fig. 3. Note, all problem spaces reached their allotted computational time without converging. In fact $\mathscr{P}_{f.rnd}$ and $\mathscr{P}_{j.rndn}$ needed restarting during the optimisation process. Furthermore, the evolution of the decision variables **w** are depicted in Fig. 4. As one can see, the relative contribution for each weight clearly differs between problem spaces. Note, that in the case of $\mathscr{P}_{j.rndn}$ (cf. Fig. 4b), CMA-ES restarts around generation 1,000 and quickly converges back to its previous fitness. However, lateral relation of weights has completely changed, implying that there are many optimal combinations of weights to be used. This can be expected due to the fact some features in Table 2 are a linear combination of others, e.g. $\phi_3 = \phi_1 + \phi_2$.

5.3 Scalability

As a benchmark, the linear ordinal regression model (PREF) from [10] was created. Using the weights obtained from optimising Eq. 7 and applying them on their 6×5 training data. Their main statistics of Eq. 5 are reported in Table 4 for all training sets described in Table 1. Moreover, the best SDR from which the features in Table 2 were inspired by, are also reported for comparison, i.e., most work remaining (MWR) for all JSP problem spaces, and least work remaining (LWR) for all PFSP problem spaces.

To explore the scalability of the learning models, a similar comparison to Sect. 5.2 is made for applying the learning models on their corresponding 10×10 testing data. Results are reported in Table 5. Note, that only resulting $C_{\max}$ is reported as the optimum makespan is not known and Eq. 5 is not applicable.

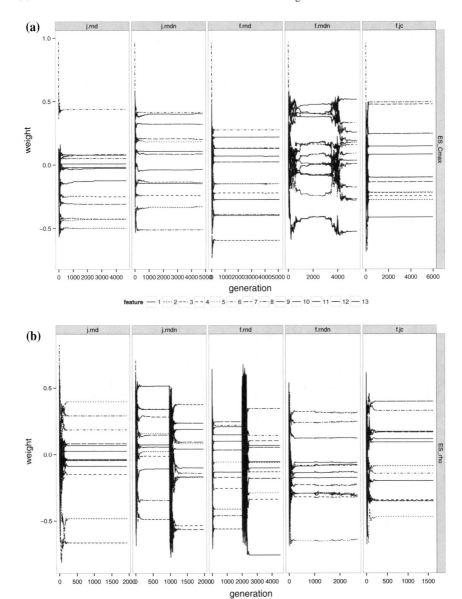

Fig. 4 Evolution of weights of features (given in Table 2) at each generation of the CMA-ES opti-misation. Note, weights are normalised such that $||\mathbf{w}|| = 1$. **a** Minimise w.r.t. Eq. 6. **b** Minimise w.r.t. Eq. 7

Table 4 Main statistics of percentage relative deviation from optimality, ρ, defined by Eq. 5 for various models, using corresponding 6×5 training data

(a) $\mathscr{P}_{j.rnd}^{6 \times 5}$

Model	Mean	Med	sd	Min	Max
$ES_{C_{max}}$	8.54	10	6	0	26
ES_ρ	8.26	10	6	0	26
PREF	10.18	11	7	0	30
MWR	16.48	16	9	0	45

(b) $\mathscr{P}_{j.rndn}^{6 \times 5}$

$ES_{C_{max}}$	8.68	11	6	0	31
ES_ρ	8.69	11	6	0	31
PREF	10.00	11	6	0	31
MWR	14.02	13	8	0	37

(c) $\mathscr{P}_{f.rnd}^{6 \times 5}$

$ES_{C_{max}}$	7.44	7	5	0	23
ES_ρ	7.48	7	5	0	34
PREF	9.87	9	7	0	38
LWR	20.05	19	10	0	71

(d) $\mathscr{P}_{f.rndn}^{6 \times 5}$

$ES_{C_{max}}$	8.09	8	2	0	11
ES_ρ	0.94	1	1	0	4
PREF	2.38	2	1	0	7
LWR	2.25	2	1	0	7

(e) $\mathscr{P}_{f.jc}^{6 \times 5}$

$ES_{C_{max}}$	0.33	0	0	0	2
ES_ρ	0.36	0	0	0	2
PREF	1.08	1	1	0	5
LWR	1.13	1	1	0	6

6 Discussion and Conclusions

Data distributions considered in this study either varied w.r.t. the processing time distributions, continuing the preliminary experiments in [10] , or w.r.t. the job ordering permutations—i.e., homogeneous machine order for PFSP versus heterogeneous machine order for JSP. From the results based on 6×5 training data given in Table 4, it's obvious that CMA-ES optimisation substantially outperforms the previous PREF methods from [10] for all problem spaces considered. Furthermore, the results hold when testing on 10×10 (cf. Table 5), suggesting the method is indeed scalable to higher dimensions.

Moreover, the study showed that the choice of objective function for evolutionary search is worth investigating. There was no statistical difference from minimising the

Table 5 Main statistics of C_{max} for various models, using corresponding 10×10 test data

(a) $\mathscr{P}_{j.rnd}^{10\times10}$

Model	Mean	Med	sd	Min	Max
$ES_{C_{max}}$	922.51	914	73	741	1173
ES_ρ	931.37	931	71	735	1167
PREF	1011.38	1004	82	809	1281
MWR	997.01	992	81	800	1273

(b) $\mathscr{P}_{j.rndn}^{10\times10}$

$ES_{C_{max}}$	855.85	857	50	719	1010
ES_ρ	855.91	856	51	719	1020
PREF	899.94	898	56	769	1130
MWR	897.39	898	56	765	1088

(c) $\mathscr{P}_{f.rnd}^{10\times10}$

$ES_{C_{max}}$	1178.73	1176	80	976	1416
ES_ρ	1181.91	1179	80	984	1404
PREF	1215.20	1212	80	1006	1450
LWR	1284.41	1286	85	1042	1495

(d) $\mathscr{P}_{f.rndn}^{10\times10}$

$ES_{C_{max}}$	1065.48	1059	32	992	1222
ES_ρ	980.11	980	8	957	1006
PREF	987.49	988	9	958	1011
LWR	986.94	987	9	959	1010

(e) $\mathscr{P}_{f.jc}^{10\times10}$

$ES_{C_{max}}$	1135.44	1134	286	582	1681
ES_ρ	1135.47	1134	286	582	1681
PREF	1136.02	1135	286	582	1685
LWR	1136.49	1141	287	581	1690

fitness function directly and its normalisation w.r.t. true optimum (cf. Eqs. (6) and (7)), save for $\mathscr{P}_{f.rndn}$. Implying, even though ES doesn't rely on optimal solutions, there are some problem spaces where it can be of great benefit. This is due to the fact that the problem instances can vary greatly within the same problem space [11]. Thus normalising the objective function would help the evolutionary search to deviate from giving too much weight for problematic problem instances.

The main drawback of using evolutionary search for learning optimal weights for Eq. 1 is how computationally expensive it is to evaluate the mean expected fitness. Even for a low problem dimension 6-job 5-machine JSP, each optimisation run reached their walltime of 288 h without converging. Now, 6×5 JSP requires 30 sequential operations where at each time step there are up to 6 jobs to choose from— i.e., its complexity is $\mathcal{O}(n^{n \cdot m})$ making it computationally infeasible to apply this framework for higher dimensions as is. However, evolutionary search only requires the rank of the candidates and therefore it is appropriate to retain a sufficiently accurate surrogate for the value function during evolution in order to reduce the number of costly true value function evaluations, such as the approach in [9]. This could reduce the computational cost of the evolutionary search considerably, making it feasible to conduct the experiments from Sect. 5 for problems of higher dimensions, e.g. with these adjustments it is possible to train on 10×10 and test on for example 14×14 to verify whether scalability holds for even higher dimensions.

References

1. Ak, B., Koc, E.: A guide for genetic algorithm based on parallel machine scheduling and flexible job-shop scheduling. Procedia—Soc. Behav. Sci. **62**, 817–823 (2012)
2. Burke, E.K., Gendreau, M., Hyde, M., Kendall, G., Ochoa, G., Ozcan, E., Qu, R.: Hyperheuristics: a survey of the state of the art. J. Oper. Res. Soc. **64**(12), 1695–1724 (2013)
3. Cheng, R., Gen, M., Tsujimura, Y.: A tutorial survey of job-shop scheduling problems using genetic algorithms-I. Representation. Comput. Ind. Eng. **30**(4), 983–997 (1996)
4. Cheng, R., Gen, M., Tsujimura, Y.: A tutorial survey of job-shop scheduling problems using genetic algorithms, part II: hybrid genetic search strategies. Comput. Ind. Eng. **36**(2), 343–364 (1999)
5. Dhingra, A., Chandna, P.: A bi-criteria M-machine SDST flow shop scheduling using modified heuristic genetic algorithm. Int. J. Eng., Sci. Technol. **2**(5), 216–225 (2010)
6. Gurobi Optimization, Inc.: Gurobi optimization (version 5.6.2) [software]. http://www.gurobi.com/ (2013)
7. Hansen, N., Ostermeier, A.: Completely derandomized self-adaptation in evolution strategies. Evol. Comput. **9**(2), 159–195 (2001)
8. Haupt, R.: A survey of priority rule-based scheduling. OR Spectr. **11**, 3–16 (1989)
9. Ingimundardottir, H., Runarsson, T.P.: Sampling strategies in ordinal regression for surrogate assisted evolutionary optimization. In: 2011 11th International Conference on Intelligent Systems Design and Applications (ISDA), pp. 1158–1163 (2011)
10. Ingimundardottir, H., Runarsson, T.P.: Supervised learning linear priority dispatch rules for job-shop scheduling. In: Coello, C. (ed.) Learning and Intelligent Optimization. Lecture Notes in Computer Science, vol. 6683, pp. 263–277. Springer, Heidelberg (2011)
11. Ingimundardottir, H., Runarsson, T.P.: Determining the characteristic of difficult job shop scheduling instances for a heuristic solution method. In: Hamadi, Y., Schoenauer, M. (eds.) Learning and Intelligent Optimization. Lecture Notes in Computer Science, pp. 408–412. Springer, Heidelberg (2012)
12. Jayamohan, M., Rajendran, C.: Development and analysis of cost-based dispatching rules for job shop scheduling. Eur. J. Oper. Res. **157**(2), 307–321 (2004)
13. Qing-dao-er ji, R., Wang, Y.: A new hybrid genetic algorithm for job shop scheduling problem. Comput. Oper. Res. **39**(10), 2291–2299 (2012)
14. Koza, J.R., Poli, R.: Genetic programming. In: Burke, E., Kendal, G. (eds.) Introductory Tutorials in Optimization and Decision Support Techniques, chap. 5. Springer (2005)
15. Nguyen, S., Zhang, M., Johnston, M., Tan, K.C.: Learning iterative dispatching rules for job shop scheduling with genetic programming. Int. J. Adv. Manuf. Technol. (2013)
16. Panwalkar, S.S., Iskander, W.: A survey of scheduling rules. Oper. Res. **25**(1), 45–61 (1977)
17. Pinedo, M.L.: Scheduling: Theory, Algorithms, and Systems, 3rd edn. Springer Publishing Company, Incorporated (2008)
18. Rice, J.R.: The algorithm selection problem. Adv. Comput. **15**, 65–118 (1976)
19. Smith-Miles, K., James, R., Giffin, J., Tu, Y.: A knowledge discovery approach to understanding relationships between scheduling problem structure and heuristic performance. In: Sttzle,

T. (ed.) Learning and Intelligent Optimization. Lecture Notes in Computer Science, vol. 5851, pp. 89–103. Springer, Heidelberg (2009)

20. Smith-Miles, K., Lopes, L.: Generalising algorithm performance in instance space: a timetabling case study. In: Coello, C. (ed.) Learning and Intelligent Optimization. Lecture Notes in Computer Science, vol. 6683, pp. 524–538. Springer, Heidelberg (2011)

21. Tay, J.C., Ho, N.B.: Evolving dispatching rules using genetic programming for solving multi-objective flexible job-shop problems. Comput. Ind. Eng. **54**(3), 453–473 (2008)

22. Tsai, J.T., Liu, T.K., Ho, W.H., Chou, J.H.: An improved genetic algorithm for job-shop scheduling problems using Taguchi-based crossover. Int. J. Adv. Manuf. Technol. **38**(9–10), 987–994 (2007)

23. Vázquez-Rodríguez, J.A., Petrovic, S.: A new dispatching rule based genetic algorithm for the multi-objective job shop problem. J. Heuristics **16**(6), 771–793 (2009)

24. Watson, J.P., Barbulescu, L., Whitley, L.D., Howe, A.E.: Contrasting structured and random permutation flow-shop scheduling problems: search-space topology and algorithm performance. INFORMS J. Comput. **14**, 98–123 (2002)

Occupational Diseases Risk Prediction by Genetic Optimization: Towards a Non-exclusive Classification Approach

Antonio di Noia, Paolo Montanari and Antonello Rizzi

Abstract This paper deals with the health risk prediction problem in workplaces through computational intelligence techniques. The available dataset has been collected from the Italian Local Health Authority (ASL) as part of the Surveillance National System. The main aim of this work is the design of a software application that can be used by occupational physicians in monitoring workers, performing a risk assessment of contracting some particular occupational diseases. The proposed algorithms, based on clustering techniques, includes a genetic optimization in order to automatically determine the weights of the adopted distance measure between patterns and the number of clusters for the final classifier's synthesis. In particular, we propose a novel approach, consisting in defining the overall classifier as an ensemble of class-specific ones, each trained to recognize patterns of risk conditions characterizing a single pathology. First results are encouraging and suggest interesting research tasks for further system development.

Keywords Occupational diseases · Risk prediction · Computational intelligence · Cluster analysis · Genetic algorithm · Non-exclusive classification · Ensemble of classifiers

A. di Noia (✉) · A. Rizzi
Department of Information Engineering, Electronics and Telecommunications,
University of Rome "La Sapienza", Via Eudossiana 18, Rome, Italy
e-mail: antonio.dinoia@uniroma1.it

A. Rizzi
e-mail: antonello.rizzi@uniroma1.it

P. Montanari
Research Department, National Institute for Insurance Against Accidents
at Work (INAIL), Rome, Italy
e-mail: p.montanari@inail.it

© Springer International Publishing Switzerland 2016
J.J. Merelo et al. (eds.), *Computational Intelligence*,
Studies in Computational Intelligence 620, DOI 10.1007/978-3-319-26393-9_5

1 Introduction

In many countries, spending for health systems has been gradually increasing in recent years, both in absolute terms and relative to Gross Domestic Product (GDP), because of the greater attention to welfare, and of the increase of life expectancy. Moreover in many countries the working life has been prolonged, and consequently exposure to risk factors has been extended, increasing the possibility of techno-pathies development.

Employee health care is gaining attention by both private and public companies, as well by OHS (Occupational Health and Safety) organizations worldwide. In fact, part of the public costs dedicated to healthcare can be reduced by monitoring and controlling workplaces hazards. In this scenario, an interesting challenge is to apply data mining, classification methods and knowledge discovery techniques to perform occupational health risk assessment. To this aim, several studies show that the application of computational intelligence techniques can lead to reveal the existence of structures in the data difficult to detect with other approaches. For example, in [1] have been developed a decision support system for employee healthcare, while in [2] have been applied clustering techniques to medical data to predict the likelihood of diseases. In [3] artificial neural networks have been applied for occupational diseases incidence forecast.

The goal of the work depicted in this paper is the development of a software application, based on computational intelligence techniques, for predicting the likelihood of contracting a disease as a function of some characteristics of both the worker and the working environment. In the following we describe the classification systems developed to this aim, showing the results yielded on a specific occupational diseases database. This database contains data collected over a decade by the Local Health Authority of the Italian Lombardy region. As a first processing step, the available dataset was filtered by considering only the most common diseases, identifying a set of features potentially related to the classification task at hand. A suited classification system relying on cluster analysis is the core procedure of the machine learning engine. In order to automatically determine both the parameters (features' weights) of the dissimilarity measure between patterns and to identify the best structural complexity of the classification model (number of clusters), a genetic algorithm has been employed to synthetize the best performing classifier.

2 Data Processing

The data set has been extracted from the archive of occupational diseases collected by the Local Health Authority (ASL) as part of the National System of Surveillance "MalProf", managed by the National Institute for Insurance against Accidents at Work (INAIL). The data set contains records for each pathology collected from ASL, storing information on registry of the worker, on his work history and his pathologies. For each worker more than one record may be present in the archive (a single record for each pathology).

In order to develop and test the whole prediction system, a first data set has been defined by considering only the cases of the Lombardy Italian region recorded in the period 1999–2009. Moreover, in order to simplify pattern's structure, only records related to workers with a single pathology and an occupational history consisting of a single working activity have been considered. This data set has been cleaned removing ambiguous records and inconsistent or missing data. This first preprocessing step yielded a data set of 3427 records as shown in Table 1; as a further filtering, the records associated with diseases scarcely represented (i.e. below 5 % threshold) were not considered, yielding the final data set of 2722 records, covering about 80 % of cases, highlighted with colored background in Table 1.

The final available data set has been partitioned into three subsets by random stratification: the training set (50 % of the total number of available patterns, denoted with STR), the validation set (25 %, SVAL) and the test set (the remaining 25 %, STEST). Table 2 shows the distribution of diseases and their labels as integer number codes.

The similarity between the subjects was evaluated through a distance function based on six features (Table 3), 3 numerical and 3 categorical, identified by both a preliminary analysis of data and a priori knowledge in the field.

Worker's profession is coded by a pair of characters based on the Italian version of the classification system ISCO. The International Standard Classification of Occupations (ISCO) is a tool for organizing jobs into a clearly defined set of groups

Table 1 Pathologies statistics in the considered data base

Disease	N. of records	Cumulative N. of records	Freq.	Cumulative Freq.
Hearing loss	1493	1493	0.436	0.436
Spinal diseases	334	1827	0.097	0.533
Musculoskeletal disorders (excluding spinal diseases)	288	2115	0.084	0.617
Tumors of the pleura and peritoneum	232	2347	0.068	0.685
Carpal tunnel syndrome	199	2546	0.058	0.743
Skin diseases	176	2722	0.051	0.794
Disorders of the ear (except hearing loss)	137	2859	0.040	0.834
Mental illness	98	2957	0.029	0.863
Diseases of the respiratory system	76	3033	0.022	0.885
Other diseases	394	3427	0.115	1

Table 2 Pathologies sorted by raw frequencies in descending order

Pathologies	Training set	Validation set	Test set
1—Hearing loss	747	373	373
	54.89 %	54.77 %	54.85 %
2—Spinal diseases	167	84	83
	12.27 %	12.33 %	12.21 %
3—Musculoskeletal disorders	144	72	72
	10.58 %	10.57 %	10.59 %
4—Tumors of the pleura and peritoneum	116	58	58
	8.52 %	8.52 %	8.53 %
5—Carpal tunnel	99	50	50
	7.27 %	7.34 %	7.35 %
6—Skin diseases	88	44	44
	6.47 %	6.47 %	6.47 %
Total	1361	681	680
	100 %	100 %	100 %

Table 3 Features

Code	Meaning	Data Type
x_1	Age of the worker at the time of disease assessment (years)	Numerical
x_2	Duration of the working period (months)	Numerical
x_3	Age at the beginning of the working period (years)	Numerical
x_4	Gender	Categorical
x_5	Profession carried out by the worker	Categorical
x_6	Company's economic activity	Categorical

according to the tasks and duties undertaken in the job. The economic activity of the company is coded by a pair of characters based on the Italian version of the NACE classification system. NACE (Nomenclature des Activités Économiques dans la Communauté Européenne) is a European industry standard classification system similar in function to Standard Industry Classification (SIC) and North American Industry Classification System (NAICS) for classifying business activities.

3 The Adopted Classification Systems

In order to design an algorithm able to evaluate the probability of contracting an occupational disease as a function of some characteristic of the worker, of his work history and of his work environment, the risk prediction problem has been reformulated as a classification problem. The classification system is a clustering based one, trained in a supervised fashion to discover clusters in S_{TR}. All algorithms were coded in C++ language.

3.1   Basic Algorithm

The core procedure for the synthesis of a classification model consists in clustering S_{TR} by the well–known k-means algorithm [4]. To this aim, an ad hoc dissimilarity measure δ between patterns was defined as a convex linear combination of inner dissimilarity measures δ_i between homologues features:

$$\delta(u, v) = \sum_{i=1}^{N} p_i \delta_i(u, v) \tag{1}$$

where N is the number of the features (six in our case) and $p_i \in \mathfrak{R}$; $p_i \in [0, 1]$ is the relative weight of the i-th feature.

The $\delta_i(u, v)$ distance between patterns u and v relative to the i-th feature have been defined differently on the basis of the considered feature type, which can be continuous or categorical (discrete nominal) values:

- for age (in years) and the duration of the activity (in months), δ_i is the Euclidean distance normalized in the unitary interval [0,1];
- for gender and economic activity of the company, in the case of concordance $\delta_i = 0$, otherwise $\delta_i = 1$ (simple match);
- for the job of the individual, in the case of concordance of both characters $\delta_i = 0$, in the case of concordance of only the first character $\delta_i = \frac{1}{2}$, otherwise $\delta_i = 1$.

The overall classification system has been designed to automatically determine the weights p_i of the dissimilarity measure (1) and the optimal number of clusters K, in order to maximize the classification accuracy:

$$f_1 = accuracy = \frac{1}{|S|} \sum_{x \in S} h(\omega_x, \omega_{Kx}) \tag{2}$$

where:
S is the labelled pattern set on which is computed the accuracy;
$\Omega = \{hearing\ loss,\ spinal\ diseases,\ musculoskeletal\ disorders,\ tumors\ of\ the\ pleura\ and\ peritoneum,\ carpal\ tunnel\ syndrome,\ skin\ diseases\}$ is the considered label set;
$\omega_x \in \Omega$ is the pathology of worker $x \in S$ (ground true class label);
$\omega_{Kx} \in \Omega$ is the label assigned by the classification model to x;

$$h(\omega_x, \omega_{Kx}) = 1\ if\ \omega_x = \omega_{Kx};$$

$$h(\omega_x, \omega_{Kx}) = 0\ if\ \omega_x \neq \omega_{Kx};$$

In order to perform this optimization task, we have developed a suited implementation of a genetic algorithm [5]. The generic individual of the population subject to evolution by genetic operators is formed of two data structures (sections) [6] for a total of 7 parameters to be optimized:

1. a vector of 6 real numbers in the range [0, 1], corresponding to the weights associated with the features in the distance function δ;
2. an integer between 2 and a maximum value fixed in the system parameters, corresponding to the number of clusters to be used for training set clustering.

From one generation to the next, each individual in the GA is evaluated by a fitness function defined as the accuracy (2), computed on S_{VAL}. The selection is simulated using a roulette wheel operator. The crossover and mutation affect the entire individual, formed by six weights and the number of clusters.

The individuals of the initial population of the GA are generated as random samples. For each individual, a k-means clustering procedure is performed on the training set with distance weights fixed in the first section of the individual's genetic code and setting the number of clusters as the integer number stored in the second section. Once obtained a partition of the S_{TR}, each cluster is assigned with a unique label, defined as the most frequent pathology in the cluster. Successively the fitness is computed as the classification accuracy on the validation set, according to (2).

Reproduction, crossover and mutation are applied to the individuals of the GA to evolve the population, until a stop criterion based on a maximum number of generations is met. The algorithm is summarized in Table 4.

Table 4 Summary of the basic algorithm

Input parameters
– Maximum number of clusters: $Kmax$
– Number of population's individuals in the GA: Pop
– Number of generations of GA: $nGeneration$.
1. Reading data from S_{TR} and S_{VAL}
2. Initialization *(Generation = 0)*
For $j = 1$ to Pop
• Random assignment of weights p_i of the 6 features and of the value $K \leq Kmax$
• Clustering of the elements of S_{TR} into K clusters using the distance function (Eq. 1) in the individual j
• Evaluation of the fitness function (2) or (3) on S_{VAL}
3. For $q = 1$ to $nGeneration$
• Application of elitism
• Repeat
– Selection of individuals of the old population by roulette wheel operator
– Crossover between pairs of the selected individuals
– Mutation with a low probability on each element
– Clustering of S_{TR} in K clusters using the distance function (1) with the parameters encoded in the individual
– Evaluation of the fitness function (2) or (3) on S_{VAL}

The distribution of pathologies in the data set shows that class labels (diseases) are not well balanced, and this could distort the values of fitness by giving excessive importance to the most frequent pathologies. Therefore, it is introduced a fitness function variant aiming to equally weight all misclassifications, regardless of their number. The new fitness (Eq. 3) is given by the weighted accuracy, i.e. the mean value of the percentages of correct answers for each pathology:

$$f_2 = accuracy_{weighted} = \frac{1}{|\Omega|} \sum_{\omega \in \Omega} \frac{1}{|S_\omega|} \sum_{x \in S} h(\omega_x, \omega_{Kx}) \tag{3}$$

where:
S_ω is the subset of S of all elements associated with pathology $\omega \in \Omega$ ($S_\omega \subset S$). Tests were performed with both fitness functions, and the results have been compared. In the following, we will refer to this version as BA (Basic Algorithm).

3.2 CBA: A Variant of the Basic Algorithm

The basic algorithm leads to the formation of clusters containing more than one disease. The label associated with the cluster coincides with the most frequent pathology in it. This procedure cannot assure the presence of at least one cluster for each class. To make sure that all the pathologies are represented in the final classification model, a second version of the proposed classification system has been designed.

For this purpose, the training set S_{TR} has been partitioned into six subsets, one for pathology. The new algorithm runs six cluster analyses in parallel, one for each of the six subsets of S_{TR}. As a consequence, each cluster will contain patterns associated with a unique class label and will consequently be directly labeled (see Fig. 1). The union of the six sets of labeled clusters originated will be directly employed for the classification model definition.

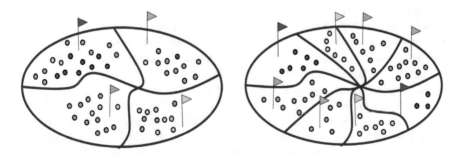

Fig. 1 Examples of clusters labelling by BA (*left*) and CBA (*right*) class assignment rule

The way a generic individual in the GA is coded has been adapted to the new algorithm. In particular, the second part of the individual no longer contains a single integer, but six distinct integers, each representing the number of clusters to be determined in cluster analysis performed in parallel on each subset of the training set (one for each class label). The initialization step of the first generation of the GA is similar to the BA initialization step. As for the previous version, we considered both the fitness functions f_1 and f_2 computed on S_{VAL} (Eqs. (2) and (3)) for individual fitness evaluation. In the following, we will refer to this version as CBA (Class-aware Basic Algorithm).

3.3   The NCBA Algorithm

Both BA and CBA are designed to face an exclusive classification problem, where the final classification model is intended to assign a unique class label to the input pattern, where classes are considered as mutually exclusive. However, as in the case of any clinical diagnostic system, the presence of risk factors for a particular disease does not exclude logically that the same worker could develop another pathology as well. For this reason, we have developed a new classification approach where class labels are considered as non-exclusive. To this aim, a specific classification model is trained for each class (pathology) included in the data set at hand. Once synthetized, the classification model M_i will act as a recognizer for the i-th pathology. The synthesis of M_i is based on a re-labeling procedure of the original training set. Specifically, considered the i-th class label ω_i, each pattern in S_{tr} originally belonging to a class different from ω_i will be associated with a common code (for instance "0"), while ω_i will be re-encoded with a different one (for example "1"). Let $S_{tr}^{(i)}$ be the re-encoded training set. M_i will be synthetized by the CBA variant, starting from $S_{tr}^{(i)}$. Thus, each cluster will contain patterns associated with a unique class label and will consequently be directly labeled. All the classification models (six in our case) will be embedded in an ensemble of classifiers as an overall diagnostic tool, conceived as a set of classifiers working in parallel on the same input pattern, able to recognize more than one possible pathology at the same time.

Note that this approach allows the introduction of new pathologies, without the necessity to train again the classifier on the whole available dataset. For this reason, this last version is characterized by much better scalability properties, in terms of computational cost of the training procedure, as well as the requirements for a parallel implementation on a distributed system.

The generic individual of the population of the GA has been adapted to the new algorithm; in particular, the second part of the individual contains two integers, the first representing the number of clusters to use in the cluster analysis of sick subjects, the second integer represents the number of clusters to use in the cluster analysis of healthy subjects. The two cluster analysis are performed in parallel, as in CBA. The initialization step of the first generation of the GA is similar to the

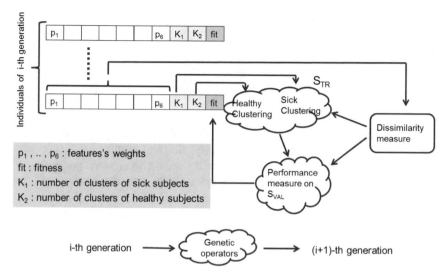

Fig. 2 The NCBA algorithm

previous algorithms. As concerns this last version, we considered only the fitness functions f_2 (Eq. 3) computed on S_{VAL} for individual fitness evaluation.

The whole training algorithm is depicted in Fig. 2.

4   Results

Let us consider a generic diagnostic test conceived to screen people for a disease. The test outcome can be positive (predicting that the person is affected by the considered disease) or negative (predicting that the person is healthy). The test result for each subject may or may not match the subject's actual status. We can have the following outcomes:

- True positive: sick people correctly diagnosed as sick
- False positive: healthy people incorrectly identified as sick
- True negative: healthy people correctly identified as healthy
- False negative: sick people incorrectly identified as healthy

Given the content of the dataset, formed only by workers affected by at least one pathology, for each disease we consider as healthy the workers not affected from that particular pathology. Table 5 shows an example of the usual representation of these values by the confusion table in the case of the pathology "1-hearing loss" using BA with the fitness function f_1. The columns "Positive to test" and "Negative to test" of Confusion tables contain the number of workers that the test predicts respectively as sick (i.e. affected by the disease in question) or healthy (i.e. affected

Table 5 Confusion table for pathology "1-hearing loss"—BA using f_1

	Positive to test	Negative to test
Actual positive	True positives	False negatives
	351	22
Actual negative	False positives	True negatives
	124	183

by other diseases). The rows "Actual positive" and "Actual negative" contain the number of those who actually are, respectively, sick and healthy.

The confusion tables allow a better understanding than mere proportion of correct guesses (accuracy). Especially when dealing with diagnostic tools, the average accuracy is not a reliable metric for the real performance of a classifier, because it will yield misleading results if the data set is unbalanced. For example, if there were 95 sick-labelled patterns and only 5 healthy ones in the data set, the classifier could be biased easily into classifying all the samples as sick. The overall accuracy would be 95 %, but in practice the classifier would have a 100 % recognition rate for the sick class and a 0 % recognition rate for the wealthy class. For these reasons, we reported for all experiments the overall confusion table containing the average values for all classes.

We have conducted 5 series of experiments, the first two with BA using both the fitness functions f_1 and f_2, the third and fourth series with CBA using the fitness f_1 and f_2, the fifth series related to NCBA with the fitness f_2.

All experiments were conducted using the GA by evolving a population of 100 individuals for 50 generations. All performances reported in the following tables are computed on the test set. Data in all the following tables refer to the best individual found by the genetic optimization.

As concerns the first test (BA with f_1), the maximum number of clusters was fixed to 20. In Table 6 the first six columns represents the confusion tables for the disease specified in the column header, while the last one contains the average values of the first six columns.

The results of the second experiment, employing BA with fitness f_2, are shown in Table 7. Also in this test the maximum number of clusters was fixed to 20.

The number of clusters of the best individual after the last generation is 20, of which 13 are labeled as "1—hearing loss", 2 as "2—spinal diseases", 1 as "3—musculoskeletal disorders", 2 as " 4—tumors of the pleura and peritoneum", 1 as "5—carpal tunnel" and 1 as " 6—skin diseases."

Table 6 Summarized data of the confusion tables—BA with f_1

	Pathology						Average values
	1	2	3	4	5	6	
True positives	351	36	0	32	23	13	75.8
False positives	124	56	0	8	30	7	37.5
False negatives	22	47	72	26	27	31	37.5
True negatives	183	541	608	614	600	629	529.2

Table 7 Summarized data of the confusion tables—BA with f_2

	Pathology						Average values
	1	2	3	4	5	6	
True positives	338	35	14	32	21	12	75.3
False positives	115	29	30	17	34	3	38.0
False negatives	35	48	58	26	29	32	38.0
True negatives	192	568	578	605	596	633	528.7

Table 8 Summarized data of the confusion tables—CBA with f_1

	Pathology						Average values
	1	2	3	4	5	6	
True positives	312	42	12	35	1	19	70.2
False positives	127	59	19	15	20	19	43.2
False negatives	61	41	60	23	49	25	43.2
True negatives	180	538	589	607	610	617	523.5

In the third experiment, based on CBA with fitness f_1, the maximum number of clusters was fixed to 20 for each pathology. The total number of clusters in the best individual is 36, with the following class distribution: 20 are labelled as "1—hearing loss", 6 as "2—spinal diseases", 2 as "3—musculoskeletal disorders", 4 as "4—tumors of the pleura and peritoneum", 2 as "5—carpal tunnel" and 2 as "6—diseases of the skin". In Table 8 are summarized the data of the six tables of confusion (one for disease).

In the fourth experiment CBA with fitness f_2 has been tested. As in the third experiment, the maximum number of clusters was fixed to 20 for each pathology. In correspondence of the best individual after the last generation, the total number of clusters was 71, with the following class distribution: 18 are labeled as "1—hearing loss", 10 as "2—spinal diseases", 7 as "3—musculoskeletal disorders", 11 as "4—tumors of the pleura and peritoneum", 15 as "5—carpal tunnel", 10 as "6—skin diseases". The results for this experiment are shown in Table 9.

The fifth experiment was based on the NCBA algorithm with fitness f_2. This algorithm provides a distinct classifier for each disease. For each one the maximum

Table 9 Summarized performances of the confusion tables—CBA with f_2

	Pathology						Average values
	1	2	3	4	5	6	
True positives	210	44	17	35	30	25	60.2
False positives	48	61	45	30	89	46	53.2
False negatives	163	39	55	23	20	19	53.2
True negatives	259	536	563	592	541	590	513.5

Table 10 Summarized performances of the confusion tables—NCBA algorithm with f_2

	Pathology						Average values
	1	2	3	4	5	6	
True positives	314	62	56	54	44	38	94.7
False positives	84	88	204	73	116	110	112.5
False negatives	59	21	16	4	6	6	18.7
True negatives	223	509	404	549	514	526	454.2

Table 11 Chromosome of GAs

	BA, f_1	BA, f_2	CBA, f_1	CBA, f_2	NCBA, f2 hearing loss	NCBA, f2 tumors of the pleura
Feature x_1	1.000	1.000	0.870	0.660	1.000	1.000
Feature x_2	0.041	0.134	0.894	0.709	0.352	0.455
Feature x_3	0.078	0.346	0.569	1.000	0.514	0.269
Feature x_4	0.592	0.519	1.000	0.196	0.450	0.147
Feature x_5	0.189	0.220	0.280	0.726	0.237	0.911
Feature x_6	0.265	0.076	0.096	0.141	0.118	0.107
N. clusters	10	20	–	–	–	–
N. clusters pathology 1	–	–	20	18	–	–
N. clusters pathology 2	–	–	6	10	–	–
N. clusters pathology 3	–	–	2	7	–	–
N. clusters pathology 4	–	–	4	11	–	–
N. clusters pathology 5	–	–	2	15	–	–
N. clusters pathology 6	–	–	2	10	–	–
N. clusters sick	–	–	–	–	48	47
N. clusters healthy	–	–	–	–	44	41

number of clusters was fixed to 50 for sick subjects and 50 for healthy subjects. The results are shown in Table 10.

Table 11 shows the genetic code of the best individual produced by the GA for each experiment. The first six parameters encode the weight of the features (normalized values) and the other parameters encode the clusters number. For the NCBA algorithm, in Table 11 are shown only the results of the two best performing classifiers: "1—hearing loss" and "4—tumors of the pleura and peritoneum".

Table 12 Diagnostic Tests indicators	Sensitivity = True positive/Actual positive
	Specificity = True negative/Actual negative
	Positive predictive value = True positive/Positive to test
	Negative predictive value = True negative/Negative to test

Another significant tool for performance analysis, commonly used in the evaluation of diagnostic tests, consists in the use of *sensitivity, specificity, positive predictive value* and *negative predictive value*, defined in Table 12.

In Table 13a–f are shown the diagnostic test's indicators relative to the six considered diseases, while in Table 14 the averages values are summarized for a better insight of comparison between the adopted algorithms.

As we can see in the Table 13a–f, the first experiment (BA using f1) shows for the subset "hearing losses" (the most frequent class in the considered dataset) a very high sensitivity value (0.941) and the specificity presents a value close to 0.6. For all the other pathologies the sensitivity is about 0.5, except for "3—musculoskeletal disorders" that presents a value of 0 because the pathology was never predicted by the algorithm, and the specificity is close to 1. These first results show that function f1 privileges the most frequent pathology. In the second experiment (BA with f2), the sensitivity no longer has null values. Specificity and sensitivity have values similar to the ones in the previous experiment.

The third experiment (CBA with f_1) shows performance values in general slightly worse compared to BA. However, there is an improvement for the sensitivity of some pathologies.

The fourth experiment (CBA with f_2) shows that the use of the second version of the fitness function f_2 compared to f_1 has led to an improvement of the sensitivity, except for the group "hearing losses". Regarding the specificity, for all pathologies the values are greater than 0.84.

The fifth experiment (NCBA algorithm) shows performance values in general slightly better compared to the other algorithms. For all pathologies the sensitivity is greater than 0.74 and the specificity is greater than 0.66.

Table 14 (average values on all pathologies of diagnostic test's indicators) allows a quick comparison between the performance of the algorithms. This table shows how the NCBA algorithm presents the highest sensitivity. Regarding the specificity and the negative and positive predictive values, we have substantially similar behaviours for the five experiments. The high values, close to unity in all experiments, for specificity and negative predictive values can be considered very interesting. As regard to execution time, the CBA and NCBA algorithms have better performance by reducing the time to a third compared to BA, because clustering procedures are performed on smaller data sets. The performances are stable over multiple runs, assuring a good reliability to the results. Note that in general for the groups "hearing loss" (the largest group) and "tumors of the pleura and peritoneum" (more severe disease) the results are better than for other diseases, including the sensitivity and the positive predictive value.

Table 13 Diagnostic test's indicators

a Diagnostic test's indicators relative to "1—hearing loss"

	BA, f1	BA, f2	CBA, f1	CBA, f2	NCBA, f2
Sensitivity	0.941	0.906	0.836	0.563	0.842
Specificity	0.596	0.625	0.586	0.844	0.726
Negative predictive value	0.893	0.846	0.747	0.614	0.791
Positive predictive value	0.739	0.746	0.711	0.814	0.789
Average values	0.792	0.781	0.720	0.709	0.787

b Diagnostic test's indicators relative to "2—spinal diseases"

	BA, f1	BA, f2	CBA, f1	CBA, f2	NCBA, f2
Sensitivity	0.434	0.422	0.506	0.530	0.747
Specificity	0.906	0.951	0.901	0.898	0.853
Negative predictive value	0.920	0.922	0.929	0.932	0.960
Positive predictive value	0.391	0.547	0.416	0.419	0.413
Average values	0.663	0.711	0.688	0.695	0.743

c Diagnostic test's indicators relative to "3—musculoskeletal disorders"

	BA, f1	BA, f2	CBA, f1	CBA, f2	NCBA, f2
Sensitivity	0.000	0.194	0.167	0.236	0.778
Specificity	1.000	0.951	0.969	0.926	0.664
Negative predictive value	0.894	0.909	0.908	0.911	0.962
Positive predictive value	0.000	0.318	0.387	0.274	0.215
Average values	0.474	0.593	0.608	0.587	0.655

d Diagnostic test's indicators relative to "4—tumors of the pleura and peritoneum"

	BA, f1	BA, f2	CBA, f1	CBA, f2	NCBA, f2
Sensitivity	0.552	0.552	0.603	0.603	0.931
Specificity	0.987	0.973	0.976	0.952	0.883
Negative predictive value	0.959	0.959	0.963	0.963	0.993
Positive predictive value	0.800	0.653	0.700	0.538	0.425
Average values	0.825	0.784	0.811	0.764	0.808

e Diagnostic test's indicators relative to "5—carpal tunnel"

	BA, f1	BA, f2	CBA, f1	CBA, f2	NCBA, f2
Sensitivity	0.460	0.420	0.020	0.600	0.880
Specificity	0.952	0.946	0.968	0.859	0.816
Negative predictive value	0.957	0.954	0.926	0.964	0.988
Positive predictive value	0.434	0.382	0.048	0.252	0.275
Average values	0.701	0.675	0.490	0.669	0.740

f Diagnostic test's indicators relative to "6—skin diseases"

	BA, f1	BA, f2	CBA, f1	CBA, f2	NCBA, f2
Sensitivity	0.295	0.273	0.432	0.568	0.864
Specificity	0.989	0.995	0.970	0.928	0.827
Negative predictive value	0.953	0.952	0.961	0.969	0.989
Positive predictive value	0.650	0.800	0.500	0.352	0.257
Average values	0.722	0.755	0.716	0.704	0.734

Table 14 Diagnostic test's indicators—average values

	BA, f1	BA, f2	CBA, f1	CBA, f2	NCBA, f2
Sensitivity	0.447	0.461	0.427	0.517	0.840
Specificity	0.905	0.907	0.895	0.901	0.795
Negative predictive value	0.929	0.923	0.906	0.892	0.947
Positive predictive value	0.503	0.574	0.460	0.442	0.396
Average values	0.696	0.716	0.672	0.688	0.744

5 Conclusions

In this chapter we compare three different classification systems designed to face the health risk prediction problem in workplaces, aiming in developing a software application to help reducing costs in performing clinical trials on all the interested workers. First results are encouraging and suggest further improvements. In particular, it is important to underline that good negative predictive values can be considered a prerequisite for the practical use of the above classification systems in a suited automatic screening procedure, since a negative classification for a given worker is sufficient to reliably ascertain his health status. By examining the features weights (Table 11), no explicit ranking can be deduced, since there is no accordance between the considered algorithms. However, in all the experiments, only the economic activity of the company (feature x_6) seems less important than the other features, suggesting to replace this feature with others more related with the classification problem at hand.

References

1. Mukherjee, C., Gupta, K., Nallusamy, R.: A decision support system for employee healthcare. In: Third International Conference on Services in Emerging Markets (2012)
2. Paul, R., Md. Latiful Hoque, A.S.: Clustering medical data to predict the likelihood of diseases. In: IEEE—Fifth International Conference on the Digital Information Management (ICDIM) (2010)
3. Huang, Z., Yu, D., Zhao, J.: Application of neural networks with linear and nonlinear weights in occupational disease incidence forecast. In: Circuits and systems. IEEE APCCAS 2000 (2000)
4. Jane, A.K., Dubes, R.C.: Algorithms for Clustering Data. Prentice-Hall, Englewood Cliffs (1988)
5. Sastry, K., Goldberg, D., Kendall, G.: Genetic algorithms. In: Search Methodologies. Springer, New York (2005)
6. Savinov, A.A.: Mining possibilistic set-valued rules by generating prime disjunctions. In PKDD'99, 3rd European Conference on Principles and Practice of Knowledge Discovery in Databases, vol. 1704, pp. 536–541. Springer (1999)

A Statistical Approach to Dealing with Noisy Fitness in Evolutionary Algorithms

J.J. Merelo, Zeineb Chelly, Antonio Mora, Antonio Fernández-Ares,
Anna I. Esparcia-Alcázar, Carlos Cotta, P. de las Cuevas and Nuria Rico

Abstract In most computer games as in life, the outcome of a match is uncertain due to several reasons: the characters or assets appear in different initial positions or the response of the player, even if programmed, is not deterministic; different matches will yield different scores. That is a problem when optimizing a game-playing engine: its fitness will be noisy, and if we use an evolutionary algorithm it will have to deal with it. This is not straightforward since there is an inherent uncertainty in the true value of the fitness of an individual, or rather whether one

J.J. Merelo (✉) · A. Mora · A. Fernández-Ares
Department of Computer Architecture and Technology, University of Granada,
ETSIIT, and CITIC, 18071 Granada, Spain
e-mail: jmerelo@geneura.ugr.es
URL: http://citic.ugr.es

A.Mora
e-mail: amorag@geneura.ugr.es

A. Fernández-Ares
e-mail: antares@geneura.ugr.es

Z. Chelly
Laboratoire de Recherche Opérationelle de Décision et de Contrôle de Processus,
Institut Supérieur de Gestion, Le Bardo, Tunisia
e-mail: zeinebchelly@yahoo.fr

C. Cotta
Departamento de Lenguajes y Ciencias de la Computaciôn, Universidad de Málaga,
Málaga, Spain
e-mail: ccottap@lcc.uma.es

N. Rico
Departamento de Estadística e Investigación Operativa, Universidad de Granada,
Granada, Spain
e-mail: nrico@ugr.es

A.I. Esparcia-Alcázar · P. de las Cuevas
Software Production Methods Research Centre, Universitat Politècnica de València,
Valencia, Spain
e-mail: aesparcia@ieee.org

P. de las Cuevas
e-mail: paloma@geneura.ugr.es

© Springer International Publishing Switzerland 2016
J.J. Merelo et al. (eds.), *Computational Intelligence*,
Studies in Computational Intelligence 620, DOI 10.1007/978-3-319-26393-9_6

79

chromosome is better than another, thus making it preferable for selection. Several methods based on implicit or explicit average or changes in the selection of individuals for the next generation have been proposed in the past, but they involve a substantial redesign of the algorithm and the software used to solve the problem. In this paper we propose new methods based on incremental computation (memory-based) or fitness average or, additionally, using statistical tests to impose a partial order on the population; this partial order is considered to assign a fitness value to every individual which can be used straightforwardly in any selection function. Tests using several hard combinatorial optimization problems show that, despite an increased computation time with respect to the other methods, both memory-based methods have a higher success rate than *implicit* averaging methods that do not use memory; however, there is not a clear advantage in success rate or algorithmic terms of one method over the other.

Keywords Evolutionary algorithms · Noisy optimization problems · Dynamic problems

1 Introduction

In our research on the optimization of the behavior of bots or game strategies, we have frequently found that the fitness of a bot is *noisy*, in the sense that repeated evaluations will yield different values [1] which is a problem since fitness is the measure used to select individuals for reproduction. If we look at it in a more general setting, noise in the fitness of individuals in the context of an evolutionary algorithm has different origins. It can be inherent to the individual that is evaluated; for instance, in [1] a game-playing bot (autonomous agent) that includes a set of application rates is optimized. This results in different actions in different runs, and obviously different success rates and then fitness. Even comparisons with other individuals can be affected: given exactly the same pair of individuals, the chance of one beating the other can vary in a wide range. In other cases like the one presented in the MADE environment, where whole worlds are evolved [2] the same kind of noisy environment will happen. When using evolutionary algorithms to optimize stochastic methods such as neural networks [3] using evolutionary algorithms the measure that is usually taken as fitness, the success rate, will also be noisy since different training schedules will result in slightly different success rates.

The examples mentioned above are actually included in one of the four categories where uncertainties in fitness are found, fitness functions with intrinsic noise. These four types include also, according to [4] approximated fitness functions (originated by, for instance, surrogate models); robust functions, where the main focus lies on finding values with high tolerance to change in initial evaluation conditions, and finally dynamic fitness functions, where the *inherent* value of the function changes with time. Our main interest will be in the first type, since it is the one that we have actually met in the past and has led to the development of this work.

At any rate, in this paper we will not be dealing with actual problems; we will try to simulate the effect of noise by adding to the fitness function Gaussian noise centered in 0 and $\sigma = 1, 2, 4$. We will deal mainly with combinatorial optimization functions with noise added having the same shape and amplitude, that we actually have found in problems so far. In fact, from the point of view of dealing with fitness, these are the main features of noise we will be interested in.

This paper is an extension of [5]. The main problem found in those set of experiments was the amount of evaluations needed to find a solution to the problem; while an individual was in the population, it was re-evaluated (re-sampled). However, statistically a limited amount of evaluations is usually enough to make meaningful, and statistically significant, comparisons. That is why in this paper we have done two kind of improvements: improve speed, because implementation always matters [6], and limit the amount of evaluations while checking if the results achieved (mainly success rate) remain the same.

The rest of the paper is structured as follows: in the next section we describe the state of the art in the treatment of noise in fitness functions. The method we propose in this paper, called Wilcoxon Tournament, will be shown in Sect. 4; experiments are described and results shown in Sect. 5; finally its implications are discussed in the last section of the paper.

2 State of the Art

The most comprehensive review of the state of the art in evolutionary algorithms in *uncertain* environments was done by [4], although recent papers such as [7–9] include a brief update of the state of the art. In that first survey of evolutionary optimization by Jin and Branke in uncertain environments that uncertainty is categorized into noise, robustness, fitness approximation and time-varying fitness functions, and then, different options for dealing with it are proposed. In principle, the approach presented in this paper was designed to deal with the first kind of uncertainty, noise in fitness evaluation, although it could be argued that there is uncertainty in the true fitness as stated in the third category. In any case it could be applied to other types of noise, since it depends more on the shape and intensity of noise than the origin.

In the same situation, a noisy fitness evaluation, several solutions have been proposed and explained in the survey [4]. These will be explained next. For scientists not concerned on solving the problem of noise, but on a straightforward solution of the optimization problem without modification of existing tools and methodologies, a usual approach is just to disregard the fact that the fitness is noisy and use whatever value is returned by a single evaluation or after re-evaluation each generation. This was the option in our previous research in games [1, 10, 11] and evolution of neural networks [3, 12] and leads, if the population is large enough, to an *implicit averaging* as mentioned in [4]. In fact, selection used in evolutionary algorithms is also stochastic, so noise in fitness evaluation will have the same effect as randomness in selection or a higher mutation rate, which might make the evolution process easier and not harder in some particular cases [7]. In fact, Miller and Goldberg proved that an infinite population would not be affected by noise [13] and Jun-Hua and Ming studied the effect of noise in convergence rates [14], proving that an elitist genetic algorithm finds at least one solution, although with a lowered convergence rate. But real populations are finite, so the usual approach to dealing with fitness with a degree of randomness is to increase the population size to a value bigger than would be needed in a non-noisy environment. In fact, it has been recently proved that using *sex*, that is, crossover, is able to deal successfully with noise [15], while an evolutionary algorithm based on mutation would suffer a considerable degradation of performance. However, crossover is part of the standard kit of evolutionary algorithms, so using it and increasing the population size has the advantage that no special provision or change to the implementation has to be made, just different values of the standard parameters.

Another more theoretically sound way is using a statistical central tendency indicator, which is usually the *average*. This strategy is called *explicit averaging* by Jin and Branke and is used, for instance, in [14]. Averaging decreases the variance of fitness but the problem is that it is not clear in advance what would be the sample size used for averaging [16]. Most authors use several measures of fitness for each new individual [17], although other averaging strategies have also been proposed, like averaging over the neighbourhood of the individual or using *resampling*, that is, more measures of fitness in a number which is decided heuristically [18]. This assumes that there is, effectively, an average of the fitness values which is true for Gaussian random noise and other distributions such as Gamma or Cauchy but not necessarily for all distributions. To the best of our knowledge, other measures like the median which might be more adequate for certain noise models have not been tested; the median always exists, while the average might not exist for non-centrally distributed variables. Besides, most models keep the number of evaluations fixed and independent of its value, which might result in bad individuals being evaluated many times before being discarded; some authors have proposed *resampling*, that is, re-evaluate the individuals a number of times to increase the precision in fitness [19, 20], which will effectively increase the number of evaluations and thus slow down the search. In any case, using average is also a small change to the overall algorithm framework, requiring only using as new fitness function the average of several evaluations. We will try to address this in the model presented in this paper.

These two approaches that are focused on the evaluation process might be complemented with changes to the selection process. For instance, using a threshold [20, 21] that is related to the noise characteristics to avoid making comparisons of individuals that might, in fact, be very similar or statistically the same; this is usually called *threshold selection* and can be applied either to explicit or implicit averaging fitness functions. The algorithms used for solution, themselves, can be also tested, with some authors proposing, instead of taking more measures, testing different solvers [22], some of which might be more affected by noise than others. However, recent papers have proved that sampling might be ineffective [9] in some types of evolutionary algorithms, adding running time without an additional benefit in terms of performance. This is one lead we will use in the current paper.

Any of these approaches do have the problem of statistical representation of the *true* fitness, even more so if there is not such a thing, but several measures that represent, *as a set* the fitness of an individual. This is what we are going to use in this paper, where we present a method that uses resampling via an individual memory and use either explicit averaging or statistical tests like the non-parametric Wilcoxon test. First we will examine and try to find the shape of the noise that actually appears in games; then we will check in this paper what is the influence on the quality of results of these two strategies and which one, if any, is the best when working in noisy environments.

3 Noise in Games: An Analysis

In order to measure the nature of noise, we are going to use the data that is available to us, fitness evaluations in the Planet Wars game, that has been used as a framework for evolving strategies, e.g., [1], and other game contents, e.g., [23]. In this game when one player *invades* a planet the outcome depends deterministically on the number of *ships* each player has; however, players are positioned randomly with the constraint that they should be far enough from each other; other than that, any planet in the game can be an initial position. The planetary map is also random, chosen from 100 possibilities. Besides, in the strategy used in that game by our bot [24], actions are not deterministic, since every player is defined by a set of probabilities to take one course of action.

In order to have enough measures to model the noise, an evolutionary algorithm with standard parameters was run with the main objective of measuring the behavior of fitness. A sample of ten individuals from generation 1 and another 10 from generation 50 were extracted. The fitness of each individual was measured 100 times. The main intention was also to see how noise evolved with time. Intuitively we thought that, since the players become better with evolution, the noise and thus the standard deviation would decrease. However, what we found is shown in Fig. 1, which shows a plot of the standard deviation in both generations and illustrates the fact that the spread of fitness values is *bigger* as the evolution proceeds, going from around 0.15 to around 0.20. This result might be a bit misleading since the average values of the fitness increase at the same time, but it implies that the noise level might be around 20 % of the *signal* in these kind of problems.

By taking these measures, we were also interested in checking whether, in fact, the normal distribution is the best fit for the fitness measures, thus making averages of fitness appropriate for modeling the whole set of fitness measures. We tested three distributions: Gamma, Weibull and normal (Gaussian) distribution after doing an initial test that included (and discarded) Cauchy and Exponential. All this analysis was done using the library `fitrdist` in R, and data as well as scripts needed to do it are publicly available in the GitHub repository. After trying to fit these three distribution to data in generation 1 and 50, we analyzed goodness-of-fit using the

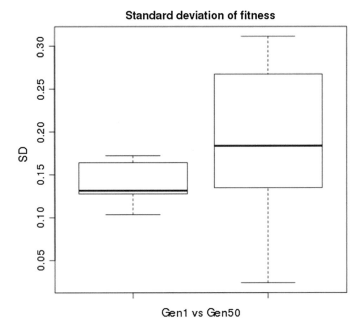

Fig. 1 Boxplot of standard deviation of noise fitted to a normal distribution, *left* for the first generation and *right* for the 50th. Fitness averages around 1

same package and the `gofstat` function. This function yields several measures of goodness, including the Akaike Information Criterion and the Kolmogorov-Smirnov statistic.

What we found was that, in all cases, Gamma is the distribution that better fits the data. That does not mean that the noise effectively follows this distribution, but that it is, of all distributions, the one that best fits the data. In fact, just a few individuals have a good fit (to 95 % accuracy using the Kolmogorov-Smirnov), and none of them in generation 50. The fit for an individual in the first generation that does follow that distribution (individual 8) is shown in Fig. 2. The main difference between Gamma and Gaussian is that Gamma is skewed, instead of being centered about a particular value.

This figure also shows that, even if it is skewed, its skewness is not too high which makes it close to the standard distribution (which is considered a good approximation if $k > 10$). However, it is interesting to note the skewness inherent in the distribution as a feature of noise, at least in this particular game.

Some interesting facts that can be deduced from these measures is that in general, fitness is skewed and its standard deviation has a high value, up to 20 % in some cases. Besides, it follows a gamma distribution, which, if we want to model noise accurately, should be the one used. However, we are rather interested in the overall shape of noise so since the skewness parameter of the gamma distribution is rather

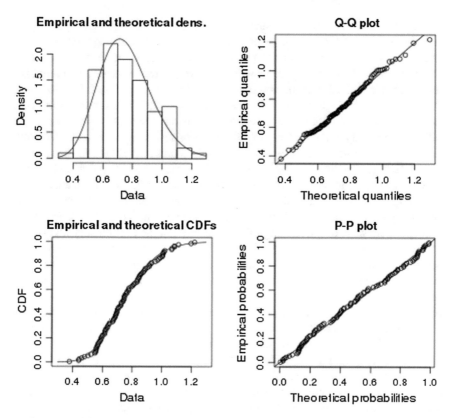

Fig. 2 Fit of the fitness value of an individual in the first generation of the evolutionary algorithm to a gamma distribution, showing an histogram in the *top left corner*, CDFs in the *bottom left corner* and quantile-quantile and percentile-percentile plots in the *right hand side*

high we will use, in this paper, a Gaussian noise to simulate it; Gaussian is a good approximation of Gamma for high values of the skewness parameter. This will be used in the experiments shown in Sects. 5 and 4.

4 Fitness Memory and Statistical Significant Differences

As indicated in the previous section, most explicit averaging methods use several measures to compute fitness as an average, with resampling, that is, additional measures, in case comparisons are not statistically significant. In this paper we will introduce a fitness *memory*, which amounts to a resampling every generation an individual survives. An individual is born with a fitness memory of a single value, with memory size increasing with *survival* time. This is actually a combination of an implicit and an explicit evaluation strategy: *younger* individuals are rejected outright if their fitness computed after a single evaluation is not enough to participate in the pool, while *older* ones use several measures to compute average fitness, which means that averages will be a more precise representation of actual value. As evolution proceeds, the best individuals will, effectively, have an underlying non-noisy best value. We will call this method *Extended Temporal Average* or ETA. However, since average is a single value, selection methods might, in fact, pick as the best individuals some that are not if the comparison is not statistically significant; this will happen mainly in the first and middle stages of search, which might effectively eliminate them from the pool or not adequately represent individuals that constitute, in fact, good solutions. That is why we introduce an additional feature: using Wilcoxon test [25] for comparing not the average, but all fitness values attached to an individual. This second method introduces a partial order in the population pool: two individuals might be different (one better than the other) or not [26]. There are many possible ways of introducing this partial order in the evolutionary algorithm; however, what we have done is to pair individuals a certain number of times (10, by default) and have every individual score a point every time it is better than the other in the couple; its score will be decreased by one if it is the worse one. An individual that is better that all its couples will have a fitness of 20; one whose comparisons are never significant according to the Wilcoxon test will score exactly 10, the same as if it wins as many times as it loses, and the one that always loses will score 0. We will call this method Partial order Wilcoxon-test, or WPO for short.

Initial tests, programmed in Perl using `Algorithm::Evolutionary` [27] (available with an open source license[1]) were made with these two types of algorithms and the Trap function [26], showing the best results for the WPO method and both of them being better that the implicit average method that uses a single evaluation per individual, although they needed more time and memory. Since it does not need to perform averages or make additional fitness measures every generation, it is twice as fast as the next method, the one that uses explicit average fitness. An exploration of memory sizes[2] (shown in Fig. 3 for a typical run) showed that they are distributed unevenly but, in general, there is no single memory size overcoming all the population. Besides, distribution of fitness[3] shows a distribution with most values concentrated along the middle (that is, fitness equal to 10) or individuals that cannot be compared with any other, together with a few with the highest fitness and many with the lowest fitness. Besides showing that using the partial order for individual selection is a valid strategy, it also shows that a too greedy selection method would eliminate many individuals that might, in fact, have a high fitness. This will be taken into account when assigning parameter values to the evolutionary algorithm that will be presented next.

---

[1]http://git.io/a-e.

[2]Published at http://jj.github.io/Algorithm-Evolutionary/graphs/memory/.

[3]http://jj.github.io/Algorithm-Evolutionary/graphs/fitness-histo/.

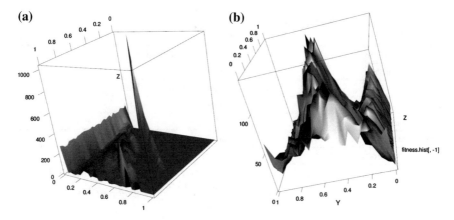

Fig. 3 (*Left*) 3D plot of the distribution of memory sizes for a single execution of the Wilcoxon-test based partial order. (*Right*) 3D plot of the distribution of fitness values along time for the WPO method on the Trap function. **a** Memory size. **b** Fitness distribution

5 Results

ETA and WPO have been tested using two well-known benchmarks, the deceptive bimodal Trap [28] function and the Massively Multimodal Deceptive Problem [29] (MMDP). We chose to use just these two functions because they have different fitness landscapes, are usually difficult for an evolutionary algorithm and have been extensively used for testing other kind of operators and algorithms. In Sect. 5.1 we will show results with no limitations on the number of evaluations; then we will show results limiting the number of evaluations in Sect. 5.2.

5.1 Noisy Fitness, Unlimited Evaluations

Several methods were tested: a baseline algorithm without noise to establish the time and number of evaluations needed to find the solution, a 0-memory (implicit average) method that uses noisy fitness without making any special arrangement, ETA and WPO. Evolutionary algorithm parameters and code for all tests were the same, except in one particular case: we used 2-tournament with 50 % replacement, 20 % mutation and 80 % crossover, $p = 1024$ and stopping when the best was found or number of evaluations reached. This was 200 K for the Trap, which used 40 as chromosome size, and 1 M for MMDP, which used 60 as chromosome size; these parameters are shown in Table 1. We have also used an additive Gaussian noise centered in zero and different σ, which is independent of the range of variation of the fitness values. By default, noise will follow a normal distribution with center in 0 and $\sigma = 1, 2$.

Table 1 Common evolutionary algorithm parameters

Parameter	Value
Chromosome length	40 (Trap) 60 (MMDP)
Population size	1024
Selection	2 tournament selection
Replacement rate	50 %
Mutation rate	20 %
Crossover rate	80 %
Max evaluations	200 K (Trap) 1 Million (MMDP)
Stopping criterion	Non-noisy best found or maximum number of evaluations reached

All tests use the `Algorithm::Evolutionary` library, and the scripts are published, as above, in the GitHub repository, together with raw results using the serialization language YAML and processed results as R scripts and `.csv` files. The evolutionary algorithm code used in all cases is exactly the same except for WPO, which, since it uses the whole population to evaluate fitness, needed a special reproduction and replacement library. This also means that the replacement method is not exactly the same: while WPO every generation replaces 50 % of the individuals, the rest evaluateS new individuals before replacement and eliminateS the worst 512 (50 % of the original population). Replacement is, thus, less greedy in the WPO case, but we do not think that this will have a big influence on the results (although it might account for the bigger number of evaluations obtained in some cases), besides, it just needed a small modification of code and was thus preferred for that reason. All values shown are the result of 30 independent runs.

The results for different noise levels are shown in Fig. 4. The boxplot on the left hand side compares the number of evaluations for the baseline method and the three methods with $\sigma = 1$. The implicit average method (labelled as 0-memory) is only slightly worse than the baseline value of around 12 K evaluations, with the ETA and WPO methods yielding very similar values which are actually worse than the 0-memory method. However, the scenario on the right, which shows how the number of evaluations scales with the noise level, is somewhat different. While the 0-memory method still has the smallest number of evaluations *for successful runs*, the success rate degrades very fast, with roughly the same and slightly less than 100 % for $\sigma = 2$ but falling down to 63 % for 0-memory and around 80 % for ETA and WPO (86 and 80 %). That is, best success rate is shown by the ETA method, but the best number of evaluations for roughly the same method is achieved by WPO.

These results also show that performance degrades quickly with problem difficulty and the degree of noise, that is why we discarded the 0-memory method due to its high degree of failure (a high percentage of the runs did not find the solution) with noise = 10 % max fitness and evaluated ETA and WPO over another problem, MMDP with similar absolute σ, with the difference that, in this case, $\sigma = 2$ would

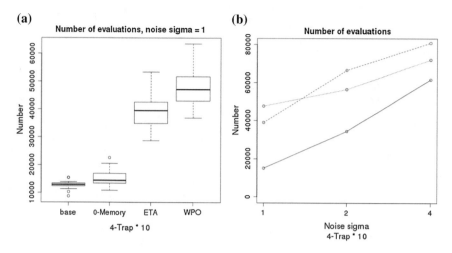

Fig. 4 (*Left*) Comparison of number of evaluations for the 4-Trap x 10 function and the rest of the algorithms with a noise σ equal to 1. (*Right*). Plot of average number of evaluations for different methods: 0 memory (*black, solid*), ETA (*red, dashed*), WPO (*blue, dot-dashed*). **a** Memory size. **b** Fitness distribution

be 20 % of the max value, which is close to the one observed experimentally, as explained in the Sect. 3. The evolutionary algorithm for MMDP used exactly the same parameters as for the Trap function above, except the max number of evaluations, which was boosted to one million. Initial tests with the 0-memory method yielded a very low degree of success, which left only the two methods analyzed in this paper for testing with MMDP. Success level was in all cases around 90 % and very similar in all experiments; the number of evaluations is more affected by noise and shown in Fig. 5. In fact, WPO and ETA have a very similar number of evaluations. It is statistically indistinguishable for $\sigma = 2$, and different only at the 10 % level (p-value = 0.09668) for $\sigma = 1$, however, if we take the time needed to reach solutions into account, ETA is much faster since it does not apply 10 * 1024 statistical tests every generation. However, WPO is more robust, with a lower standard deviation, in general, at least for high noise levels. However, both methods obtain good results with a much higher success rate than the implicit fitness (0-memory) method. Besides, ETA and WPO incorporate explicit fitness evaluation naturally into the population *resampling* only surviving individuals. This accounts for a predictable behavior of the algorithm, since the number of evaluations per generation is exactly the population size, which is important for optimization processes with a limited budget.

5.2 MMDP with Limited Evaluations

As indicated in the state of the art *resampling*, that is, performing new evaluations, is only necessary in the case statistical significance is not met. If two individuals are

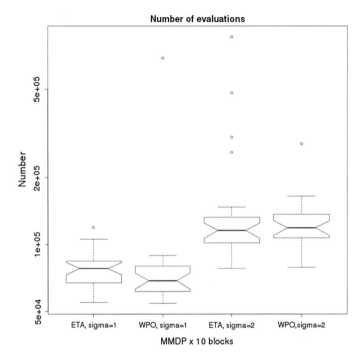

Fig. 5 Number of evaluations for successful runs ETA and WPO needed for solving the MMDP problem with 10 blocks and different noise levels, $\sigma = 1, 2$

different enough, the comparison will be significant even if they have only one fitness value. That is why some authors [9] have proved that resampling is not needed. However, as indicated above, an increasing number of evaluations will be needed as generations proceed. This is due to the fact that, as shown in the noise study in Sect. 3, σ increases with time and it becomes increasingly difficult to take apart different inherent values of the fitness. That is the main reason that has led us to performed resampling every generation in this work. However, it is quite clear that this resampling should not proceed unchecked beyond what is statistically reasonable, hence we have performed experiments limiting the maximum number of evaluations to 30, and evaluated its impact in performance. We have repeated the experiments adding lines of code that skip making a new evaluation if the vector of fitness measures has reached length 30. Code is, as usual, available in the repository https://github.com/JJ/algorithm-evolutionary-examples. Experiment results and log are under the directory ECTA-book. Results are shown in Figs. 6 and 7.

Please note that the unlabelled boxplots in Figs. 6 and 7 have the same values as the ones shown in Fig. 5. In fact, we should look at the columns labelled *lted*, for *limited*. First, the number of evaluations in Fig. 6 does not significantly change. It is essentially the same for ETA and slightly less for WPO (median 68610 for unlimited evaluations, 66550 for limited) but the difference is not significant; it is not significant

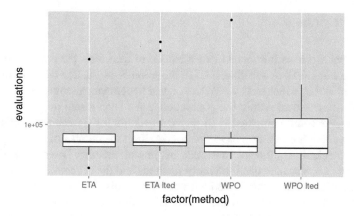

Fig. 6 Number of evaluations for successful runs ETA and WPO, with and without limitation of evaluations needed for solving the MMDP problem with 10 blocks and $\sigma = 1$. Please note that y axis is algorithmc, for clarity

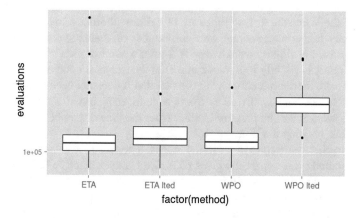

Fig. 7 Number of evaluations for successful runs ETA and WPO, with and without limitation of evaluations needed for solving the MMDP problem with 10 blocks and $\sigma = 2$. Please note that, for the sake of clarity, the y axis is logarithmic

either in success rate, which is essentially the same. However, with a higher noise level, shown in Fig. 7, number of evaluations is *higher* and in this case the difference is significant (the limited version averages 231400 evaluations while the unlimited averages 126700).

It should be noted that the average number of evaluations *increases* even if the number of evaluations per individual *decreases*. This means that, in some cases, this limitation makes *worse* individuals pass to the next generation, decreasing the quality of evolutionary search, and making it slower. In general, our conclusion here is that the number of evaluations in WPO should be kept unlimited or, at least, limited to a high value. It might increase slightly the performance of the algorithm at low noise levels, but it will result in *worse* performance for higher noise levels.

6 Conclusions

In this paper we have introduced two methods to deal with the problem of noisy fitness functions. The two methods, ETA, based on re-evaluation of surviving individuals and WPO, which uses the Wilcoxon test to compare a sample of individuals and partial-order them within the population, have been tested over two different fitness functions and compared with implicit average (or 0-memory) methods, as well as among themselves; we have also tried to limit the number of evaluations and generally checked its influence in results. In general, memory-based methods have much higher success rate than 0-memory methods and the difference increases with the noise level, with 0-memory methods crashing at noise levels close to 20 % while ETA and WPO maintain a high success level. Limiting the number of evaluations, in general, has an impact that is very lightly positive or negative.

It is difficult to choose between the two proposed methods, ETA and WPO. However, ETA is much faster since it avoids costly statistical analysis. In a profiling of the WPO method we have discovered that the vast majority of program time is spent computing the Wilcoxon test, so times are of a different order of magnitude. It also has a slightly higher success rate, and the number of evaluations it needs to find the solution is only slightly worse; even if from the point of view of the evolutionary algorithm it is slightly less robust and slightly worse, it compensates the time needed to make more evaluations with the fact that it does not need to perform statistical tests to select new individuals. So, in general, using a central and fast statistic such as the average will be preferred, either from the time or number of evaluations point of view, to using the Wilcoxon test.

However, this research is in its initial stages. The fact that we are using a centrally distributed noise apparently gives ETA an advantage since, in fact, comparing the mean of two individuals will be essentially the same as doing a statistical comparison, since when the number of measures is enough, statistical significance will be reached. In fact, with a small difference ETA might select as better an individual whose fitness is actually the same, something that would be correctly spotted by WPO, but, in fact, since there is an average selective pressure this is not going to matter in the long run. Using higher selective pressures might make a difference, but this is something that will have to be tested.

It might matter in different situations, for instance in numerical optimization problems and also when noise follows a uniform distribution; behavior might in this case be similar to when noise levels are higher. These are scenarios that are left for future research, and destined to find out in which situations WPO is better than ETA and the other way round.

Besides exploring noise in different problems and modelling its distribution, we will explore different parameters in the model. The first one is the number of comparisons in WPO. Initial explorations have proved that changing it from 5 to 32 does not yield a significant difference. Looking for a way to speed up this method would also be important since it would make its speed closer to ETA. It might be an improvement to use Wilcoxon at the selection stage, not as we do know at the fitness computaton stage. All these research questions open new research avenues that will be explored in the future, along with testing with fuzzy selection techniques and using clustering for selecting tournament sets.

Acknowledgments This work has been supported in part by project EphemeCH (TIN2014-56494-C4-3-P and TIN2014-56494-C4-1-P) and DNEMESIS (P10-TIC-6083). The authors would like to thank the FEDER of European Union for financial support via project "Sistema de Información y Predicción de bajo coste y autónomo para conocer el Estado de las Carreteras en tiempo real mediante dispositivos distribuidos"(SIPEsCa) of the"Programa Operativo FEDER de Andalucía 2007-2013". We also thank all Agency of Public Works of Andalusia Regional Government staff and researchers for their dedication and professionalism.
Our research group is committed to Open Science and the writing and development of this paper has been carried out in GitHub at this address https://github.com/JJ/noisy-fitness-eas. We encourage you to visit, comment and to do all kind of suggestions or feature requests.
We would like also to thank Marc Schoenauer for the suggestion that has been put to test in this paper.

Agencia de Obra Pública de la Junta de Andalucía
CONSEJERÍA DE FOMENTO Y VIVIENDA

Andalucía
se mueve con Europa

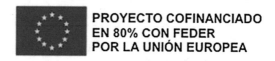
PROYECTO COFINANCIADO
EN 80% CON FEDER
POR LA UNIÓN EUROPEA

References

1. Mora, A.M., Fernández-Ares, A., Merelo-Guervós, J.J., García-Sánchez, P., Fernandes, C.M.: Effect of noisy fitness in real-time strategy games player behaviour optimisation using evolutionary algorithms. J. Comput. Sci. Technol. **27**, 1007–1023 (2012)
2. García-Ortega, R.H., García-Sánchez, P., Merelo, J.J.: Emerging archetypes in massive artificial societies for literary purposes using genetic algorithms. ArXiv e-prints (2014). http://adsabs.harvard.edu/abs/2014arXiv1403.3084G
3. Castillo, P.A., González, J., Merelo-Guervós, J.J., Prieto, A., Rivas, V., Romero, G.: G-Prop-III: global optimization of multilayer perceptrons using an evolutionary algorithm. In: GECCO-99: Proceedings of the Genetic and Evolutionary Computation Conference, p. 942 (1999)
4. Jin, Y., Branke, J.: Evolutionary optimization in uncertain environments—a survey. IEEE Trans. Evol. Comput. **9**, 303–317 (2005). (cited By (since 1996) 576)
5. Merelo, J.J., Castillo, P.A., Mora, A., Fernández-Ares, A., Esparcia-Alcázar, A.I., Cotta, C., Rico, N.: Studying and tackling noisy fitness in evolutionary design of game characters. In: Rosa, A., Merelo, J. J., Filipe, J. (eds.) ECTA 2014—Proceedings of the International Conference on Evolutionary Computation Theory and Applications, pp. 76–85 (2014)
6. Merelo-Guervós, J.J., Romero, G., García-Arenas, M., Castillo, P.A., Mora, A.M., Jiménez-Laredo, J.L.: Implementation matters: programming best practices for evolutionary algorithms. [30] 333–340

 7. Qian, C., Yu, Y., Zhou, Z.H.: Analyzing evolutionary optimization in noisy environments. CoRR (2013). abs/1311.4987
 8. Bhattacharya, M., Islam, R., Mahmood, A.: Uncertainty and evolutionary optimization: a novel approach. In: 2014 IEEE 9th Conference on Industrial Electronics and Applications (ICIEA), pp. 988–993 (2014)
 9. Qian, C., Yu, Y., Jin, Y., Zhou, Z.H.: On the effectiveness of sampling for evolutionary optimization in noisy environments. In: Bartz-Beielstein, T., Branke, J., Filipic, B., Smith, J. (eds.) Parallel Problem Solving from Nature PPSN XIII. Volume 8672 of Lecture Notes in Computer Science, pp. 302–311. Springer (2014)
10. Mora, A.M., Montoya, R., Merelo, J.J., Sánchez, P.G., Castillo, P.A., Laredo, J.L.J., Martínez, A.I., Espacia, A.: Evolving bots ai in unreal. In: di Chio C., et al. (ed.) Applications of Evolutionary Computing, Part I. Volume 6024 of Lecture Notes in Computer Science, pp. 170–179. Springer, Istanbul (2010)
11. Liberatore, F., Mora, A.M., Castillo, P.A., Merelo-Guervós, J.J.: Evolving evil: optimizing flocking strategies through genetic algorithms for the ghost team in the game of Ms. Pac-Man. In: Camacho, D., Gómez-Martín, M.A., González-Calero, P.A. (eds.) Proceedings 1st Congreso de la Sociedad Española para las Ciencias del Videojuego, CoSECivi 2014, Barcelona, Spain, June 24, 2014. Volume 1196 of CEUR Workshop Proceedings, pp. 111–116. CEUR-WS.org (2014)
12. Merelo-Guervós, J.J., Prieto, A., Morán, F.: 4. In: Optimization of Classifiers Using Genetic Algorithms, pp. 91–108. MIT press (2001). ISBN: 0262162016. http://geneura.ugr.es/pub/papers/g-lvq-book.ps.gz
13. Miller, B.L., Goldberg, D.E.: Genetic algorithms, selection schemes, and the varying effects of noise. Evol. Comput. **4**, 113–131 (1996)
14. Jun-hua, L., Ming, L.: An analysis on convergence and convergence rate estimate of elitist genetic algorithms in noisy environments. Opt. Int. J. Light Electron Opt. **124**, 6780–6785 (2013)
15. Friedrich, T., Kötzing, T., Krejca, M., Sutton, A.M.: The Benefit of sex in noisy evolutionary search. ArXiv e-prints (2015)
16. Aizawa, A.N., Wah, B.W.: Scheduling of genetic algorithms in a noisy environment. Evol. Comput. **2**, 97–122 (1994)
17. Costa, A., Vargas, P., Tinós, R.: Using explicit averaging fitness for studying the behaviour of rats in a maze. Adv. Artif. Life ECAL **12**, 940–946 (2013)
18. Liu, J., St-Pierre, D.L., Teytaud, O.: A mathematically derived number of resamplings for noisy optimization. In: Proceedings of the 2014 Conference Companion on Genetic and Evolutionary Computation Companion. GECCO Comp '14, New York, pp. 61–62. ACM, NY (2014)
19. Rada-Vilela, J., Johnston, M., Zhang, M.: Population statistics for particle swarm optimization: Resampling methods in noisy optimization problems. In: Swarm and Evolutionary Computation (2014, in press)
20. Rakshit, P., Konar, A., Nagar, A.: Artificial bee colony induced multi-objective optimization in presence of noise. In: 2014 IEEE Congress on Evolutionary Computation (CEC), pp. 3176–3183 (2014)
21. Rudolph, G.: A partial order approach to noisy fitness functions. Proc. IEEE Conf. Evol. Comput. ICEC **1**, 318–325 (2001)
22. Cauwet, M.L., Liu, J., Teytaud, O., et al.: Algorithm portfolios for noisy optimization: compare solvers early. In: Learning and Intelligent Optimization Conference (2014)
23. Lara-Cabrera, R., Cotta, C., Fernández-Leiva, A.J.: On balance and dynamism in procedural content generation with self-adaptive evolutionary algorithms. Nat. Comput. **13**, 157–168 (2014)
24. Fernández-Ares, A., Mora, A.M., Guervós, J.J.M., García-Sánchez, P., Fernandes, C.M.: Optimizing strategy parameters in a game bot [30] 325–332
25. Wilcoxon, F.: Individual comparisons by ranking methods. Biometrics Bull. **1**, 80–83 (1945)
26. Merelo-Guervós, J.J.: Using a Wilcoxon-test based partial order for selection in evolutionary algorithms with noisy fitness. Technical Report, GeNeura Group, University of Granada (2014). http://dx.doi.org/10.6084/m9.figshare.974598

27. Merelo-Guervós, J.J., Castillo, P.A., Alba, E.: Algorithm::Evolutionary, a flexible Perl module for evolutionary computation. Soft Comput. **14**, 1091–1109 (2010). http://sl.ugr.es/000K
28. Deb, K., Goldberg, D.E.: Analyzing deception in trap functions. FOGA **2**, 98–108 (1992)
29. Goldberg, D.E., Deb, K., Horn, J.: Massive multimodality, deception, and genetic algorithms. In: Männer, R., Manderick, B. (eds.) Parallel Problem Solving from Nature, vol. 2, pp. 37–48. Elsevier, Amsterdam (1992)
30. Cabestany, J., Rojas, I., Caparrós, G.J., (eds.): Advances in Computational Intelligence—11th International Work-Conference on Artificial Neural Networks, IWANN 2011, Torremolinos-Málaga, Spain, June 8-10, 2011, Proceedings, Part II. IWANN (2). Volume 6692 of Lecture Notes in Computer Science. Springer (2011)

Particle Swarm Optimization with Dynamic Topology and Conservation of Evaluations

Carlos M. Fernandes, Juan L.J. Laredo, Juan Julián Merelo, Carlos Cotta and Agostinho C. Rosa

Abstract This paper investigates a Particle Swarm with dynamic topology and a conservation of evaluations strategy. The population is structured on a 2-dimensional grid of nodes, through which the particles interact and move according to simple rules. As a result of this structure, each particle's neighbourhood degree is time-varying. If at given time step a particle p has no neighbours except itself, p is not evaluated until it establishes at least one link to another particle. A set of experiments demonstrates that the dynamics imposed by the structure provides a consistent and stable behaviour throughout the test set when compared to standard topologies, while the conservation of evaluations significantly reduces the convergence speed of the algorithm. The working mechanisms of the proposed structure are very simple and, except for the size of the grid, they do not require parameters and tuning.

Keywords Particle swarm optimization · Population structure · Dynamic topologies · Swarm intelligence

C.M. Fernandes (✉) · A.C. Rosa
Institute for Systems and Robotics (ISR/IST), LARSyS, Instituto Superior Técnico, University of Lisbon, Lisbon, Portugal
e-mail: cfernandes@laseeb.org

A.C. Rosa
e-mail: acrosa@laseeb.orgjuanlu

C.M. Fernandes · J.J. Merelo
Departamento de Arquitectura y Tecnología de Computadores, University of Granada, Granada, Spain
e-mail: jmerelo@geneura.ugr.es

J.L.J. Laredo
LITIS, University of Le Havre, Le Havre, France
e-mail: juanlu.jimenez@gmail.com

C. Cotta
Lenguages y Ciencias de la Computacion, University of Malaga, Malaga, Spain
e-mail: ccottap@lcc.uma.es

© Springer International Publishing Switzerland 2016
J.J. Merelo et al. (eds.), *Computational Intelligence*,
Studies in Computational Intelligence 620, DOI 10.1007/978-3-319-26393-9_7

1 Introduction

The Particle Swarm Optimization (PSO) algorithm is a population-based meta-heuristic for binary and real-valued function optimization inspired by the swarm and social behavior of organisms in bird flocks and fish schools [4]. The optimization is performed by a swarm of candidate solutions, called particles, which move around the problem's search space guided by mathematical rules that define their velocity and position at each time step. Each particle's velocity vector is influenced by its best known position and by the best known positions of its neighbors. The neighborhood of each particle—and consequently the flow of information through the population—is defined a priori by the population topology.

The reason why the swarm is interconnected is the core of the algorithm: the particles communicate so that they acquire information on the regions explored by other particles. In fact, it has been claimed that the uniqueness of the PSO algorithm lies in the interactions of the particles [5]. As expected, the population topology deeply affects the balance between exploration and exploitation and the convergence speed and accuracy of the algorithm.

The population can be structured on any possible topology, from sparse to dense (or even fully connected) graphs, with different degrees of connectivity and clustering. The classical and most used population structures are the *lbest* (which connects the individuals to a local neighbourhood) and the *gbest* (in which each particle is connected to every other individual). These topologies are well-studied and the major conclusions are that *gbest* is fast but is frequently trapped in local optima, while *lbest* is slower but converges more often to the neighborhood of the global optima.

Since the first experiments on *lbest* and *gbest* structures, researchers have tried to design networks that hold the best traits given by each structure [9]. Some studies also try to understand what makes a good structure. For instance, Kennedy and Mendes [5] investigate several types of topologies and recommend the use of a lattice with von Neumann neighbourhood (which results in a connectivity degree between that of *lbest* and *gbest*).

Recently, dynamic structures have been introduced in PSO for improving the algorithm's adaptability to different fitness landscapes and overcome the rigidity of static structures, like [7], for instance. Fernandes et al. [1] try a different approach and propose a dynamic and partially connected von Neumann structure with Brownian motion. In this paper, we use the same model but a strategy for the conservation of function evaluations [8] is introduced in order to make the most of the underlying structure and reduce convergence speed. A formal description of the dynamic network is given here, opening the way for more sophisticated dynamics.

In the proposed topology, n particles are placed in a 2-dimensional m-nodes grid where $m > n$. Every time-step, each individual checks its von Neumann neighborhood and, as in the standard PSO, updates its velocity and position using the information given by the neighbours. However, while the connectivity degree of the von Neumann topology is $k = 5$ the degree of the proposed topology is variable in the range $1 \leq k \leq 5$. Furthermore, the structure is dynamic: in each time-step, every particle updates its position on the grid (which is a different concept from the position of the particle on the fitness landscape) according to a pre-defined rule that selects the destination node. The movement rule, which is implemented locally and without any knowledge on the global state of the system, can be based on stigmergy [2] or Brownian motion.

As stated above, the connectivity degree k of each particle in each time-step is variable and lies in the range $1 \leq k \leq 5$. Depending on the size of the grid, there may be particles with $k = 1$. These particles without neighbors (except the particle itself) do not learn from any local neighbourhood at that specific iteration. Therefore, it is expected that they continue to follow their previous trajectory in the fitness land-scape. Taking into account these premises, the algorithm proposed in this study does not evaluate the position of the particles when $k = 1$. Regardless of the loss of informant intrinsic to a *conservation of evaluations* policy, we hypothesize that the strategy is particularly suited for the proposed dynamic topology (in which the particles are sometimes isolated from the flow of information) and the number of function evaluations required for meeting the stop criteria can be significantly reduced. Furthermore, it is the structure of the population and the position of the particles at a specific time-step that decides the application of the conservation rule and not any extra parameter or pre-defined decision rule.

A classical PSO experimental setup is used for the tests and the results demonstrate that the proposed algorithm consistently improves the speed of convergence of the standard von Neumann structure without degrading the quality of solutions. The experiments also demonstrate that the introduction of the conservation strategy reduces significantly the convergence speed without affecting the quality of the final solutions.

The remaining of the paper is organized as follows. Section 2 describes PSO and gives an overview on population structures for PSOs. Section 3 gives a formal description of the proposed structure. Section 4 describes the experiments and discusses the results and, finally, Sect. 5 concludes the paper and outlines future research.

2 Background Review

PSO is described by a simple set of equations that define the velocity and position of each particle. The position of the ith particle is given by $\vec{X}_i = (x_{i,1}, x_{i,2}, \ldots x_{1,D})$, where D is the dimension of the search space. The velocity is given by $\vec{V}_i = (v_{i,1}, v_{i,2}, \ldots v_{1,D})$. The particles are evaluated with a fitness function $f(\vec{X}_i)$ and then their positions and velocities are updated by:

$$v_{i,d}(t) = v_{i,d}(t-1) + c_1 r_1 \left(p_{i,d} - x_{i,d}(t-1) \right) + c_2 r_2 \left(p_{g,d} - x_{i,d}(t-1) \right) \qquad (1)$$

$$x_{i,d}(t) = x_{i,d}(t-1) + v_{i,d}(t) \qquad (2)$$

were p_i is the best solution found so far by particle i and p_g is the best solution found so far by the neighborhood. Parameters r_1 and r_2 are random numbers uniformly distributed in the range [0, 1] and c_1 and c_2 are acceleration coefficients that tune the relative influence of each term of the formula. The first term is known as the *cognitive* part, since it relies on the particle's own experience. The last term is the *social* part, since it describes the influence of the community in the trajectory of the particle.

In order to prevent particles from stepping out of the limits of the search space, the positions $x_{i,d}(t)$ are limited by constants that, in general, correspond to the domain of the problem: $x_{i,d}(t) \in [-Xmax, Xmax]$. Velocity may also be limited within a range in order to prevent the *explosion* of the velocity vector: $v_{i,d}(t) \in [-Vmax, Vmax]$.

For achieving a better balancing between local and global search, Shi an Eberhart [12] added the inertia weight ω as a multiplying factor of the first term of Eq. 1. This paper uses PSOs with inertia weight.

The neighbourhood of the particle defines the value of p_g and is a key factor in the performance of PSO. Most of the PSOs use one of two simple sociometric principles for defining the neighbourhood network. One connects all the members of the swarm to one another, and it is called *gbest*, were g stands for *global*. The degree of connectivity of *gbest* is $k = n$, where n is the number of particles. Since all the particles are connected to every other and information spreads easily through the network, the *gbest* topology is known to converge fast but unreliably (it often converges to local optima).

The other standard configuration, called *lbest* (where l stands for local), creates a neighbourhood that comprises the particle itself and its k nearest neighbors. The most common *lbest* topology is the ring structure, in which the particles are arranged in a ring structure (resulting in a degree of connectivity $k = 3$, including the particle). The *lbest* converges slower than the *gbest* structure because information spreads slower through the network but for the same reason it is less prone to converge prematurely to local optima. In-between the ring structure with $k = 3$ and the *gbest* with $k = n$ there are several types of structure, each one with its advantages on a certain type of fitness landscapes. Choosing a proper structure depends on the target problem and also on the objectives or tolerance of the optimization process.

Kennedy and Mendes [5] published an exhaustive study on population structures for PSOs. They tested several types of structures, including the *lbest*, *gbest* and von Neumann configuration with radius 1 (also kown as *L5* neighborhood). They also tested populations arranged in randomly generated graphs. The authors conclude that when the configurations are ranked by the performance the structures with $k = 5$ (like the *L5*) perform better, but when ranked according to the number of iterations needed to meet the criteria, configurations with higher degree of connectivity perform better. These results are consistent with the premise that low connectivity favors robustness, while higher connectivity favors convergence speed (at the expense of reliability). Amongst the large set of graphs tested in [5], the Von Neumann with radius 1 configuration performed more consistently and the authors recommend its use.

Alternative topologies that combine standard structures' characteristics or introduce some kind of dynamics in the connections have been also proposed. Parsopoulos and Vrahatis [9] describe the unified PSO (UPSO), which combines the *gbest* and *lbest* configurations. Equation 1 is modified in order to include a term with p_g and a term with p_i and a parameter balances the weight of each term. The authors argue that the proposed scheme exploits the good properties of *gbest* and *lbest*.

Peram et al. [10] proposed the fitness-distance-ratio-based PSO (FDR-PSO), which defines the neighbourhood of a particle as its k closest particles in the population (measured by the Euclidean distance). A selective scheme is also included: the particle selects nearby particles that have also visited a position of higher fitness. The algorithm is compared to a standard PSO and the authors claim that FDR-PSO performs better on several test functions. However, the FDR-PSO is compared only to a *gbest* configuration, which is known to converge frequently to local optima in the majority of the functions of the test set.

More recently, a comprehensive-learning PSO (CLPSO) was proposed [7]. Its learning strategy abandons the global best information and introduces a complex and dynamic scheme that uses all other particles' past best information. CLPSO can significantly improve the performance of the original PSO on multimodal problems. Finally, Hseigh et al. [3] use a PSO with varying swarm size and solution-sharing that, like in [7], uses the past best information from every particle.

A different approach is given in 1. The authors describe a structure that is based on a grid of m nodes (with $m > n$) on which the particles move and interact. In this structure, a particle, at a given time-step, may have no neighbours except itself. The isolated particles will continue to follow its previous trajectory, based on their current information, until they find another particle in the neighbourhood. Therefore, we intend to investigate if the loss of information caused by not evaluating these particles is overcome by the payoff in the convergence speed.

Common ways of addressing the computational cost of evaluating solutions in hard real-world problems are function approximation [6], fitness inheritance [11] and conservation of evaluations [8]. Due to the underlying structure of the proposed algorithm, we have tested a conservation policy similar to the GREEN-PSO proposed by Majercik [8]. However, in our algorithm the decision on evaluating or not is defined by the position of the particle in the grid (isolated particles are not evaluated) while in the GREEN-PSO the decision is probabilistic and the likelihood of conserving a solution is controlled by a parameter.

The following section gives a formal description of the proposed network and presents the transition rules that define the model for dynamic population structures.

3 Partially Connected Structures

Let us consider a rectangular grid G of size $q \times s \geq \mu$, where μ is the size of the population of any population-based metaheuristics or model. Each node G_{uv} of the grid is a tuple $\langle \eta_{uv}, \zeta_{u,v} \rangle$, where $\eta_{uv} \in \{1, \ldots, \mu\} \cup \{\bullet\}$ and $\zeta_{uv} \in (D \times \mathbb{N}) \cup \{\bullet\}$ for some domain D. The value η_{uv} indicates the index of the individual that occupies the position $\langle u, v \rangle$ in the grid. If $\eta_{uv} = \bullet$ then the corresponding position is empty. However, that same position may still have information, namely a mark (or clue) ζ_{uv}. If $\zeta_{uv} = \bullet$ then the position is empty and unmarked. Please note that when $q \times s = \mu$, the topology is a static 2-dimensional lattice and when $q \times s = \mu$ and $q = s$ the topology is the standard square grid graph.

In the case of a PSO, the marks are placed by particles that occupied that position in the past and they consist of information about those particles, like their fitness ζ_{uv}^f or position in the fitness landscape, as well as a time stamp ζ_{uv}^t that indicates the iteration in which the mark was placed. The marks have a lifespan of K iterations, after which they are deleted.

Initially, $G_{uv} = (\bullet, \bullet)$ for all $\langle u, v \rangle$. Then, the particles are placed randomly on the grid (only one particle per node). Afterwards, all particles are subject to a movement phase (or grid position update), followed by a PSO phase. The process (position update and PSO phase) repeats until a stop criterion is met.

The PSO phase is the standard iteration of a PSO, comprising position and velocity update. The only difference to a static structure is that in this case a particle may find empty nodes in its neighbourhood.

In the position update phase, each individual moves to an adjacent empty node. Adjacency is defined by the Moore neighborhood of radius r, so an individual i at $\rho_g(i) = \langle u, v \rangle$ can move to an empty node $\langle u', v' \rangle$ for which $L_\infty(\langle u, v \rangle \langle u', v' \rangle) \leq r$. If empty positions are unavailable, the individual stays in the same node. Otherwise, it picks a neighboring empty node according to the marks on them. If there are no marks, the destination is chosen randomly amongst the free nodes.

With this framework, there are two possibilities for the position update phase: stimergic, whereby the individual looks for a mark that is similar to itself; and Brownian, whereby the individual selects an empty neighbor regardless of the marks. For the first option, let $\mathbb{N}\langle u, v \rangle = \{\langle u^{(1)}, v^{(1)} \rangle, \ldots \langle u^w, v^w \rangle\}$ be the collection of empty neighboring nodes and let i be the individual to move. Then, the individual attempts to move to a node whose mark is as close as possible to its own corresponding trait (fitness or position in the fitness landscape, for instance) or to an adjacent cell picked at random if there are no marks in the neighborhood. In the alternative Brownian policy, the individual moves to an adjacent empty position picked at random. In either case, the process is repeated for the whole population.

For this paper, the investigation is restricted to the Brownian structure. The algorithm is referred in the remaining of the paper has PSO-B, followed by the grid size $q \times s$. An extension of the PSO-B is constructed by introducing a conservation of function evaluations (*cfe*) strategy. If at a given time-step a particle has no neighbors, then the particle is updated but its position is not evaluated. This version of the algorithm is referred to as PSO-B*cfe*. The following section describes the results attained by the PSOs with dynamic structure and Brownian movement, with and without conservation of function evaluations and compares them to the standard topology.

4   Experiments and Results

An experimental setup was constructed with eight benchmark unimodal and mul-timodal functions that are commonly used for investigating the performance of PSO. The functions are described in Table 1. The dimension of the search space is set to $D = 30$ (except Schaffer, with $D = 2$). In order to obtain a square grid graph for the standard *von Neumann* topology, the population size n is set to 49 (which is

Table 1 Benchmarks for the experiments

Function	Mathematical representation	Range of search/Range of initialization	Stop criteria
Sphere f_1	$$f_1(\vec{x}) = \sum_{i=}^{D} x_i^2$$	$(-100, 100)^D$ $(50, 100)^D$	0.01
Rosenbrock f_2	$$f_2(\vec{x}) = \sum_{i=1}^{D-1} (100(x_{i+1} - x_i^2)^2 + (x_i - 1)^2$$	$(-100, 100)^D$ $(15, 30)^D$	100
Rastrigin f_3	$$f_3(\vec{x}) = \sum_{i=1}^{D-1} (x_i^2 - 10\cos(2\pi x_i) + 10)$$	$(-10, 10)^D$ $(2.56, 5.12)^D$	100
Griewank f_4	$$f_4(\vec{x}) = 1 + \frac{1}{4000}\sum_{i=1}^{D} x_i^2 - \prod_{i=1}^{D}\cos\left(\frac{x_i}{\sqrt{i}}\right)$$	$(-600, 600)^D$ $(300, 600)^D$	0.05
Schaffer f_5	$$f_5(\vec{x}) = 0.5 + \frac{\left(\sin\sqrt{x^2+y^2}\right)^2 - 0.5}{(1.0+0.001(x^2+y^2))^2}$$	$(-100, 100)^2$ $(15, 30)^2$	0.00001
Weierstrass f_6	$$f_6(\vec{x}) = \sum_{i=1}^{D}\left(\sum_{k=0}^{kmax}[a^k\cos(2\pi b^k(x_i+0.5))]\right)$$ $$- D\sum_{k=0}^{kmax}[a^k\cos(2\pi b^k \cdot 0.5)],$$ $$a = 0.5, b = 3, kmax = 20$$	$(-0.5, 0.5)^D$ $(0.5, 0.2)^D$	0.01
Ackley f_7	$$f_7(\vec{x}) = -20exp\left(-0.2\sqrt{\frac{1}{D}\sum_{i=1}^{D}x_i^2}\right)$$ $$- exp\left(\frac{1}{D}\sum_{i=1}^{D}\cos(2\pi x_i)\right) + 20$$ $$+ e$$	$(-32.768, 32.768)^D$ $(2.56, 5.12)^D$	0.01
Schwefel f_8	$$f_7(\vec{x}) = 418.9829 \times D - \sum_{i=1}^{D} x_i sin\left(\left|x_i\right|^{\frac{1}{2}}\right)$$	$(-500, 500)^D$ $(-500, 500)^D$	3000

Dynamic range, initialization range and stop criteria

within the typical range of PSO's swarm size). The acceleration coefficients were set to 1.494 and the inertia weight is 0.729, as in [13]. *Xmax* is defined as usual by the domain's upper limit and *Vmax* = *Xmax*. A total of 50 runs for each experiment are conducted. *Asymmetrical initialization* is used (the initialization range for each function is given in Table 1).

Two experiments were conducted. Firstly, the algorithms were run for a limited amount of function evaluations (147000 for f_1 and f_5, 49000 for the remaining) and the fitness of the best solution found was averaged over 50 runs. In the second experiment the algorithms were run for 980000 evaluations (corresponding to 20000 iterations of standard PSO with $n = 49$) or until reaching a stop criterion. For each function and each algorithm, the evaluations required to meet the criterion was recorded and averaged over the 50 runs. A success measure is defined as the number of runs in which an algorithm attains the fitness value established as the stop criterion.

Tables 1 and 2 compare PSO-B with the standard PSO (with von Neumann topology): Table 1 gives the averaged best fitness found by the swarms while Table 2 gives, for each algorithm and each function, the averaged number of iterations required to meet the criterion, and the number of runs in which the criterion was met.

The best fitness values are similar in both configurations. In fact, the differences are not statistical significant except for function f_1, for which PSO-Bcfe significantly better than PSO. (For the statistical tests comparing two algorithms, non-parametric Kolmogorov-Smirnov tests (with 0.05 level of significance) have been used.) As for the convergence speed, PSO-B is faster in every test function. The results are significantly different in f_1, f_2, f_3, f_5, f_6 and f_8. PSO-B and the standard PSO attain similar fitness values, but PSO-B is faster.

The main hypothesis of this paper is that a conservation of evaluations strategy further improves the convergence speed of the dynamic topology. Moreover, we also expect that PSO-B*cfe* performance is less affected when the size of the grid is increased. Large grid sizes result in large rates of isolated particles, deprived from social information, which reduces the convergence speed of the algorithm. By not evaluating these particles, the computational effort can be significantly reduced, hopefully without degrading the overall performance. In order to investigate these hypotheses, we have compared PSO-B and PSO-B*cfe*, while varying the size of the grid (Table 3).

Table 4 shows the average fitness values attained by PSO-B and PSO-B*cfe* with different grid sizes. Table 5 displays the average number of function evaluations required to meet the stop criteria as well as the number of successful runs. The performance according to the fitness values is very similar, with no significant differences between the algorithm in every function except f_1 (in which PSO-B*cfe* is significantly better). When considering the number of function evaluations (i.e., the convergence speed), PSO-B*cfe* is significantly better or statistically equivalent in every function.

The results confirm that PSO-B*cfe* is able to improve the convergence speed of PSO-B without degrading the accuracy of the solutions. The loss of information

Table 2 Standard PSO with von Neumann topology and Brownian PSO with 10×10 grid

	f_1	f_2	f_3	f_4	f_5	f_6	f_7	f_8
PSO	4.26e − 36	1.39e + 01	6.40e + 01	**5.61e − 03**	0.00e + 00	**4.43e − 02**	**1.12e − 15**	3.07e + 03
	±1.32e − 35	±2.39e + 01	±1.59e + 01	**±8.78e − 03**	±0.00e + 00	±1.69e − 01	±2.24e − 16	±4.92e + 02
PSO-B	**8.74e − 38**	**1.25e + 01**	**6.23e + 01**	7.73e − 03	**0.00e + 00**	4.61e − 02	**1.12e − 15**	**2.80e + 03**
	±1.29e − 37	**±2.07e + 01**	**±1.88e + 01**	±9.50e − 03	**±0.00E + 00**	±2.50e − 01	**±2.24e − 16**	**±6.85e + 02**

Best fitness values averaged over 50 runs

Table 3 Standard PSO with von Neumann topology and Brownian PSO (10×10)

	f_1	f_2	f_3	f_4	f_5	f_6	f_7	f_8
PSO	23530.78	72707.18	18424.00	22015.70	17622.36	41677.84	24420.62	21384.05
	±954.74	±92916.33	±11082.75	±1304.60	±16056.68	±1360.37	±958.81	±20450.21
	(50)	(50)	(49)	(50)	(50)	(44)	(50)	(22)
PSO-B	**22700.72**	**65769.76**	**15114.46**	**21574.70**	**10741.78**	**38976.00**	**24410.86**	**12943.90**
	±906.39	**±70232.15**	**±3939.74**	**±1107.56**	**±10658.16**	**±1654.12**	**±1533.28**	**±3422.18**
	(50)	**(50)**	**(48)**	**(50)**	**(50)**	**(42)**	**(50)**	**(31)**

Function evaluations averaged over 50 runs and success rates

Table 4 PSO-B and PSO-Bcfe

	f_1	f_2	f_3	f_4	f_5	f_6	f_7	f_8
PSO-B 8 × 8	2.71e − 40 ±4.21e − 40	7.98e + 00 ±1.11e + 01	6.57e + 01 ±1.78e + 01	7.43e − 03 ±9.02e − 03	1.17e − 03 ±3.19e − 03	1.55e − 01 ±4.04e − 01	1.17e − 15 ±2.15e − 16	2.85e + 03 ±5.05e + 02
PSO-Bcfe 8 × 8	3.20e − 40 ±8.44e − 40	1.12e + 01 ±2.02e + 01	6.29e + 01 ±1.35e + 01	7.44e − 03 ±7.96e − 03	1.94e − 04 ±1.37e − 03	3.59e − 02 ±2.15e − 01	1.20e − 15 ±2.4 − 16	2.78e + 03 ±5.48e + 02
PSO-B 10 × 10	8.74e − 38 ±1.29e − 37	1.25e + 01 ±2.07e + 01	6.23e + 01 ±1.88e + 01	7.73e − 03 ±9.50e − 03	**0.00e + 00** **±0.00e + 00**	4.61e − 02 ±2.50e − 01	**1.12e − 15** **±2.24e − 16**	2.80e + 03 ±6.85e + 02
PSO-Bcfe 10 × 10	2.43e − 39 ±4.38e − 39	**1.09e + 01** **±1.67e + 01**	6.16e + 01 ±1.60e + 01	5.61e − 03 ±7.47e − 03	**0.00e + 00** **±0.00e + 00**	1.15e − 01 ±3.30e − 01	1.19e − 15 ±2.09e − 16	2.91e + 03 ±5.37e + 02
PSO-B 15 × 15	3.47e − 33 ±4.31e − 33	1.53e + 01 ±2.65e + 01	**6.05e + 01** **±1.45e + 01**	**3.79e − 03** **±6.22e − 03**	3.89e − 04 ±1.92e − 03	3.65e − 02 ±2.15e − 01	1.14e − 15 ±2.23e − 16	**2.72e + 03** **±5.07e + 02**
PSO-Bcfe 15 × 15	**2.47e − 45** **±4.21e − 45**	1.31 e + 01 ±2.42e + 01	6.62e + 01 ±2.05e + 01	6.45e − 03 ±9.90e − 03	**0.00e + 00** **±0.00E + 00**	**3.59e − 02** **±2.15e − 01**	1.13e − 15 ±2.24e − 16	**2.72e + 03** **±5.81e + 02**

Best fitness values averaged over 50 runs

Table 5 PSO-B and PSO-Bcfe

	f_1	f_2	f_3	f_4	f_5	f_6	f_7	f_8
PSO-B 8 × 8	21070.98 ±1023.60 (50)	56472.50 ±49342.60 (50)	13224.90 ±3894.30 (48)	19959.66 ±1408.36 (50)	14369.25 ±20599.03 (46)	37385.64 ±1864.74 (36)	22112.72 ±1172.33 (50)	**10756.34** **±2866.56** (29)
PSO-Bcfe 8 × 8	21157.7 ±1092.81 (50)	96148.42 ±112675.6 (50)	13427.12 ±2259.73 (50)	19968.40 ±1103.20 (50)	10275.69 ±6952.548 (49)	33519.66 ±1522.85 (33)	21975.32 ±1079.08 (50)	11477.00 ±3318.38 (33)
PSO-B 10 × 10	22700.72 ±906.39 (50)	65769.76 ±70232.15 (50)	15114.46 ±3939.74 (48)	21574.70 ±1107.56 (50)	10741.78 ±10658.16 (50)	38976.00 ±1654.12 (42)	24310.86 ±1533.28 (50)	12943.90 ±3422.18 (31)
PSO-Bcfe 10 × 10	21796.04 ±832.54 (50)	**57704.16** **±71260.77** **(50)**	13953.04 ±3341.41 (49)	20430.34 ±1176.32 (50)	11817.06 ±15647.63 (50)	39002.15 ±2753.01 (40)	23159.40 ±1105.97 (50)	12191.81 ±2252.50 (27)
PSO-B 15 × 15	26122.88 ±950.08 (50)	74321.24 ±83535.59 (50)	20408.50 ±3692.10 (50)	24626.42 ±1406.53 (50)	11830.83 ±11576.37 (48)	45721.17 ±1471.31 (37)	27151.88 ±966.74 (50)	20961.41 ±11807.10 (37)
PSO-Bcfe 15 × 15	**19600.76** **±730.62** **(50)**	77348.06 ±91374.00 (50)	**16713.59** **±4387.78** **(46)**	**18734.94** **±1029.26** **(50)**	**10890.55** **±11624.59** **(47)**	**33519.66** **±1522.85** **(38)**	**20345.96** **±795.368** **(50)**	13614.55 ±5513.17 (38)

Best fitness values averaged over 50 runs

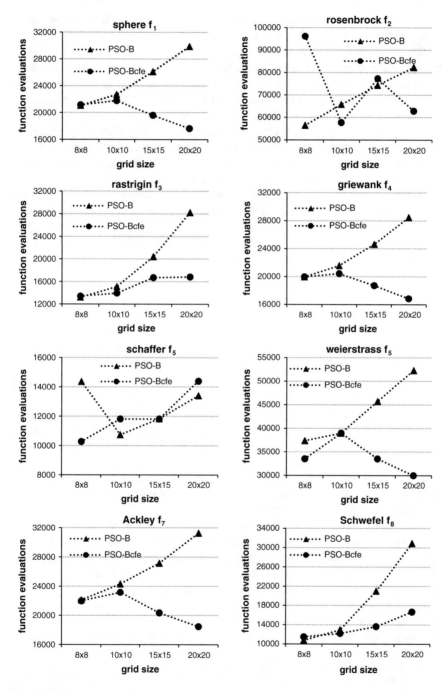

Fig. 1 PSO-B and PSO-B*cfe*. Function evaluations required to meet stop criteria when using grids with different sizes

that results from conserving evaluations is clearly overcome by the benefits of reducing the computational cost per iteration.

In the case of f_1, PSO-B*cfe* also significantly improves the quality of the solutions, namely with larger grids. The proposed scheme seems to be particularly efficient in unimodal landscapes, but further tests are required in order to confirm this hypothesis and understand what mechanisms make PSO-B*cfe* so efficient in finding more precise solutions for the sphere function.

The differences in the convergence speed of the algorithm are particularly noticeable when the grid is larger. While PSO-B's speed tends to decrease when the grid size increases, the behavior of PSO-B*cfe*, is much more stable, and in some functions it is even faster when the grid is expanded.

Figure 1 graphically depicts the above referred observations. When the grid size grows from 8×8 to 20×20, PSO-B's convergence speed degrades consistently, except in function f_5, where the behavior is more irregular. PSO-Bcfe, on the other hand, is sometimes faster with larger grids. When its convergence speed decreases with size (f_8, for instance), it scales better than PSO-B.

5 Conclusions

This paper proposes a general scheme for structuring dynamic populations for the Particle Swarm Optimization (PSO) algorithm. The particles are placed on a grid of nodes where the number of nodes is larger than the swarm size. The particles move on the grid according to simple rules and the network of information is defined by the particle's position on the grid and its neighborhood (*von Neumann* vicinity is considered here). If isolated (i.e., no neighbors except itself), the particle is updated but its position is not evaluated. This strategy may result in some loss of information, but the results show that the payoff in convergence speed overcomes the loss of information: the convergence speed is increased in the entire test set, while the accuracy of the algorithm (i.e., the averaged final fitness) is not degraded by the conservation of evaluations strategy.

The proposed algorithm is tested with a Brownian motion rule and compared to the standard static topology. The conservation of evaluations strategy results in a more stable performance when varying the grid size. Removing the strategy from the proposed dynamic structure results in a drop of the convergence speed when the size of the grid increases in relation to the swarm size.

The present study is restricted to dynamic structures based on particles with Brownian motion. Future research will be focused on dynamic structures with stigmergic behavior based on the fitness and position of the particles.

Acknowledgments The first author wishes to thank FCT, *Ministério da Ciência e Tecnologia*, his Research Fellowship SFRH/BPD/668/6/2009. The work was supported by FCT PROJECT [PEst-OE/EEI/LA0009/2013], Spanish Ministry of Science and Innovation projects TIN2011-28627-C04-02 and TIN2011-28627-C04-01, Andalusian Regional Government P08-TIC-03903 and P10-TIC-6083, CEI-BioTIC UGR project CEI2013-P-14, and UL-EvoPerf project.

References

1. Fernandes, C.M., Laredo, J.L.J., Merelo, J.J., Cotta, C., Nogueras, R., Rosa, A.C.: Performance and scalability of particle Swarms with dynamic and partially connected grid topologies. In: Proceedings of the 5th International Joint Conference on Computational Intelligence IJCCI 2013, pp. 47–55 (2013)
2. Grassé, P.-P.: La reconstrucion du nid et les coordinations interindividuelles chez bellicositermes et cubitermes sp. La théorie de la stigmergie: Essai d'interpretation du comportement des termites constructeurs, Insectes Sociaux **6**, 41–80 (1959)
3. Hseigh, S.-T., Sun, T.-Y, Liu, C.-C., Tsai, S.-J.: Efficient population utilization strategy for particle swarm optimizers. IEEE Trans. Syst., Man Cybern. Part B **39**(2), 444–456 (2009)
4. Kennedy, J., Eberhart, R.: Particle swarm optimization. Proc. IEEE Int. Conf. Neural Netw. **4**, 1942–1948 (1995)
5. Kennedy, J., Mendes, R.: Population structure and particle swarm performance. In: Proceedings of the IEEE World Congress on Evolutionary Computation, pp. 1671–1676 (2002)
6. Landa-Becerra, R., Santana-Quintero, L.V., Coello Coello, C.A.: Knowledge incorporation in multi-objective evolutionary algorithms. In: Multi-Objective Evolutionary Algorithms for Knowledge Discovery from Databases, pp. 23–46 (2008)
7. Liang, J.J., Qin, A.K., Suganthan, P.N., Baskar, S.: Comprehensive learning particle swarm optimizer for global optimization of multimodal functions. IEEE Trans. Evol. Comput. **10**(3), 281–296 (2006)
8. Majercik, S.: GREEN-PSO: conserving function evaluations in particle swarm optimization. In: Proceedings of the IJCCI 2013—International Joint Conference on Computational Intelligence, pp. 160–167 (2013)
9. Parsopoulos, K.E., Vrahatis, M.N.: UPSO: a unified particle swarm optimization scheme. In: Proceedings of the International Conference of Computational Methods in Sciences and Engineering (ICCMSE 2004), pp. 868–887 (2004)
10. Peram, T., Veeramachaneni, K., Mohan, C.K.: Fitness-distance-ratio based particle swarm optimization. In: Proceedings of the Swarm Intelligence Symposium SIS'03, pp. 174–181 (2003)
11. Reyes-Sierra, M., Coello Coello, C.A.: A study of techniques to improve the efficiency of a multiobjective particle swarm optimizer. In: Studies in Computational Intelligence (51), Evolutionary Computation in Dynamic and Uncertain Environments, pp. 269–296 (2007)
12. Shi, Y., Eberhart, R.C.: A Modified Particle Swarm Optimizer. In: Proceedings of IEEE 1998 International Conference on Evolutionary Computation, pp. 69–73. IEEE Press (1998)
13. Trelea, I.C.: The particle swarm optimization algorithm: convergence analysis and parameter selection. Inf. Proc. Lett. **85**, 317–325 (2003)

A Dissimilarity Learning Approach by Evolutionary Computation for Faults Recognition in Smart Grids

Enrico De Santis, Fabio Massimo Frattale Mascioli, Alireza Sadeghian and Antonello Rizzi

Abstract In a modern power grid known also as a Smart Grid (SG) its of paramount importance detecting a fault status both from the electricity operator and consumer feedback. The modern SG systems are equipped with Smart Sensors scattered within the real-world power distribution lines that are able to take a fine-grain picture of the actual power grid status gathering a huge amount of heterogeneous data. The Computational Intelligence paradigm has proven to be a useful approach in pattern recognition and analysis in facing problems related to SG. The present work deals with the challenging task of synthesizing a recognition model that learns from heterogeneous information that relates to environmental and physical grid variables collected by the Smart Sensors on MV feeders in the real-world SG that supplies the entire city of Rome, Italy. The recognition of faults is addressed by a combined approach of a multiple weighted Dissimilarity Measure, designed to cope with mixed data types like numerical data, Time Series and categorical data, and a One-Class Classification technique. For Categorical data the Semantic Distance (SD) is proposed, capable of grasping semantical information from clustered data. The faults model is obtained by a clustering algorithm (k-means) with a suitable initialization procedure capable to estimate the number of clusters k. A suited evolutionary algorithm has been designed to learn from the optimal weights of the Dissimilarity Measure defining a suitable performance measure computed by means of a cross-validation approach. In the present work a crisp classification rule on unseen test patterns is studied together with a soft decision mechanism based on a fuzzy membership function. Moreover

E. De Santis (✉) · F.M.F. Mascioli · A. Rizzi
Department of Information Engineering, Electronics, and Telecommunications,
"Sapienza" University of Rome, Via Eudossiana 18, 00184 Rome, Italy
e-mail: enrico.desantis@uniroma1.it, EDS@gmail.com

F.M.F. Mascioli
e-mail: mascioli@infocom.uniroma1.it

A. Rizzi
e-mail: antonello.rizzi@uniroma1.it

A. Sadeghian
Department of Computer Science, Ryerson University,
Toronto, ON M5B 2K3, Canada
e-mail: asadeghi@ryerson.ca

© Springer International Publishing Switzerland 2016
J.J. Merelo et al. (eds.), *Computational Intelligence*,
Studies in Computational Intelligence 620, DOI 10.1007/978-3-319-26393-9_8

a favorable discrimination performance between faults and standard working condition of the (One-Class) classifier will be presented comparing the SD with the well-known Simple Matching (SM) Distance for categorical data.

Keywords Evolutionary optimization · One class classification · Faults recognition · Concept learning

1 Introduction

The Smart Grid (SG) is one of the best technological breakthrough concerning efficient and sustainable management of power grids. According to the definition of the Smart Grid European Technology Platform a SG should "*intelligently integrate the actions of all the connected users, generators, consumers and those that do both, in order to efficiently deliver sustainable economic and secure electricity supply*" [1]. To reach that global goal the key word is the "*integration*" of technologies and research fields to add value to the power grid. The SG can be considered an evolution rather than a "revolution" [2] with improvements in monitoring and control tasks, in communications, in optimization, in self-healing technologies and in the integration of the sustainable energy generation. This evolution process is possible if it will be reinforced by the symbiotic exchange with Information Communications Technologies (ICTs), that, with secure network technologies and powerful computer systems, will provide the "*nervous system*" and the "*brain*" of the actual power grid. Smart Sensors are the fundamental driving technology that together with wired and wireless network communications and cloud systems are able to take a fine grained picture not only of the power grid state but also of the surrounding environment. At this level of abstraction, the SG ecosystem acts like a Complex System with an inherent non-linear and time-varying behavior emerging from heterogeneous elements with high degree of interaction, exchanging energy and information. Computational Intelligence (CI) techniques can face complex problems [3] and is a natural way to "inject" intelligence in artificial computing systems taking inspiration from the nature and providing capabilities like monitoring, control, decision making and adaptations [4].

An important key issue in SGs is the Decision Support System (DSS), which is an expert system that provides decision support for the commanding and dispatching system of the power grid. The information provided by the DSS can be used for Condition Based Maintenance (CBM) in the power grid [5]. Collecting heterogeneous measurements in modern SG systems is of paramount importance. As an instance, the available measurements can be used for dealing with various important pattern recognition and data mining problems on SGs, such as event classification [6], or diagnostic systems for cables and accessories [7]. On the basis of the specific type of considered data, different problem types could be formulated. In [8] authors have established a relationship between environmental features and fault causes. A fault cause classifier based on the linear discriminant analysis (LDA) is proposed in [9]. Information regarding weather conditions, longitude-latitude information, and measurements of physical quantities (e.g., currents and voltages) related to the power grid have been taken into account. The One-Class Quarter-Sphere SVM algorithm is proposed [10] for faults classification in the power grid. The reported experimental evaluation is however performed on synthetically generated data only. This paper addresses this topic, facing the challenging problem of faults prediction and recognition on a real distribution network, in order to report in real time possible defects, before failures can occur, or as an off-line decision making aid, within the corporate strategic management procedures. The data set provided by ACEA Distribuzione S.p.a (ACEA is the company managing the electrical network feeding the whole province of Rome, Italy) collects all the information considered by company's field experts as related to the events of a particular type of faults, namely Localized Faults (LF). This paper follows our previous work [11] where the posed problem of faults recognition and prediction is framed as an unsupervised learning problem approached with the One Class Classification (OCC) paradigm [12] because of the availability only of positive or target instances (faults patterns). This modeling problem can be faced by synthesizing reasonable decision regions relying on a k-means clustering procedure in which the parameters of a suited dissimilarity measure and the boundaries of decision regions are optimized by a Genetic Algorithm, such that unseen target test patterns are recognized properly as faults or not. This paper focuses on two important issues: (*i*) the initialization of k-means with an automatic procedure in order to find the optimal number k of clusters; (*ii*) to find a more reliable dissimilarity measure for the categorical features of the faults patterns; to this aim, the Semantic Distance (SD) is adopted, addressing the problem of better grasping the semantic content of a well-formed cluster. The paper is organized as follows.

Section 2 deals with a bird's eye view on the "ACEA Smart Grid Pilot Project" of which this study is based. A brief review of the faults patterns is given in Sect. 3.1, while in Sect. 3.2 will be introduced the OCC system for fault recognition. In Sect. 3.3 is described the k-means initialization algorithm. In Sect. 3.4 is presented the weighted dissimilarity measure and the proposed SD for categorical features. A novel approach to associate a reliability measure to an hard decision about a test pattern based on a fuzzy membership function is presented in Sect. 4. In Sect. 5 it is shown and discussed the experimental results in terms of classifications performances comparing the well-known Simple Matching measure with the proposed Semantic Distance for categorical attributes. Finally, in the Sect. 6, conclusions are drawn.

2 The ACEA Smart Grid Project

The following work is a branch of a general project, the "ACEA Smart Grid project" [13]. The aim is to develop an automated tool for diagnostic, recognition and analysis of fault states in the powergrid managed by the ACEA distribution company. The process flow diagram is depicted in Fig. 1 where it is shown how raw data coming

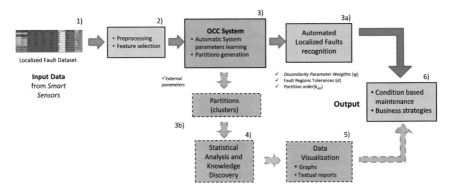

Fig. 1 Process flow diagram describing the "ACEA Smart Grid Project". Those last two post-processing stages (*gray boxes*), belonging to the work packages set of the overall project, are not discussed in this paper

from the SSs is transformed into meaningful information in order to support business strategies. After a preliminary preprocessing stage on data, operated together with the ACEA experts, the faults dataset is used as input for the herein presented OCC which by means of an evolutionary strategy is in charge of learning typical fault scenarios. A clustering technique is adopted to define the model of the proposed OCC. The synthesized partition is used also for post-processing purposes, such data analysis and visualization (the lower branch in Fig. 1).

3 The One Class-Classification Approach for Faults Detection

3.1 The Fault Patterns

The ACEA power grid is constituted of backbones of uniform section exerting radially with the possibility of counter-supply if a branch is out of order. Each backbone of the power grid is supplied by two distinct Primary Stations (PS) and each half-line is protected against faults through the breakers. The underlined SG is equipped with Secondary Stations (SSs) located on Medium Voltage (MV) lines, each ones fed by a PS able to collect faults data. A fault is related to the failure of the electrical insulation (e.g., cables insulation) that compromises the correct functioning of (part of) the grid. Therefore, a LF is actually a fault in which a physical element of the grid is permanently damaged causing long outages. LFs must be distinguished from both: (i) "short outages" that are brief interruptions lasting more than 1 s and less than 3 min; (ii) "transient outages" in which the interruptions don't exceed 1 s. The last ones can be caused, for example, by a transient fault of a cable's electrical insulation of very brief duration not causing a blackout.

The proposed one-class classifier is trained and tested on a dataset composed by 1180 LFs patterns structured in 21 different features. The features belong to different data types: categorical (nominal), quantitative (i.e., data belonging to a normed space) and times series (TSs). The last ones describes the sequence of short outages that are automatically registered by the protection systems as soon as they occur. LFs on MV feeders are characterized by heterogeneous data, including weather conditions, spatio-temporal data (i.e., longitude-latitude pairs and time), physical data related to the state of power grid and its electric equipments (e.g., measured currents and voltages), and finally meteorological data. The whole database was provided by ACEA and contains data concerning a temporal period of 3 years across 2009–2011. This database was validated, by cleaning it from human errors and by completing in an appropriate way missing data. A detailed description of the considered features is provided in Table 1.

3.2 The OCC Classifier

The main idea in order to build a model of LF patterns in the considered SG is to use a clustering technique. In this work a modified version of k-means is proposed, capable to find a suitable partition $P = \{C_1, C_2, \ldots, C_k\}$ of data set and to determine at the same time the optimal number of clusters k. The main assumption is that similar status of the SG have similar chances of generating a LF, reflecting the cluster model. The OCC System is designed to find a proper decision region, namely the "faults space", F, relying on the positions of target patterns denoting the LFs. A (one-class) classification problem instance is defined as a triple of disjoint sets, namely training set (S_{tr}), validation set (S_{vs}), and test set (S_{ts}), all containing fault pattern instances. Given a specific parameters setting, a classification model instance is synthesized on S_{tr} and it is validated on S_{vs}. Finally, performance measures are computed on S_{ts}. As depicted in the functional model (see Fig. 2) this paradigm is objectified by designing the OCC classifier as the composition of three modules wrapped in an optimization block. In order to synthesize the LF region, the learning procedure is leaded: (1) by the clustering module that operate an hard partition of S_{tr}; (2) by the validation module operating on S_{vs}, designed to refine the LF boundaries; the decision rule (that leads the task of the *patterns assignment*) is based on the proximity of the LF pattern at hand to the clusters representatives. Thus the core of the OCC system is the dissimilarity measure $d : F \times F \rightarrow \mathbb{R}^+$, reported in Sect. 3.4, that depends on a weighting parameter vector w. For this reason the proposed (One-Class) classifier fully belongs to the Metric Learning framework [14]. The decision regions $B(C_i)$ are derived from a "cluster extent" measure $\delta(C_i)$ characterizing the C_i clusters and summed to a tolerance parameter σ (thus $B(C_i)=\delta(C_i)+ \sigma$) that together to the dissimilarity weights belongs to the search space for the optimization algorithm. Here C_i is the ith cluster ($i = 1, 2, \ldots, k$) and $\delta(C_i)$ is the average intra-cluster dissimilarity. The overall decision regions is the union of the k hyperspheres, that is:

Table 1 Considered features representing a fault pattern

Feature	Data type	Description
(1) Day start		Day in which the LF was detected
(2) Time start		Time stamp (minutes) in which the LF was detected
(12) Current out of bounds	Quantitative (Integer)	The maximum operating current of the backbone is less than or equal to 60 % of the threshold "out of bounds", typically established at 90 % of capacity
(11) # Secondary Stations (SSs)		Number of out of service secondary stations due to the LF
(3) Primary Station (PS) code		Unique backbone identifier
(4) Protection tripped		Type of intervention of the protective device
(5) Voltage line	Categorical (String)	Nominal voltage of the backbone
(6) Type of element		Element that caused the damage
(17) Cable section		Section of the cable, if applicable
(7) Location element		Element positioning (aerial or underground)
(8) Material		Constituent material element (CU, AL)
(9) Primary station fault distance		Distance between the primary station and the geographical location of the LF
(10) Median point		Fault location calculated as median point between two secondary stations
(13) Max. temperature		Maximum registered temperature
(14) Min. temperature	Quantitative (Real)	Minimum registered temperature
(15) Delta temperature		Difference between the maximum and minimum temperature
(16) Rain		Millimeters of rainfall in a period of 2 h preceding the LF
(18) Backbone Electric Current		Extracted feature from Time Series of electric current values that flows in a given backbone of the considered power grid. It is the difference between the average of the current's value, in two consecutive temporal windows of 12 h each one, before the fault
(19) Interruptions (breaker)		Sequence of opening events of the *breakers* in the primary station
(20) Petersen alarms	TS (Integers sequence)	Sequence of alarms detected by the device called "Petersen's coil" due to loss of electrical insulation on the power line
(21) Saving interventions		Sequences of decisive interventions of the Petersen's coil which have prevented the LF

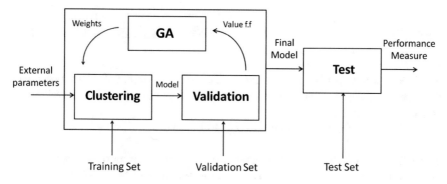

Fig. 2 Block diagram depicting the optimized classification model synthesis

$$B = \bigcup_{i=1}^{k} B(C_i) \tag{1}$$

In Fig. 3 is depicted one hypersphere model.

In this work in addition to the weights w and the σ parameters the search space is completed by a γ parameter controlling the proposed k-means initialization algorithm (see Sect. 3.3). Finally it is defined the search space, constituted by the all model's parameters, as $p = [w, \sigma, \gamma]$.

In this work the representative of the cluster, denoted as $c_i = R(C_i)$, is the Min-SOD [15]. So for each cluster C_i the representative one will be chosen as the pattern that belongs to the considered cluster and for which the sum of distances from the other patterns of the cluster has the lower value. A cluster representative c_i can be considered as a prototype of a *typical fault scenario* individuated in S_{tr}. The decision rule to establish if a test pattern is a target pattern or not is performed computing its overall dissimilarity measure d from the representatives of all clusters C_i and verifying if it falls in the decision region (see Fig. 3) built up on the nearest cluster. A standard Genetic Algorithm is used in the learning phase in order to minimize the

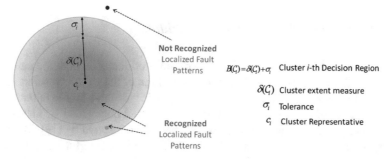

Fig. 3 Cluster decision region and its characterizing parameters

Fig. 4 Composition of the
chromosome

trade-off between a performance measure (we specify the nature of such a measure in the experiments section) on S_{vs} and the threshold σ' value by means of the following Objective Function:

$$f(p) = \alpha A(S_{vs}) + (1 - \alpha)\,\sigma', \tag{2}$$

where $\alpha \in [0, 1]$ is an external parameter controlling the importance in minimizing $A(S_{vs})$ versus σ'. In other words, α is a meta-parameter by which it is possible to control the relative importance in minimizing $A(S_{vs})$ or in minimizing the overall faults decision region extent. σ' is the threshold value normalized with respect to the diagonal D of the hypercube (see Sect. 3.4) of the overall space: $\sigma' = \sigma/D$. As concerns the chromosome coding (see Fig. 4), each individual of the population consists in the weights w_s, $s = (1, 2, \ldots, N_w)$ associated to each feature, where N_w is the number of the considered features, the value σ that is the threshold added at each "cluster extent" measure during the validation phase and the γ parameter mentioned above. The overall number of genes in an individual's chromosome is therefore $l = N_w + 2$. The functional dependencies between the discussed parameters, in the proposed OCC system, are:

$$\begin{cases} k_{opt} = k_{opt}(w, \gamma) \\ \sigma_i = \sigma_i(k_{opt}, w) \\ A = A(w, \sigma_i), \end{cases} \tag{3}$$

where k_{opt} is the optimum number of clusters found by the k-means initialization algorithm described in the next section and A is the performance measure of the proposed (One-Class) classifier. The subscript index i covers the general case, not studied here, in which can be instantiated distinct thresholds values for different clusters.

3.3 The k-Means Initialization Algorithm

It is well known that the k-means behavior depends critically on both the number k of clusters, given as a fixed input, and on the position of the k initial clusters representatives. In the literature there are a wide range of algorithms for the initialization of the centroids of the k-means, each with its pros and cons [16–18]. The initialization criterion of centroids, here proposed, was initial inspired by [19]. The work is based on the idea to choose as centroids, the patterns that are furthest from each other. The provided version of the algorithm takes into account also the presences of *outliers*. To verify if the candidate centroid is an outlier we designed a simple decision rule defined by parameters: a, an integer value, and b, a real valued number ranging $[0, 1]$. The parameter a indicates the minimum number of patterns that must enter in the circumference with center the pattern candidate as centroid and radius given by $b * d_{Pmax}$, where d_{Pmax} is the distance between the furthest pattern in the whole dataset. Hence if within the distance $b * d_{Pmax}$ there are more than a patterns then the candidate centroid is not an outlier. Other inputs to the overall algorithm are a scale parameter γ, the Dissimilarity Matrix (DM) **D** and the number of initial centroids k_{ini}. The algorithm tries to calculate the best positions of the centroids and their final number k_{opt} possibly decreasing the provided initial number (k_{ini}). The main steps are the following:

Algorithm input: The initial number of centroids k_{ini}, the dissimilarity matrix **D**, the $\gamma \in [0, 1]$ parameter, a, b, d_{Pmax}.

Algorithm output: the k_{opt} centroids.

Choose a random pattern p_i among those available in S_{tr} and compute the pattern p_j furthest away from it.
while *centroid* $==$ *not found* **do**
    if p_j *is not an outlier* **then**
        choose p_j as the first centroid;
      centroid $=$ found;
    else
        choose as p_j the next pattern among those furthest away from p_i;
      centroid $=$ not found;
    end
end
Choose as second centroid the pattern p_a furthest away from p_j.
while *centroid* $==$ *not found* **do**
    if p_a *is not an outlier* **then**
        choose it as the as second centroid;
      centroid $=$ found;
    else
        choose as p_a the next pattern furthest away from p_j;
      centroid $=$ not found
    end
end
while $k < k_{ini}$ **do**
    choose as a possible centroid the pattern p_n whose sum of the distances to the other centroids, found earlier, is maximum;
    if p_n *is not an outlier* **then**
        choose it as the other centroid; $k = k + 1$;
    else
        choose the next pattern whose sum of the distances to the other centroids found earlier, is maximum;
    end
end
Calculate $d_{Cmax} = d(p_j, p_a)$ as the distance between the first two centroids. Given the external parameter $\gamma \in [0, 1]$
for $i = 1; i < k; i++$ **do**

```
for j = 1; j < k; j++ do
    if d(p_i, p_j) ≤ γ * d_Cmax then
        delete randomly one of the two considered centroid, k = k_ini − 1;
    end
end
end
```

return the $k_{opt} = k$ centroids.

The k-means with the proposed representatives initialization can be seen as an hybrid between a k-clustering and a free clustering algorithm where, once fixed an initial number of centroids, it returns an optimal number of centroids less or equal to the initial ones.

3.4 The Weighted Custom Dissimilarity Measure

The dissimilarity function between two patterns is of paramount importance in data driven modeling applications. Given two patterns x and y the wighted dissimilarity measure adopted in the proposed classifier is:

$$d(\underline{x}, \underline{y}; \underline{W}) = \sqrt{(\underline{x} \ominus \underline{y})\underline{W}\underline{W}^T(\underline{x} \ominus \underline{y})^T)} \qquad (4)$$

$$d(\underline{x}, \underline{y}; \underline{W}) \in [0, D]$$

where $(x \ominus y)$ is a *Component-Wise* dissimilarity measure, i.e. a row vector containing the specific differences between homologues features. W is a diagonal square matrix of dimension $N_w \times N_w$, in which N_w is the number of weights. In Eq. 4 the maximum value for d is the diagonal of the hypercube, that is: $D = \sqrt{\sum_{i=1}^{N_w} w_i^2}$, where w_i are the features weights. The inner specific dissimilaritiy functions differ each other depending on the nature of each feature as explained in the following.

Quantitative (*real*). Given two normalized quantitative values v_i, v_j the distance between them is the absolute difference: $d_{i,j} = |v_i - v_j|$.

As regards the features "Day start" and "Time start" the distance is calculated through the circular difference. The value of these features is an integer number between 1 and 365 (*total days in 1 year*) for the former and between 1 and 1140 (*total minutes in 1 day*) for the latter. The circular distance between two numbers is defined as the minimum value between the calculated distance in a clockwise direction and the other calculated in counter clockwise (Fig. 5).

Categorical (*nominal*). Categorical attributes, also referred to as nominal attributes, are attributes without a semantically valid ordering (see Table 1 for the data treated as nominal). Let's define c_i and c_j the values of the categorical feature for the patterns ith and jth, respectively. A one well-suited solution to compute a dissimilarity measure for categorical features is the Simple-Matching (SM) distance:

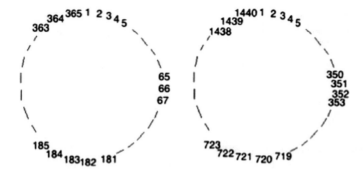

Fig. 5 Sketch of circular domains for "Day start" and "Time start" features

$$d_{i,j} = \begin{cases} 1 & \text{if } c_i \neq c_j \\ 0 & \text{if } c_i = c_j. \end{cases} \tag{5}$$

When measuring pattern-cluster dissimilarities (i.e. in the assignment of a pattern to a clusters) the Semantic Distance (SD) introduced in Sect. 3.4.1 is used.

Times Series. TSs are characterized by a non-uniform sampling since they represent sequences of asynchronous events. As a consequence, usually they don't share the same length. TSs are represented as real valued vectors containing the differences between short outages timestamps and the LF timestamp considered as a common reference. These values are normalized in the range [0, 1], dividing the values obtained by the total number of seconds in the temporal window considered. In order to measure the distance between two different TSs *(different in values and size)*, we use the Dynamic Time Warping (DTW) [20].

3.4.1 Semantic Distance for Categorical Data

The task of calculating a good similarity measure between categorical objects is challenging because of the difficulties to establish meaningful relations between them. The distance between two objects computed with the simple matching similarity measure (Eq. 5) is either 0 or 1. This often results in clusters with weak intra-similarity [21] and this may result in a loss of semantic content in a partition generated by a clustering algorithm. As concerns k-modes [22] algorithm, in the literature several frequency-based dissimilarity measures between categorical object are proposed [23, 24]. The proposed dissimilarity measure for categorical objects is a frequency-based dissimilarity measure and follows the work [25] in which a features weighted k-modes algorithm is studied, where the weights are related to the frequency value of a category in a given cluster.

Let $N_{i,j}$ be the number of instances of the ith value of the considered categorical feature F_{cc} in the cluster jth (C_j) and let's define $N_{max,j} = max(N_{1,j}, \dots N_{n,j})$, where n

Fig. 6 In this cluster the Yellow feature value is completely missing

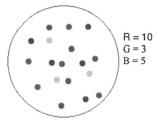

R = 10
G = 3
B = 5

is the number of the different values of F_{cc} present in C_j. We can finally define the SD between a categorical feature of the pattern P_h ($F_{cc}P_h$) and the cluster C_j as:

$$d_{F_{cc}P_h,C_j} = 1 - W_{i,j}, \quad with \ d_{F_{cc}P_h,C_j} \in [0, 1] \tag{6}$$

where $W_{i,j} = \frac{N_{i,j}}{N_{max,j}}$ is the fraction of values of the ith category of the considered categorical feature in the jth cluster with respect to the number of values of the most frequent category.

The SD takes into account the statistical information of a given cluster and it is used like a pattern-cluster dissimilarity measure. Unlike the SM distance, the SD can span in the real valued range [0, 1]. Note that this distance is characterized by the statistical properties of the specific cluster under consideration. The SD can be intended as a "local metric", since each cluster is characterized by its own statistic distribution of categorical values and thus it is characterized by its own weights that can change from one cluster to another.

For example, let us consider a categorical feature coding for one of four possible colors (red, green, blue or yellow) and let us consider the cluster depicted in Fig. 6. By means of Eq. (6) it is possible to compute the values of $W_{i,L}$ for each color (nominal attribute value), represented or not in the considered cluster and then the SD:

$$\begin{cases} W_{R,L} = \frac{10}{10}, W_{G,L} = \frac{3}{10}, W_{B,L} = \frac{5}{10} \\ if \ Color \notin C_L => W_{Color,L} = 0 \end{cases}$$

- if the value of $F_{cc}P_h$ is red: $d_{F_{cc}P_h,C_L} = 1 - \frac{10}{10} = 0$
- if the value of $F_{cc}P_h$ is green: $d_{F_{cc}P_h,C_L} = 1 - \frac{3}{10} = \frac{7}{10}$
- if the value of $F_{cc}P_h$ is blue: $d_{F_{cc}P_h,C_L} = 1 - \frac{5}{10} = \frac{1}{2}$
- if the value of $F_{cc}P_h$ is yellow: $d_{F_{cc}P_h,C_L} = 1 - \frac{0}{10} = 1$

4 Reliability Evaluation

Beside the Boolean decision rule regarding if a new test pattern is a fault or not (see Sect. 3.2), it is operatively reasonable to provide the user an additional measure that quantifies the reliability of a decision. For this purpose we equip each cluster C_i with a suitable membership function denoted in the following as $\mu_{C_i}(\cdot)$. In other words it generates a fuzzy set over C_i. The membership function is able to quantify the uncertainty (expressed by the membership degree in $[0, 1]$) of a decision about the recognition of a test pattern. Membership values close to either 0 or 1 denote "certain" and thus reliable hard decisions. If the membership degree is close to 0.5 there is no clear distinction about the hard decision. For this purpose, we used a parametric sigmoid model for $\mu_{C_i}(\cdot)$, which is defined as follows:

$$\mu_{C_i}(x) = \frac{1}{1 + \exp((d(c_i, x) - b_i)/a_i)}, \tag{7}$$

where $a_i, b_i \geq 0$ are two parameters specific to C_i, and $d(\cdot, \cdot)$ is the dissimilarity measure (4). The shape parameters of the herein sigmoidal function are: (i) a_i that controls the steepness, the lower the value, the faster the rate of change; (ii) b_i that is used to center the function in the input domain. If a cluster (that models a typical fault situation found in the training set) is very compact, then it describes a very specific fault scenario. Therefore, no significant variations should be accepted to consider test patterns as members of this cluster. The converse is also true. If a cluster is characterized by a wide extent, then we might be more tolerant in the evaluation of the membership. Accordingly, the parameter a_i is set equal to $\delta(C_i)$. On the other hand, we can define $b_i = \delta(C_i) + \sigma_i/2$. This allows us to position the part of the sigmoid that changes faster right in-between the area of the decision region determined by the dissimilarity values falling in $[B(C_i) - \sigma_i, B(C_i)]$.

5 Experimental Results

5.1 Test on k-Means Initialization Algorithm

The proposed initialization algorithm has been tested on a toy problem, where patterns are generated from three distinct Gaussian distributions, as depicted in Fig. 7. Setting the initial number of centroids $K_{ini} = 10$, the proposed algorithm converges to an optimal number of clusters equal to 3 (see also Fig. 8).

5.2 Tests on ACEA Dataset

In this section we report on the first tests of the proposed (One-Class) classifier system on real data. The synthesized classification model should be able to correctly recognize fault patterns and at the same time avoid raising wrong alarm signals, recognizing faults system's measurements corresponding to normal operating conditions. The proposed (One-Class) classifier generates both hard and soft decisions on each test pattern. In cases of a hard decision we evaluate the recognition performance

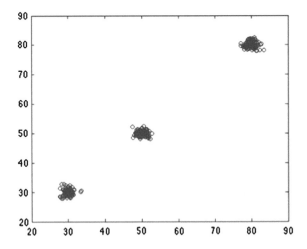

Fig. 7 Patterns distribution in the considered toy problem

(a)   **(b)**

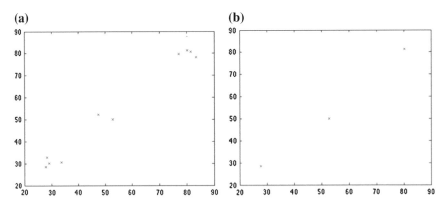

Fig. 8 **a** Centroids found before the close representatives removal step, with $k_{ini} = 10$. **b** The final optimal centroids ($K_{opt} = 3$)

by means of the Confusion Matrix extrapolating the False Positive Rate (FPR), the True Positive Rate (TPR) and the Accuracy (A) [26]. On the other hand, in soft decision cases the correctness of the classifier is quantified by computing the area under the Receiver Operating Characteristic (ROC) curve (AUC) [26] generated by considering the membership degree as a suitable "score" assigned by the classifier to the test pattern. The performance measure (2) that guides the evolution of the model parameters by means of the GA is the accuracy elaborated by the Confusion Matrix. The search space for the γ parameter that controls the number of retrieved cluster is selected in the real valued interval [0.1, 1], while the initial number of cluster k_{ini} is fixed to 15. The adopted GA performs stochastic uniform selection, Gaussian mutation and scattered crossover (with crossover fraction of 0.7). It implements a form of elitism that imports the two fittest individuals in the next generation; the population size is kept constant throughout the generations and equal to 50 individuals. The stop criterion is defined by considering a maximum number of iterations (200) and checking the variations of the best individual fitness.

Since non-faults patterns (negative instances) are not available in the ACEA dataset, in order to properly measure system performances, negative instances are formed by randomly generating, with a uniform distribution, each feature value characterizing a fault pattern (see Sect. 3.1 for details on the features). In close cooperation with the ACEA experts, following their precious advice the LF model is trained on the features 1–4 and 6–18 (described in Table 1). The training set is composed of 532 fault patterns; in the validation set we have 470 fault patterns and 500 non-fault patterns; finally for the test set we have 82 fault patterns and 300 non-fault patterns.

Table 2 reports the results of six simulations carried out for three distinct values of the alpha parameter ($\alpha = [0.3, 0.5, 0.7]$) and for each one it is compared to the performance achieved using the Simple Matching (SM) distance and the Statistic Distance (SD). According to the definition in [26] where *"the AUC of a classifier is equivalent to the probability that the classifier will rank a randomly chosen positive instance higher than a randomly chosen negative instance"* the test performed with $\alpha = 0.3$ and using the SD for categorical variables (Sim. 1) achieves clearly the best result with an AUC value of 92.1 %. Moreover, as we claim the classifier shows the better true positive rate (TPR = 96.3 %) and the lower false alarms rate measured through the false positive rate (FPR = 35.5 %). Also the Accuracy (A) is good in the previous case, meaning that the hard classification works well. Results from Table 2 show also how the classifier performs better in terms of **AUC** in every case in which is used the SD. Hence in general terms the Sim. 1 achieves a favorable Decision Region that characterizes a much better faults patterns, with a much more limited extension in the whole input domain, thus avoiding to cover non-faults patterns. To confirm this interpretation we have computed an index that measures the compactness and separability of clusters (the lower the better), namely the Davies-Bouldin index [27], on the training set partitions (**DB** in Table 2). For Sim. 1 the DB index is 9.53 compared to the reported worst case in terms of the **AUC** in which the DB is 14.61. Figure 9 shows the first two components of the Principal Component Analysis (PCA) com-

Table 2 Results of the simulations obtained with the Semantic Distance (SD) and Simple Matching (SM) for several values of α parameter in the fitness function (Eq. (2))

Sim.	Distance	α	k_i	k_{opt}	FPR	TPR	AUC	γ	A	DB
1	SD	0.3	15	3	35.5 %	96.3 %	92.1 %	0.5201	75.3 %	9.53
2	SM	0.3	15	9	44.3 %	83.2 %	71.3 %	0.5011	64.6 %	14.61
3	SD	0.5	15	6	41.4 %	92.7 %	85.2 %	0.4041	67.8 %	12.80
4	SM	0.5	15	7	46.1 %	96.3 %	82.1 %	0.4113	67.8 %	10.04
5	SD	0.7	15	8	42.6 %	96.3 %	84.8 %	0.4898	70.2 %	15.13
6	SM	0.7	15	8	60.9 %	98.8 %	78.1 %	0.4905	58.7 %	14.13

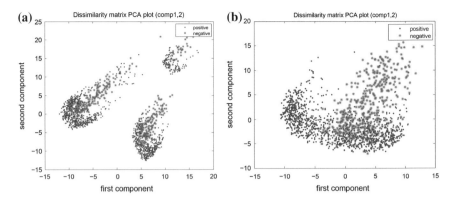

Fig. 9 The first two components of the PCA elaborated over the Dissimilarity Matrix **D** constructed from the ACEA dataset containing both fault and non-fault patterns. Figure b is obtained with the DM weighted with the dissimilarity weights $\underline{w}$ carried out from Sim. 1 (see Table 2). **a** Unitary weights, **b** non-unitary weights

puted on the overall ACEA dataset in the case of: (a) Dissimilarity Matrix (DM) weighted with unitary weights, (b) DM weighted with weights retrieved from Sim. 1. Within the Sim 1 settings the OCC accomplishes the best performance retrieving 3 clusters (visible in Fig. 9a). The PCA computed in the weighted case shows how the OCC transforms the (PCA of the) dissimilarity space separating better the fault patterns (positive instances) from the non-fault patterns (negative instances).

6 Conclusions

In this paper we propose a MV lines faults recognition system as the core element of a Condition Based Maintenance procedure to be employed in the electric energy distribution network of Rome, Italy, managed by ACEA Distribuzione S.p.A. By relying on the OCC approach, the faults decision region is synthesized by partitioning the available samples of the training set. A suited pattern dissimilarity measure has been defined in order to deal with different features data types. The adopted clustering procedure is a modified version of k-means, with a novel procedure for centroids initialization. A genetic algorithm is in charge to find the optimal value of the dissimilarity measure weights, as well as two parameters controlling the initial centroids positioning and the fault decision region extent, respectively. According to our tests, the new proposed method for k-means initialization shows a good reliability in finding automatically the best number of clusters and the best positions of the centroids. Furthermore, the proposed SD for categorical features subspaces performs better than the plain SM distance when used to define a pattern to cluster dissimilarity measure. Since faults decision region is synthesized starting from each cluster decision region, this measure has a key role in defining a proper inductive inference engine, and thus in improving the generalization capability of the recognition system. Future works will be focused on the possibility to evaluate other clustering algorithms and different global optimization schemes. Lastly, tests results performed on real data make us confident about further systems developments possibility, towards a final commissioning into the Rome electric energy distribution network.

Acknowledgments The authors wish to thank ACEA Distribuzione S.p.A. for providing the faults data and for their useful support during the OCC system design and test phases. Special thanks to Ing. Stefano Liotta, Chief Network Operation Division, and to Ing. Silvio Alessandroni, Chief Electric Power Distribution Remote Control Division.

References

1. The SmartGrids European Technology Platform (2013). http://www.smartgrids.eu/ETPSmartGrids
2. International Energy Outlook 2011—Energy Information Administration (2013). http://www.eia.gov/forecasts/ieo/index.cfm
3. Venayagamoorthy, G.K.: Dynamic, stochastic, computational, and scalable technologies for smart grids. IEEE Comput. Intell. Mag. **6**, 22–35 (2011)
4. De Santis, E., Rizzi, A., Sadeghian, A., Frattale M.F.M.: Genetic optimization of a fuzzy control system for energy flow management in micro-grids. In: IEEE Joint IFSA World Congress and NAFIPS Annual Meeting. **2013**, 418–423 (2013)
5. Raheja, D., Llinas, J., Nagi, R., Romanowski, C.: Data fusion/data mining-based architecture for condition-based maintenance. Int. J. Prod. Res. **44**, 2869–2887 (2006)
6. Afzal, M., Pothamsetty, V.: Analytics for distributed smart grid sensing. In: Innovative Smart Grid Technologies (ISGT), 2012 IEEE PES, pp. 1–7 (2012)
7. Rizzi, A., Mascioli, F.M.F., Baldini, F., Mazzetti, C., Bartnikas, R.: Genetic optimization of a PD diagnostic system for cable accessories. IEEE Trans. Power Delivery **24**, 1728–1738 (2009)
8. Guikema, S.D., Davidson, R.A., Haibin, L.: Statistical models of the effects of tree trimming on power system outages. IEEE Trans. Power Delivery **21**, 1549–1557 (2006)
9. Cai, Y., Chow, M.Y.: Exploratory analysis of massive data for distribution fault diagnosis in smart grids. In: Power Energy Society General Meeting, 2009. PES '09. IEEE, pp. 1–6 (2009)
10. Shahid, N., Aleem, S., Naqvi, I., Zaffar, N.: Support vector machine based fault detection amp; classification in smart grids. In: Globecom Workshops (GC Wkshps), 2012 IEEE, pp. 1526–1531 (2012)
11. De Santis, E., Rizzi, A., Livi, L., Sadeghian, A., Frattale M.F.M.: Fault recognition in smart grids by a one-class classification approach. In: 2014 IEEE World Congress on Computational Intelligence, IEEE (2014)
12. Khan, S.S., Madden, M.G.: A survey of recent trends in one class classification. In: Coyle, L., Freyne, J. (eds.) Artificial Intelligence and Cognitive Science. Lecture Notes in Computer Science, vol. 6206, pp. 188–197. Springer, Heidelberg (2010)
13. The ACEA smart grid pilot project (in italian) http://www.autorita.energia.it/allegati/operatori/elettricita/smartgrid/V%20Rel%20smart%20ACEA%20D.pdf
14. Bellet, A., Habrard, A., Sebban, M.: A survey on metric learning for feature vectors and structured data. CoRR (2013). abs/1306.6709
15. Del Vescovo, G., Livi, L., Frattale M.F.M., Rizzi, A.: On the problem of modeling structured data with the MinSOD representative. Int. J. Comput. Theory Eng. **6**, 9–14 (2014)
16. Dan Pelleg, A.M.: X-means: extending k-means with efficient estimation of the number of clusters. In: Proceedings of the Seventeenth International Conference on Machine Learning, pp. 727–734. Morgan Kaufmann, San Francisco (2000)

17. Tibshirani, R., Walther, G., Hastie, T.: Estimating the number of clusters in a data set via the gap statistic. J. R. Stat. Soc. Ser. B (Stat. Methodol.) **63**, 411–423 (2001)
18. Laszlo, M., Mukherjee, S.: A genetic algorithm using hyper-quadtrees for low-dimensional k-means clustering. IEEE Trans. Pattern Anal. Mach. Intell. **28**, 533–543 (2006)
19. Barakbah, A., Kiyoki, Y.: A pillar algorithm for k-means optimization by distance maximization for initial centroid designation. 61–68 (2009)
20. Müller, M.: Dynamic time warping. In: Information Retrieval for Music and Motion, pp. 69–84. Springer, Heidelberg (2007)
21. Ng, M.K., Junjie, M., Joshua, L., Huang, Z., He, Z.: On the impact of dissimilarity measure in k-modes clustering algorithm. IEEE Trans. Pattern Anal. Mach. Intell. (2007)
22. Huang, Z.: Extensions to the k-means algorithm for clustering large data sets with categorical values (1998)
23. Cheng, V., Li, C.H., Kwok, J.T., Li, C.K.: Dissimilarity learning for nominal data. Pattern Recogn. **37**, 1471–1477 (2004)
24. Quang, L., Bao, H.: A conditional probability distribution-based dissimilarity measure for categorial data. In: Dai, H., Srikant, R., Zhang, C. (eds.) Advances in Knowledge Discovery and Data Mining. Lecture Notes in Computer Science, vol. 3056, pp. 580–589. Springer, Berlin Heidelberg (2004)
25. He, Z., Xu, X., Deng, S.: Attribute value weighting in k-modes clustering. Expert Syst. Appl. **38**, 15365–15369 (2011)
26. Fawcett, T.: An introduction to ROC analysis. Pattern Recogn. Lett. **27**, 861–874 (2006)
27. Davies, D.L., Bouldin, D.W.: A cluster separation measure. IEEE Trans. Pattern Anal. Mach. Intell. **PAMI-1**, 224–227 (1979)

Noise Sensitivity of an Information Granules Filtering Procedure by Genetic Optimization for Inexact Sequential Pattern Mining

Enrico Maiorino, Francesca Possemato, Valerio Modugno
and Antonello Rizzi

Abstract One of the most essential challenges in Data Mining and Knowledge Discovery is the development of effective tools able to find regularities in data. In order to highlight and to extract interesting knowledge from the data at hand, a key problem is frequent pattern mining, i.e. to discover frequent substructures hidden in the available data. In many interesting application fields, data are often represented and stored as sequences over time or space of generic objects. Due to the presence of noise and uncertainties in data, searching for frequent subsequences must employ approximate matching techniques, such as edit distances. A common procedure to identify recurrent patterns in noisy data is based on clustering algorithms relying on some edit distance between subsequences. However, this plain approach can produce many spurious patterns due to multiple pattern matchings on close positions in the same sequence excerpt. In this paper, we present a method to overcome this drawback by applying an optimization-based step lter that identifies the most descriptive patterns among those found by the clustering process, and allows to return more compact and easily interpretable clusters. We evaluate the mining systems performances on synthetic data in two separate cases, corresponding respectively to two different (simulated) sources of noise. In both cases, our method performs well in retrieving the original patterns with acceptable information loss.

Keywords Granular modeling · Sequence data mining · Inexact sequence matching · Frequent subsequences extraction · Evolutionary computation

---

E. Maiorino (✉) · F. Possemato · A. Rizzi
Department of Information Engineering, Electronics and Telecommunications (DIET),
SAPIENZA University of Rome,Via Eudossiana 18, 00184 Rome, Italy
e-mail: enrico.maiorino@uniroma1.it

V. Modugno
Dipartimento di Ingegneria Informatica, Automatica e Gestionale (DIAG), SAPIENZA
University of Rome, Via Ariosto 25, 00185 Rome, Italy
e-mail: valerio.modugno@uniroma1.it

© Springer International Publishing Switzerland 2016
J.J. Merelo et al. (eds.), *Computational Intelligence*,
Studies in Computational Intelligence 620, DOI 10.1007/978-3-319-26393-9_9

1 Introduction

Nowadays, sequence data mining is a very interesting field of research that is going to be central in the next years due to the growth of the so called "Big Data" challenge. Moreover, available data in different application fields consist in sequences (for example over time or space) of generic objects. Generally speaking, given a set of sequences defined over a particular domain, a data mining problem consists in searching for possible frequent subsequences (patterns), relying on inexact matching procedures. In this work we propose a possible solution for the so called *approximate subsequence mining* problem, in which we admit some noise in the matching process. As an instance, in computational biology, searching for recurrent patterns is a critical task in the study of DNA, aiming to identify some genetic mutations or to classify proteins according to some structural properties. Sometimes the process of pattern extraction returns sequences that differ from the others in a few positions. Consequently, the choice of an adequate dissimilarity measure becomes a critical issue when we are designing an algorithm able to deal with this kind of problems. Handling sequences of objects is another challenging aspect, especially when the data mining task is defined over a *structured* domain of sequences [1, 2] Thinking data mining algorithms as a building block of a wider system facing a classification task, a reasonable way to treat complex sequential data is to map sequences to $\mathbb{R}^d$ vectors by means of some *feature extraction* procedures in order to use classification techniques that deal with real valued vectors as input data [3–7]. The Granular Computing (GrC) approach [8] offers a valuable framework to fill the gap between the input sequence domain and the features space $\mathbb{R}^d$ and relies on the so-called *information granules* that play the role of *indistinguishable* features at a particular level of abstraction adopted for system description. The main objective of Granular modeling consists in finding the correct level of information granulation that best describes the input data [9].

2 Frequent Substructures Mining and Matching Problem

The problem of *sequential patterns* mining was first introduced by Agrawal and Srikant [10] in a specific context: starting from a dataset of sequences of customer transactions, the objective consists in mining for *sequential patterns* in such dataset. In a dataset of sequences of customer transactions, the general object α_i of each sequence consists of the following fields: customer-id, transaction-time and the set of items purchased in the transaction. Agrawal et al. [10] introduce for the first time the notion of *itemset* as a non-empty set of items. This problem is often viewed as the discovery of "association rules", that is strictly dependent on the task of mining frequent itemsets. In [11], the authors propose the very first algorithm able to generate significant association rules between items in databases. Manager of supermarkets as well as e-commerce websites have to make decisions about which products to put on sale, how to design coupons and customize the offers in order to maximize their profits. This problem raises the need to analyze past transactions and predict future behaviors.

All the studies in this field are based on the notion of *market-basket* model of data [12]. It is used to describe relationship between *items* and *baskets*, also called "transactions". Each basket consists in an itemset and it is assumed that the number of items in a basket is much smaller than the total number of items.

The *market-basket* model (also known as a priori-like) asserts that each itemset cannot be frequent if its items are not frequent or equivalently any super-pattern of infrequent patterns cannot be frequent. Using this principle Agrawal and Srikant proposed the AprioriAll algorithm in [10]. Their approach aims to extract frequent sequential patterns and is based on a candidate generation and test paradigm. Note that during the mining procedure, candidate frequent sequential patterns can be obtained only by joining shorter frequent sequential patterns. An example of a sequential pattern is "5 % of customers bought {Apple, Orange, Flour, Coffe} in one transaction, followed by {Coffee, Sugar} in a later transaction". The weakness of the algorithm is that a huge set of candidate sequences are generated requiring an enormous amount of memory and many repeated database scans. This behavior gets worse with increasing size of sequences in the database.

In [13] a new algorithm named GSP (Generalized Sequential Patterns) is introduced. The authors propose a breadth-first search and bottom-up method to obtain the frequent sequential pattern. Moreover, they introduce a time constraint that fixes the minimum and maximum delay between adjacent elements in the candidate patterns and the possibility for items to be present in a set of transactions in a fixed time window. GSP overcomes the performances of the APrioriAll algorithm [10] reducing the number of candidate sequential patterns. However, all a priori-like sequential pattern mining methods tend to behave badly with large datasets, because they may generate a large set of candidate subsequences. Moreover, for such algorithms, multiple scans of the database are needed, one for each length of the candidate patterns and this becomes very time consuming for mining long patterns. Finally, another problem occurs with long sequential patterns: a combinatorial number of subsequences are generated and tested.

To overcome these problems, in [14], a new algorithm named SPADE is introduced. The authors use a similar approach of GSP, however they use a vertical data format and divide the mining problem into smaller sub-problems reducing significantly the number of database scans required. In [15, 16] the authors introduce two algorithms FreeSpan and PrefixSpan. They are based on a completely different approach than APrioriAll and GSP: the pattern-growth approach for mining sequential patterns in large datasets. Each time new sequential patterns are generated, the whole dataset of sequences is projected into a set of smallest projected datasets using the extracted sequential patterns and bigger sequential patterns are grown in each projected dataset analyzing only locally frequent fragments. PrefixSpan introduces new techniques to reduce the size of the projected datasets.

All presented works describe search techniques for mining non-contiguous sequences of objects. However, these approaches are not ideal when the objective is to extract frequent sequential patterns, in which the contiguity of the component objects plays a fundamental role in the information extraction.

In particular, in computation biology, even though techniques for mining sequential *noncontiguous* patterns have many uses, they are not appropriate for many applications. Computational biology community has developed a lot of methods for detecting frequent patterns, that in this field are called *motifs*. Moreover, working with real-world data, the presence of some *noise* must be taken into account in the designing of the matching procedure [17–20]. In many fields and particularly in a biological context, patterns should have long lengths and high supports, but standard sequential pattern mining approaches tend to discover a large amount of "low quality" patterns, i.e. patterns having either short lengths or low supports. It is easy to observe that genome sequences contain errors, so it is unlikely that long subsequences generated from the same origin will be exactly identical. Moreover, the increase of the minimum number of occurrences of a subsequence in a database, in case of exact matching, obliges to accept shorter and shorter subsequences, with the possibility to obtain a massive quantity of data with a less specific meaning. In such cases, exact matches techniques can give only short and trivial patterns. So, by allowing some mismatches, it is possible to discover valuable sequential patterns, with longer length and higher approximate supports. Some works [17, 18] use Hamming distances to search for recurrent motifs in data. Other works employ suffix tree data structure [21], suffix array to store and organize the search space [22] or use a GrC framework for the extraction of frequent patterns in data [23].

The algorithm presented in [24] uses a suffix-three data structure to mine frequent approximate contiguous subsequences (also called substrings). The procedure follows a "break-down-and-build-up" strategy. The "break-down" step aims at searching, by means of a suffix-tree based algorithm, for the longest subsequences which repeat, with an exact match, in the whole database. These subsequences represent the initial sequences (called *strands*), which will be iteratively assembled into longer strands by using a local search algorithm. The "build-up" step groups the obtained strands, forming the set from which all approximate subsequences will be identified.

The algorithm [25] uses a similar approach as [24], but taking into account the quality of sequential patterns. Good quality patterns can be obtained by balancing pattern length and pattern support. Short patterns are undesirable, particularly when sequences are long, since the meaning is less specific. Patterns with low supports are not desirable too, since they can be trivial and may not describe general phenomena. Thus, the algorithm is biased toward the search for longer subsequences, characterized by a sufficient frequency. It makes use of a suffix array to store and organize in a lexicographic order the search space (i.e., the set of subsequences). The search on such a suffix array follows a *prefix extension* approach, meaning that frequent subsequences are individuated analyzing the prefixes of the input sequences, tolerating inexactness during the evaluation.

Computational biology community has developed a lot of algorithms for mining frequent motifs using the Hamming distance as similarity measure. YMF [17] is based on the computation of the statistical significance of each motif, but its performances decrease as the complexity of motifs increases. Weeder [18] is a suffix-tree-based algorithm and is faster than YMF, because it considers only certain types of mismatches for the motifs, however it can not be used for different types of motifs. Another algorithm, MITRA [19], is a mismatch-tree-based approach and uses heuristics to prune the space of possible motifs.

Analysis and interpretation of time series is another challenging problem that many authors try to solve [26, 27]. Some works consider the problem of mining for motifs in time series databases in several applications: from the analysis of stock prices to the study of the ECG in medicine, to the analysis of measures from sensors. In particular in [28] it is showed how to discretize a time series, in order to obtain a sequence of symbols, defined over a fixed alphabet and use well known motif mining algorithms. However, in the discretization process, a lot of information is lost. Moreover, this algorithm uses exact matching procedures for mining patterns and is unusable in real cases with noisy data. In Chiu et al. [29] present another algorithm, based on [20], that considers the presence of noise in data. However, also in this case, a simple model of mismatches is considered. In [30] the algorithm FLAME is presented. It consists in a suffix-tree-based technique and can be used also with time series data sets, by converting such data into a sequence of symbols, discretizing the numeric data. All these approaches suffer from the loss of information during the discretization procedure.

In the following, we present a clustering-based subsequences mining algorithm that can be used with general sequence databases, choosing a suited similarity measure, depending on the particular application. Moreover, most methods focus only on the recurrence of patterns in data without taking into account the concept of "information redundancy", or, in other words, the existence of overlapping among retrieved patterns [31]. Frequent pattern mining with approximate match is a challenging problem starting from the definition itself: even if one ignores small redundant patterns, there might be a huge number of large frequent redundant patterns. This problem should be taken in consideration, in a way that only some representatives of such patterns should survive after the mining process.

3 The Proposed Algorithm

In this work we present a new approximate subsequence mining algorithm called FRL-GRADIS (Filtered Reinforcement Learning-based GRanular Approach for DIscrete Sequences) [32] aiming to reduce the information redundancy of RL-GRADIS [33] by executing an optimization-based refinement process on the extracted patterns. In particular, this paper introduces the following contributions:

1. our approach finds the patterns that maximize the knowledge about the process that generates the sequences;
2. we employ a dissimilarity measure that can extract patterns despite the presence of noise and possible corruptions of the patterns themselves;
3. our method can be applied on every kind of sequence of objects, given a properly defined similarity or dissimilarity function defined in the objects domain;
4. the filtering operation produces results that can be interpreted more easily by application's field experts;
5. considering this procedure as an inner module of a more complex classification system, it allows to further reduce the dimension of the feature space, thus better addressing the *curse of dimensionality* problem.

This paper consists of three parts. In the first part we provide some useful definitions and a proper notation; in the second part we present FRL-GRADIS as a two-step procedure, consisting of a subsequences extraction step and a subsequences filtering step. Finally, in the third part, we report the results obtained by applying the algorithm to synthetic data, showing a good overall performance in most cases.

4 Problem Definition

Let $D = \{\alpha_i\}$ be a *domain* of *objects* α_i. The objects represent the atomic units of information. A sequence S is an ordered list of n objects that can be represented by the set of pairs

$$S = \{(i \to \beta_i) \mid i = 1, \dots, n; \ \beta_i \in D\},$$

where the integer i is the order index of the object β_i within the sequence S. S can also be expressed with the compact notation

$$S \equiv \langle \beta_1, \beta_2, \dots, \beta_n \rangle$$

A sequence database *SDB* is a set of sequences S_i of variable lengths n_i. For example, the DNA sequence $S = \langle G, T, C, A, A, T, G, T, C \rangle$ is defined over the domain of the four amino acids $D = \{A, C, G, T\}$.

A sequence $S_1 = \langle \beta_1', \beta_2', \dots, \beta_{n_1}' \rangle$ is a *subsequence* of a sequence $S_2 = \langle \beta_1'', \beta_2'', \dots, \beta_{n_2}'' \rangle$ if $n_1 \leq n_2$ and $S_1 \subseteq S_2$. The *position* $\pi_{S_2}(S_1)$ of the subsequence S_1 with respect to the sequence S_2 corresponds to the order index of its first element (in this case the order index of the object β_1') within the sequence S_2. The subsequence S_1 is also said to be *connected* if

$$\beta_j' = \beta_{j+k}'' \quad \forall j = 1, \dots, n_1$$

where $k = \pi_{S_2}(S_1)$. Two subsequences S_1 and S_2 of a sequence S are *overlapping* if

$$S_1 \cap S_2 \neq \emptyset.$$

In the example described above, the complete notation for the sequence $S = \langle G, T, C, A, A, T, G, T, C \rangle$ is

$$S = \{(1 \rightarrow G), (2 \rightarrow T), (3 \rightarrow C), \ldots\}$$

and a possible connected subsequence $S_1 = \langle A, T, G \rangle$ corresponds to the set

$$S_1 = \{(5 \rightarrow A), (6 \rightarrow T), (7 \rightarrow G)\}.$$

Notice that the objects of the subsequence S_1 inherit the order indices from the containing sequence S, so that they are univocally referred to their original positions in S. From now on we will focus only on connected subsequences, therefore the connection property will be implicitly assumed.

4.1 Pattern Coverage

The objective of this algorithm is to find a set of frequent subsequences of objects named as *patterns*. A pattern Ω is a subsequence of objects $\langle \omega_1, \omega_2, \ldots, \omega_{|\Omega|} \rangle$, with $\omega_i \in D$, that is more likely to occur within the dataset SDB. Patterns are unknown *a priori* and represent the underlying information of the dataset records. Moreover, each sequence is subject to noise whose effects include the addition, substitution and deletion of objects in a random uncorrelated fashion and this makes the recognition of recurrent subsequences more challenging.

Given a sequence $S \in SDB$ and a set of patterns $\Gamma = \{\Omega_1, \ldots, \Omega_m\}$, we want to determine a quality criterion for the description of S in terms of the pattern set Γ. A connected subsequence $C \subseteq S$ is said to be **covered** by a pattern $\Omega \in \Gamma$ iff $d(C, \Omega) \leq \delta$, where $d(\cdot, \cdot)$ is a properly defined distance function and δ is a fixed tolerance (Fig. 1). The **coverage** $C_\Omega^{(\delta)}(S)$ of the pattern Ω over the sequence S is the union set of all non-overlapping connected subsequences covered by the pattern. We can write,

$$C_\Omega^{(\delta)}(S) = \bigcup_i \left[C_i \subseteq S \text{ s.t. } d(C_i, \Omega) \leq \delta \wedge C_i \cap C_j = \emptyset \; \forall \, i \neq j \right]. \tag{1}$$

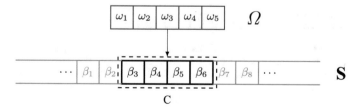

Fig. 1 Coverage of the pattern Ω over the subsequence $C \subseteq S$ with tolerance δ. *Black boxes* and *gray boxes* represent respectively the covered and the uncovered objects of the sequence S. Notice that if $\delta > 0$ the sequences Ω and C need not to be of the same length

Formally, this set is still not well defined until we expand on the meaning of the property

$$C_i \cap C_j = \emptyset, \tag{2}$$

which is the requirement for the covered subsequences to be non-overlapping. Indeed, we need to include additional rules on how to deal with these overlappings when they occur. To understand better, let us recall the example of the DNA sequences presented above, where the dissimilarity measure between two sequences is the Levenshtein distance. The set of all covered subsequences C_i (in this context referred to as *candidates*) by the pattern Ω over the sequence S will consist only of sequences with values of length between $|\Omega| - \delta$ and $|\Omega| + \delta$. Indeed, these bounds correspond respectively to the extreme cases of deleting and adding δ objects to the subsequence. In case of two overlapping candidates C_i and C_j, in order to satisfy the property (2) of the coverage $C_{\Omega}^{(\delta)}(S)$, we have to define a rule to decide which subsequence belongs to the set $C_{\Omega}^{(\delta)}(S)$ and which does not. Candidates with smaller distances from the searched pattern Ω are chosen over overlapping candidates with higher distances. If the two overlapping candidates have the same distance the first starting from the left is chosen, but if also their starting position is the same the shorter one (i.e. smaller length value) has the precedence.

A coverage example in the context of the DNA sequences is shown in Fig. 2. The coverage of the pattern $\Omega = \langle A, G, G, T \rangle$ over the sequence S is $C_{\Omega}^{(\delta)}(S) = \langle A, C, G, T \rangle \cup \langle G, G, T \rangle \cup \langle A, C, G, G, T \rangle$.

Similarly, the compound coverage of the pattern set Γ is defined as

$$C_{\Gamma}^{(\delta)}(S) = \bigcup_{\Omega \in \Gamma} C_{\Omega}^{(\delta)}(S). \tag{3}$$

It is important to notice that, in this case, this set can include overlapping subsequences only if they belong to coverages of different patterns (i.e. it is assumed that different patterns can overlap). For example consider the case shown in Fig. 3. The coverage $C_{\{\Omega_1, \Omega_2\}}^{(\delta)}(S)$ for the patterns $\Omega_1 = \langle A, G, G, T \rangle$ and $\Omega_2 = \langle G, T, C \rangle$ is equal to $C_{\{\Omega_1, \Omega_2\}}^{(\delta)}(S) = \langle A, G, G, T, C \rangle$.

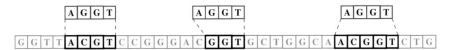

Fig. 2 Coverage examples in the case of DNA sequences. The searched pattern $\langle A, G, G, T \rangle$ is found 3 times with tolerance $\delta \leq 1$ using the Levenshtein distance. The three occurrences show all the edit operations allowed by the considered edit distance, respectively objects substitution, deletion and insertion

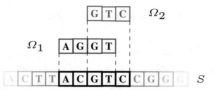

Fig. 3 Example of the compound coverage of multiple symbols, where the symbols $\langle G, T, C \rangle$ and $\langle A, G, G, T \rangle$ have Levenshtein distances from the corresponding subsequences equal to 0 and 1, respectively. Notice that different symbols can cover overlapping subsequences, while competing coverages of the same symbol are not allowed and only the most similar subsequence is chosen

5 The Mining Algorithm

In this section, we describe FRL-GRADIS, as a clustering-based sequence mining algorithm. It is able to discover clusters of connected subsequences of variable lengths that are frequent in a sequence dataset, using an inexact matching procedure. FRL-GRADIS consists in two main steps:

- the *symbols alphabet extraction*, which addresses the problem of finding the most frequent subsequences within a SDB. It is performed by means of the clustering algorithm RL-GRADIS [33] that identifies frequent subsequences as representatives of dense clusters of similar subsequences. These representatives are referred to as *symbols* and the pattern set as the *alphabet*. The clustering procedure relies on a properly defined edit distance between the subsequences (e.g. Levenshtein distance, DTW, etc.). However, this approach alone has the drawback of extracting many superfluous symbols which generally dilute the pattern set and deteriorate the *interpretability* of the produced pattern set.
- the *alphabet filtering* step deals with the problem stated above. The objective is to filter out all the spurious or redundant symbols contained in the alphabet produced by the symbols extraction step. To accomplish this goal we employ a heuristic approach based on evolutionary optimization over a validation SDB.

One of the distinctive features of this algorithm is its generality with respect to the kind of data contained in the input sequence database (e.g., sequences of real numbers or characters as well as sequences of complex data structures). Indeed, both steps outlined above take advantage of a dissimilarity-based approach, with the dissimilarity function being a whatever complex measure between two ordered sequences, not necessarily metric.

In the following, we first describe the main aspects of the symbols alphabet extraction procedure, then we present the new filtering method. For more details on the symbols alphabet construction we refer the reader to [33].

5.1 Frequent Subsequences Identification

Consider the input training dataset of sequences $\mathcal{T} = \{S_1, S_2, \ldots, S_{|\mathcal{T}|}\}$ and a properly defined dissimilarity measure $d : \mathcal{T} \times \mathcal{T} \rightarrow \mathbb{R}$ between two objects of the training dataset (e.g., Levenshtein distance for strings of characters). The goal of the subsequences extraction step is the identification of a finite set of symbols $\mathcal{A}_e = \{\Omega_1, \Omega_2, \ldots, \Omega_{|\mathcal{A}_e|}\}$,[1] computed using the distance $d(\cdot, \cdot)$ in a free clustering procedure. The algorithm we chose to accomplish this task is RL-GRADIS which is based on the well-known Basic Sequential Algorithmic Scheme (BSAS) clustering algorithm [33]. Symbols are found by analysing a suited set of variable-length subsequences of $\mathcal{T}$, also called n-grams, that are generated by expanding each input sequence $S \in \mathcal{T}$. The expansion is done by listing all n-grams with lengths varying between the values $l_{\min}$ and $l_{\max}$. The parameters $l_{\min}$ and $l_{\max}$ are user-defined and are respectively the minimum and maximum admissible length for the mined patterns. The extracted n-grams are then collected into the SDB $\mathcal{N}$. At this point, the clustering procedure is executed on $\mathcal{N}$. For each cluster we compute its representative, defined by the Minimum Sum of Distances (MinSOD) technique [33, 34], as the element having the minimum total distance from the other elements of the cluster. This technique allows to represent the corresponding clusters by means of their most characteristic elements.

The quality of each cluster is measured by its *firing strength* f, where $f \in [0, 1]$. Firing strengths are used to track the dynamics describing the *updating rate* of the clusters when the input stream of subsequences $\mathcal{N}$ is analyzed. A reinforcement learning procedure is used to dynamically update the list of candidate symbols based on their firing strength. Clusters with a low rate of update (low firing strength) are discarded in an on-line fashion, along with the processing of the input data stream $\mathcal{N}$. RL-GRADIS maintains a dynamic list of candidate symbols, named *receptors*, which are the representatives of the active clusters. Each receptor's firing strength (i.e. the firing strength of its corresponding cluster) is dynamically updated by means of two additional parameters, $\alpha, \beta \in [0, 1]$. The α parameter is used as a *reinforcement weight* factor each time a cluster $\mathcal{R}$ is updated, i.e., each time a new input subsequence is added to $\mathcal{R}$. The firing strength update rule is defined as follows:

$$f(\mathcal{R}) \leftarrow f(\mathcal{R}) + \alpha(1 - f(\mathcal{R})). \tag{4}$$

The β parameter, instead, is used to model the speed of *forgetfulness* of receptors according to the following formula:

$$f(\mathcal{R}) \leftarrow (1 - \beta)f(\mathcal{R}). \tag{5}$$

---

[1]The subscript "e" stands for "extraction" as in extraction step.

The firing strength updating rules shown in Eqs. (4) and (5) are performed for each currently identified receptor, after the analysis of each input subsequence. Therefore, receptors/clusters that are not updated frequently during the analysis of $\mathcal{N}$ will likely have a low strength value and this will cause the system to remove the receptor from the list.

5.2 Subsequences Filtering

As introduced above, the output alphabet $\mathcal{A}_e$ of the clustering procedure is generally redundant and includes many spurious symbols that make the recognition of the true alphabet quite difficult.

To deal with this problem, an optimization step is performed to reduce the alphabet size, aiming at retaining only the most significant symbols, i.e. only those that best resemble the original, unknown ones. Since this procedure works like a filter, we call the output of this optimization the *filtered alphabet* $\mathcal{A}_f$ and, clearly, $\mathcal{A}_f \subset \mathcal{A}_e$ holds. Nevertheless, it is important for the filtered alphabet's size not to be smaller than the size of the true alphabet, since in this case useful information will be lost. Let $\Gamma \subset \mathcal{A}_e$ be a candidate subset of symbols of the alphabet $\mathcal{A}_e$ and $S \in \mathcal{V}$ a sequence of a *validation SDB* $\mathcal{V}$. We assume the *descriptive power* of the symbols set Γ, with respect to the sequence S, to be proportional to the quantity $|C_\Gamma^{(\delta)}(S)|$ (cfr Eq. 3), i.e. the number of objects $\beta_i \in S$ covered by the symbols set Γ. In fact, intuitively, a lower number of uncovered objects in the whole SDB by Γ symbols can be considered as a clue that Γ itself will likely contain the true alphabet. The normalized number of uncovered objects in a sequence S by a pattern set Γ corresponds to the quantity

$$P = \frac{|S| - |C_\Gamma^{(\delta)}(S)|}{|S|}, \tag{6}$$

where the operator $|\cdot|$ stands for the cardinality of the set. The term P assumes the value 0 when the sequence S is completely covered by the pattern set Γ and the value 1 when none of the symbols in Γ are present in the sequence S. Notice that $C_\Gamma^{(\delta)}(S)$ depends on the parameter δ which represents the tolerance of the system towards the corruption of symbols' occurrences caused by noise.

On the other hand, a bigger pattern set is more likely to contain spurious patterns which tend to hinder the interpretability of the obtained results, so smaller set sizes are to be preferred. This property can be described with the normalized alphabet size

$$Q = \frac{|\Gamma|}{|\mathcal{A}_e|}, \tag{7}$$

where $\mathcal{A}_e$ is the alphabet of symbols extracted by the clustering procedure described in the last section. Clearly, the cardinality of $\mathcal{A}_e$ represents an upper bound for the size of the filtered alphabet, so the term Q ranges from 0 to 1. The terms P and Q generally show opposite trends, since a bigger set of symbols is more likely to cover a bigger portion of the sequence and vice versa.

Finding a tradeoff between these two quantities corresponds to minimizing the convex objective function

$$G_S^{(\delta)}(\Gamma) = \lambda Q + (1 - \lambda) P \tag{8}$$

where $\lambda \in [0, 1]$ is a meta-parameter that weighs the relative importance between the two constributions. It is easy to verify that

$$0 \le G_S^{(\delta)}(\Gamma) \le 1. \tag{9}$$

More generally, for a validation SDB $\mathcal{V}$, the global objective function is the mean value of $G_S^{(\delta)}(\Gamma)$ over all sequences $S_i \in \mathcal{V}$, hence

$$G_{\mathcal{V}}^{(\delta)}(\Gamma) = \frac{\displaystyle\sum_{1 \le i \le |\mathcal{V}|} G_{S_i}^{(\delta)}(\Gamma)}{|\mathcal{V}|} \tag{10}$$

and the best symbols set after the optimization procedure is

$$\mathcal{A}_f = \underset{\Gamma \subset \mathcal{A}_e}{\operatorname{argmin}} \, G_S^{(\delta)}(\Gamma). \tag{11}$$

To solve the optimization problem described by Eq. (11) we employ a standard genetic algorithm, where each individual of the population is a subset Γ of the extracted alphabet $\mathcal{A}_e = \{\Omega_1, \dots, \Omega_{|\mathcal{A}_e|}\}$. The genetic code of the individual is encoded as a binary sequence E of length $|\mathcal{A}_e|$ of the form

$$E_\Gamma = \langle e_1, e_2, \dots, e_{|\mathcal{A}_e|} \rangle \tag{12}$$

with

$$e_i = \begin{cases} 1 & \text{iff } \Omega_i \in \Gamma \\ 0 & \text{otherwise} \end{cases}.$$

It is important not to mistake genetic codes with the SDB sequences described earlier, even if they are both formally defined as ordered sequences.

Given a validation dataset $\mathcal{V}$ and a fixed tolerance δ, the fitness value $F(E_\Gamma)$ of each individual E_Γ is computed as the following affine transformation of the objective function introduced in the last paragraph

$$F(E_\Gamma) = 1 - G_{\mathcal{V}}^{(\delta)}(\Gamma) \tag{13}$$

The computation is then performed with standard crossover and mutation operators between the binary sequences and the stop condition is met when the maximum fitness does not change for a fixed number N_{stall} of generations or after a given maximum number N_{max} of iterations. When the evolution stops, the filtered alphabet $\mathcal{A}_f = \widetilde{\Gamma}$ is returned, where $\widetilde{\Gamma}$ is the symbols subset corresponding to the fittest individual $E_{\widetilde{\Gamma}}$.

6 Tests and Results

In this section, we present results from different experiments that we designed to test the effectiveness and performance of FRL-GRADIS in facing problems with varying complexity.

6.1 Data Generation

We tested the capabilities of FRL-GRADIS on synthetic sequence databases composed of textual strings. For this reason, the domain of the problem is the English alphabet

$$\mathcal{D} = \{A, B, C, \dots, Z\}.$$

Modeled noise consists in all cases of random characters insertions, deletions and substitutions to the original string. For this reason a natural choice of dissimilarity measure between sequences is the Levenshtein distance, that measures the minimum number of edit steps necessary to transform one string of characters into another. We conducted two different classes of tests, which accounted for two kinds of noise, respectively *symbols noise* and *channel noise*, presented in the following paragraphs.

6.1.1 Symbols Noise

This kind of noise simulates those situations in which symbols are altered during the composition of the sequence. In fact, each instance of a symbol being added to the data sequence has a fixed probability of being mutated by one addition, deletion or modification of its objects. Moreover, a variable number of uncorrelated objects are added between contiguous instances of symbols in the sequence, to simulate the presence of irrelevant data separating actual symbols. The detailed process of data generation is described below:

1. the true symbols alphabet $\mathcal{A}_t$ is generated. This alphabet consists of N_{sym} symbols with lengths normally distributed around the mean value L_{sym}. Each character is chosen in $\mathcal{D}$ with uniform probability and repeated characters are allowed;

2. a training SDB $\mathcal{T}$ and a validation SDB $\mathcal{V}$ respectively composed of N_{tr} and N_{val}
   sequences are generated. Each of these sequences is built by concatenating N_{symseq}
   symbols chosen randomly from $\mathcal{A}_t$. Notice that generally $N_{symseq} > N_{sym}$ so there
   will be repeated symbols;
3. in each sequence, every symbol will be subject to noise with probability μ. The
   application of noise to a symbol in a sequence corresponds to the deletion, substitution or insertion of one character to that single instance of the symbol. This
   kind of noise is referred to as *intra-pattern* noise;
4. a user-defined quantity of random characters is added between instances of symbols in each sequence. This noise is called *inter-pattern* noise. Such quantity
   depends on the parameter η that corresponds to the ratio between the number
   of characters belonging to actual symbols and the total number of character of
   the sequence after the application of inter-pattern noise, that is,

$$\eta = \frac{(\# \text{ symbol characters})}{(\# \text{ total characters})}.$$

Notice that the amount of inter-pattern noise is inversely proportional to the value
of η.

6.1.2 Channel Noise

In this case we simulate a noise affecting an hypotetical channel through which the
sequence is transmitted. This kind of noise alters the objects in a uncorrelated manner, without keeping track of the separation between symbols.

In this case we generate the original SDB $\mathcal{T}$ and $\mathcal{V}$ in the same manner as
described in steps 1 and 2 of Sect. 6.1.1. The noise is then added to these datasets
by iterating through the objects of the sequence and altering each object with a fixed
probability p. The alteration consists with equal probability in either:

- the substitution of the object with another randomly chosen object;
- the deletion of the object;
- the addition of another randomly chosen object to the right of the current object
  position in the sequence.

The generated datasets $\mathcal{T}$ and $\mathcal{V}$ are then ready to be used as input of the FRL-
GRADIS procedure. Notice that the true alphabet $\mathcal{A}_t$ is unknown in real-world applications and here is used only to quantify the performance of the algorithm.

6.2 Quality Measures

We now introduce the quality measures used in the following tests to evaluate the
mining capabilities of the FRL-GRADIS algorithm. These measures are computed

for the resulting alphabets obtained from both the extraction and the filtering steps presented in Sect. 5, in order to highlight the improvement made by the filtering procedure (i.e. the improvement of FRL-GRADIS over RL-GRADIS).

The *redundance R* corresponds to the ratio between the cardinality of the alphabet $\mathcal{A}$ and the true alphabet $\mathcal{A}_t$, that is,

$$R = \frac{|\mathcal{A}|}{|\mathcal{A}_t|} \tag{14}$$

Clearly, since the filtering step selects a subset $\mathcal{A}_f$ (filtered alphabet) of the extracted alphabet $\mathcal{A}_e$, we always have that

$$R_f < R_e.$$

The redundance measures the amount of unnecessary symbols that are found by a frequent pattern mining procedure and it ranges from zero to infinite. When $R > 1$ some redundant symbols have been erroneously included in the alphabet, while when $R < 1$ some have been missed, the ideal value being $R = 1$.

It is important to notice that the redundancy depends only on the number of symbols reconstructed, but not on their similarity with respect to the original alphabet. For this purpose we also introduce the *mining error E*, defined as the mean distance between each symbol Ω_i of the true alphabet $\mathcal{A}_t$ and its best match within the alphabet $\mathcal{A}$, where the best match means the symbol with the least distance from Ω_i. In other words, considering $\mathcal{A}_t = \{\Omega_1, \dots, \Omega_{|\mathcal{A}_t|}\}$ and $\mathcal{A} = \{\tilde{\Omega}_1, \dots, \tilde{\Omega}_{|\mathcal{A}|}\}$, the mining error corresponds to

$$E = \frac{\sum_i d(\Omega_i, \tilde{\Omega}_{(i)})}{|\mathcal{A}_t|} \tag{15}$$

where

$$\tilde{\Omega}_{(i)} = \operatorname*{argmin}_{\tilde{\Omega} \in \mathcal{A}} d(\Omega_i, \tilde{\Omega}).$$

This quantity has the opposite role of the redundancy, in fact it keeps track of the general accuracy of reconstruction of the true symbols regardless of the generated alphabet size. It assumes non-negative values and the ideal value is 0. For the same reasons stated above the inequality

$$E_f \geq E_e$$

holds, so the extraction procedure's mining error constitutes a lower bound for the mining error obtainable with the filtering step.

6.3 Results

We executed the algorithm multiple times for different values of the noise parameters, to assess the different response of FRL-GRADIS to increasing amounts of noise. Most parameters have been held fixed for all the tests and they are listed in Table 1.

As a first result, we present the synthetic tests performed by adding varying quantities of symbols noise to the data sequences, performed with $\mu = 0.5$ and variable amounts of inter-pattern noise η. It means that about half of the symbols in a sequence are subject to the alteration of one character and increasing amounts of random characters are added between symbols in each sequence. The results obtained with this configuration are shown in Figs. 4 and 5.

The redundancy plot in Fig. 4 shows an apparently paradoxical trend of the extraction procedure's redundancy: with decreasing amounts of inter-pattern noise (i.e. increasing values of η) the extraction algorithm performs more poorly, leading to higher redundancies. That can be easily explainable by recalling how the clustering procedure works.

Higher amounts of inter-pattern noise mean that the frequent symbols are more likely to be separated by random strings of characters. These strings of uncorrelated characters generate very sparse clusters with negligible cardinality that are very likely to be deleted during the clustering's reinforcement step. Clusters corresponding to actual symbols, instead, are more active and compact, their bounds being clearly defined by the noise characters, and so they are more likely to survive the reinforcement step.

In case of negligible (or non-existent) inter-pattern noise, instead, different symbols are more likely to occur in frequent successions that cause the generation of many clusters corresponding to spurious symbols, obtained from the concatenation

Table 1 Fixed parameters adopted for the tests

Parameter	Value	Parameter	Value
N_{tr}	50	N_{val}	25
N_{sym}	5	N_{symseq}	10
l_{min}	4	l_{max}	12
δ	1	λ	0.5
N_{pop}	100	N_{elite}	0.1
p_{cross}	0.8	p_{mut}	0.3
N_{max}	100	N_{stall}	50

The parameter δ corresponds to the tolerance of the Levenshtein distance considered when calculating the coverage as in Eq. (1) while λ weighs the two terms of the objective function of Eq. (8). The values shown in the second part of the table refer to the genetic algorithm's parameters. N_{pop} corresponds to the population size, N_{elite} is the fraction of individuals who are guaranteed to survive and be copied to the new population in each iteration, p_{cross} and p_{mut} are respectively the crossover and mutation probabilities. The evolution terminates if N_{evol} iterations have been performed or if for a number N_{stall} of iterations the maximum fitness has not changed

Fig. 4 Plot of the redundance R of the extraction (RL-GRADIS) and filtering (FRL-GRADIS) steps with variable inter-pattern noise and $\mu = 0.5$

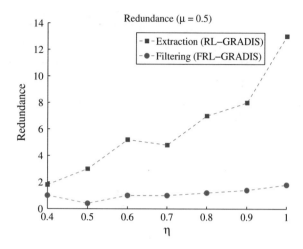

Fig. 5 Plot of the mining error E of the extraction (RL-GRADIS) and filtering (FRL-GRADIS) steps with variable inter-pattern noise and $\mu = 0.5$

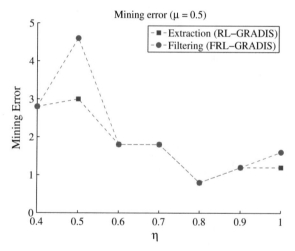

of parts of different symbols. The filtering procedure overcomes this inconvenience, as it can be seen from Fig. 4 that it is nearly not affected by the amount of inter-pattern noise. As it is evident, the filtering procedure becomes fundamental for higher values of the parameter η, where the clustering produces highly redundant alphabets that would be infeasible to handle in a real-world application. Figure 5 shows that the mining error after the filtering procedure remains mostly the same for all values of η, which means that the system is robust to the moderate alteration of the input signal.

In the second pool of tests we show the response of the system to increasing quantities of channel noise p. In Fig. 6 and 7 are shown the redundance and the mining error measured for different values of p. While FRL-GRADIS shows slightly higher mining error levels than RL-GRADIS, its redundancy is still significantly lower and,

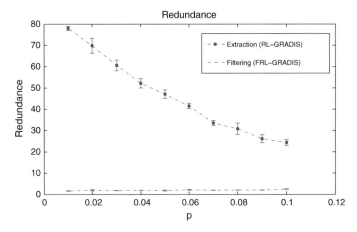

Fig. 6 Plot of the redundance R of the extraction (RL-GRADIS) and filtering (FRL-GRADIS) steps with variable channel noise. Error bars represent the standard deviation over 3 runs of the algorithm with the same parameters

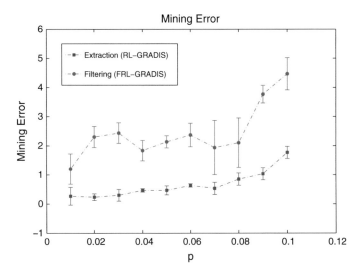

Fig. 7 Plot of the mining error E of the extraction (RL-GRADIS) and filtering (FRL-GRADIS) steps with variable channel noise. Error bars represent the standard deviation over 3 runs of the algorithm with the same parameters

to a large extent, insensitive to the quantity of noise in the system. Clearly, lower mining error levels are obtainable by setting suitable values of the parameter λ (at the expense of resulting redundancy) or stricter convergence criteria of the genetic algorithm (at the expense of convergence time).

In general, we can conclude that the system allows for a remarkable synthesis of the extracted alphabet despite of a modest additional mining error.

7 Conclusions

In this work we have presented a new approach to sequence data mining, focused on improving the interpretability of the frequent patterns found in the data. For this reason, we employed a two-steps procedure composed of a clustering algorithm, that extracts the frequent subsequences in a sequence database, and a genetic algorithm that filters the returned set to retrieve a smaller set of patterns that best describes the input data. For this purpose we introduced the concept of coverage, that helps in recognizing the true presence of a pattern within a sequence affected by noise. The experiments were performed on two cases of synthetic data affected by two different sources of noise. The results have shown a good overall performance and lay the foundations for improvements and further experiments on real data.

References

1. Possemato, F., Rizzi, A.: Automatic text categorization by a granular computing approach: facing unbalanced data sets. In: The 2013 International Joint Conference on Neural Networks (IJCNN), pp. 1–8 (2013)
2. Modugno, V., Possemato, F., Rizzi, A.: Combining piecewise linear regression and a granular computing framework for financial time series classification (2014)
3. Bianchi, F., Livi, L., Rizzi, A., Sadeghian, A.: A granular computing approach to the design of optimized graph classification systems. Soft Comput. **18**, 393–412 (2014)
4. Bianchi, F.M., Scardapane, S., Livi, L., Uncini, A., Rizzi, A.: An interpretable graph-based image classifier. In: 2014 International Joint Conference on Neural Networks (IJCNN), pp. 2339–2346. IEEE (2014)
5. Rizzi, A., Del Vescovo, G.: Automatic image classification by a granular computing approach. In: Proceedings of the 2006 16th IEEE Signal Processing Society Workshop on Machine Learning for Signal Processing, pp. 33–38 (2006)
6. Del Vescovo, G., Rizzi, A.: Automatic classification of graphs by symbolic histograms. In: IEEE International Conference on Granular Computing. GRC 2007, pp. 410–410 (2007)
7. Del Vescovo, G., Rizzi, A.: Online handwriting recognition by the symbolic histograms approach. In: IEEE International Conference on Granular Computing. GRC 2007, pp. 686–686 (2007)
8. Bargiela, A., Pedrycz, W.: Granular Computing: An Introduction. Springer (2003)
9. Livi, L., Rizzi, A., Sadeghian, A.: Granular modeling and computing approaches for intelligent analysis of non-geometric data. Appl. Soft Comput. **27**, 567–574 (2015)
10. Agrawal, R., Srikant, R.: Mining sequential patterns. In: Proceedings of the Eleventh International Conference on Data Engineering, pp. 3–14. IEEE (1995)
11. Agrawal, R., Imieliński, T., Swami, A.: Mining association rules between sets of items in large databases. In: ACM SIGMOD Record, vol. 22, pp. 207–216. ACM (1993)
12. Rajaraman, A., Ullman, J.D.: Mining of Massive Datasets. Cambridge University Press, New York (2011)
13. Srikant, R., Agrawal, R.: Mining Sequential Patterns: Generalizations and Performance Improvements. Springer (1996)
14. Zaki, M.J.: Spade: an efficient algorithm for mining frequent sequences. Mach. Learn. **42**, 31–60 (2001)
15. Han, J., Pei, J., Mortazavi-Asl, B., Chen, Q., Dayal, U., Hsu, M.C.: Freespan: frequent pattern-projected sequential pattern mining. In: Proceedings of the Sixth ACM SIGKDD International Conference on Knowledge Discovery and Data Mining, ACM, pp. 355–359 (2000)

16. Pei, J., Han, J., Mortazavi-Asl, B., Pinto, H., Chen, Q., Dayal, U., Hsu, M.C.: Prefixspan: mining sequential patterns efficiently by prefix-projected pattern growth. In: 2013 IEEE 29th International Conference on Data Engineering (ICDE), IEEE Computer Society, pp. 0215–0215 (2001)
17. Sinha, S., Tompa, M.: YMF: a program for discovery of novel transcription factor binding sites by statistical overrepresentation. Nucleic Acids Res. **31**, 3586–3588 (2003)
18. Pavesi, G., Mereghetti, P., Mauri, G., Pesole, G.: Weeder web: discovery of transcription factor binding sites in a set of sequences from co-regulated genes. Nucleic Acids Res. **32**, W199–W203 (2004)
19. Eskin, E., Pevzner, P.A.: Finding composite regulatory patterns in dna sequences. Bioinformatics **18**, S354–S363 (2002)
20. Buhler, J., Tompa, M.: Finding motifs using random projections. J. Comput. Biol. **9**, 225–242 (2002)
21. Zhu, F., Yan, X., Han, J., Yu, P.S.: Efficient discovery of frequent approximate sequential patterns. In: Seventh IEEE International Conference on Data Mining. ICDM 2007, pp. 751–756. IEEE (2007)
22. Ji, X., Bailey, J.: An efficient technique for mining approximately frequent substring patterns. In: Seventh IEEE International Conference on Data Mining Workshops. ICDM Workshops 2007, pp. 325–330. IEEE (2007)
23. Rizzi, A., Possemato, F., Livi, L., Sebastiani, A., Giuliani, A., Mascioli, F.M.F.: A dissimilarity-based classifier for generalized sequences by a granular computing approach. In: IJCNN, IEEE, pp. 1–8 (2013)
24. Zhu, F., Yan, X., Han, J., Yu, P.S.: Efficient discovery of frequent approximate sequential patterns. In: Proceedings of the 2007 Seventh IEEE International Conference on Data Mining, Washington, DC, USA, IEEE Computer Society, pp. 751–756 (2007)
25. Ji, X., Bailey, J.: An efficient technique for mining approximately frequent substring patterns. In: Proceedings of the Seventh IEEE International Conference on Data Mining Workshops. ICDMW '07, Washington, DC, USA, IEEE Computer Society, pp. 325–330 (2007)
26. Fu, A.W.C., Keogh, E., Lau, L.Y., Ratanamahatana, C.A., Wong, R.C.W.: Scaling and time warping in time series querying. VLDB J. Int. J. Very Large Data Bases **17**, 899–921 (2008)
27. Vlachos, M., Kollios, G., Gunopulos, D.: Discovering similar multidimensional trajectories. In: 18th International Conference on Data Engineering. Proceedings. IEEE, pp. 673–684 (2002)
28. Patel, P., Keogh, E., Lin, J., Lonardi, S.: Mining motifs in massive time series databases. In: 2002 IEEE International Conference on Data Mining. ICDM 2003. Proceedings. IEEE, pp. 370–377 (2002)
29. Chiu, B., Keogh, E., Lonardi, S.: Probabilistic discovery of time series motifs. In: Proceedings of the Ninth ACM SIGKDD International Conference on Knowledge Discovery and Data Mining, ACM, pp. 493–498 (2003)
30. Floratou, A., Tata, S., Patel, J.M.: Efficient and accurate discovery of patterns in sequence data sets. IEEE Trans. Knowl. Data Eng. **23**, 1154–1168 (2011)
31. Matsui, T., Uno, T., Umemori, J., Koide, T.: A new approach to string pattern mining with approximate match. In: Discovery Science, pp. 110–125. Springer (2013)
32. Maiorino, E., Possemato, F., Modugno, V., Rizzi, A.: Information granules filtering for inexact sequential pattern mining by evolutionary computation (2014)
33. Rizzi, A., Del Vescovo, G., Livi, L., Frattale Mascioli, F.M.: A new granular computing approach for sequences representation and classification. In: Proceedings of the 2012 International Joint Conference on Neural Networks, pp. 2268–2275 (2012)
34. Del Vescovo, G., Livi, L., Frattale Mascioli, M., Rizzi, A.: On the problem of modeling structured data with the minsod representative. Int. J. Comput. Theory Eng. **6**, 9–14 (2014)

A Shuffled Complex Evolution Algorithm for the Examination Timetabling Problem

Nuno Leite, Fernando Melício and Agostinho C. Rosa

Abstract In this work two instances of the examination timetabling problem are studied and solved using memetic algorithms. The first is the uncapacitated single-epoch problem instance. In the second problem instance two examination epochs are considered, with different durations. The memetic algorithm, named Shuffled Complex Evolution Algorithm, uses a population organized into sets called *complexes* which evolve independently using a recombination and local search operators. Population diversity is preserved by means of the recombination operator and a special solution update mechanism. Experimental evaluation was carried out on the public uncapacitated Toronto benchmarks (single epoch) and on the ISEL–DEETC department examination benchmark (two epochs). Results show that the algorithm is competitive on the Toronto benchmarks, attaining a new lower bound on one benchmark. In the ISEL–DEETC benchmark, the algorithm attains a lower cost when compared with the manual solution.

Keywords Examination timetabling · Shuffled complex evolution algorithm · Memetic computing · Great deluge algorithm · Toronto benchmarks · Two-epoch examination timetabling

N. Leite (✉)
Instituto Superior de Engenharia de Lisboa/ADEETC, Polytechnic Institute of Lisbon, Rua Conselheiro Emídio Navarro n.°1, 1959-007 Lisboa, Portugal
e-mail: nleite@cc.isel.ipl.pt

A.C. Rosa
Department of Bioengineering/Instituto Superior Técnico, Universidade de Lisboa, Av. Rovisco Pais, N.°1, 1049-001 Lisboa, Portugal
e-mail: acrosa@laseeb.org

N. Leite · F. Melício · A.C. Rosa
Institute for Systems and Robotics/LaSEEB, Instituto Superior Técnico, Universidade de Lisboa, Av. Rovisco Pais, N.°1 TN 6.21, 1049-001 Lisboa, Portugal

F. Melício
e-mail: fmelicio@laseeb.org

© Springer International Publishing Switzerland 2016
J.J. Merelo et al. (eds.), *Computational Intelligence*,
Studies in Computational Intelligence 620, DOI 10.1007/978-3-319-26393-9_10

1 Introduction

The Examination Timetabling Problem (ETTP) is a combinatorial optimisation problem which objective is to allocate course exams to a set of limited time slots, while respecting some *hard constraints*, such as respect maximum room capacity, guarantee room exclusiveness for given exams, guarantee that no students will sit two or more exams at the same time slot, guarantee exam ordering (e.g., larger exams must be scheduled at the beginning of the timetable), among others [19]. The ETTP is a multi-objective problem in nature as several objectives (reflecting the various interested parties, e.g., students, institution, teachers) are considered [5]. However, due to complexity reasons, the ETTP has been dealt as a single-objective problem. A second type of constraints, named *soft constraints*, are also considered but there is no obligation to observe them. The optimisation goal is usually the minimisation of the soft constraints violations.

The ETTP, as other problems (e.g. Course timetabling) belong to the general class of timetabling problems which include *Transportation* and *Sports* timetabling, *Nurse scheduling*, among others. In terms of complexity, University timetabling problems belong to the NP-complete class of problems [22]. In the past 30 years, several heuristic solution methods have been proposed to solve the ETTP. The *meta-heuristics* form the most successful methods applied to the ETTP. These are mainly divided into two classes [24]: single-solution based meta-heuristics and population-based meta-heuristics. Single-solution meta-heuristics include algorithms such as simulated annealing, tabu search, and variable neighbourhood search. Population-based meta-heuristics include genetic algorithms, ant colony optimisation, particle swarm optimisation and memetic algorithms. Particle swarm optimisation integrates the larger branch named Swarm Intelligence [15]. In Swarm Intelligence the behaviour of self-organised systems (e.g., frog swarm in a swamp, fish swarm, honeybee mating) is simulated. For a recent survey of approaches applied to the ETTP see [22]. Population-based approaches, especially hybrid methods that employ single-solution meta-heuristics in an exploitation phase, have shown to be efficient approaches for the timetabling problem. These hybrid methods are named Memetic Algorithms [21], and integrate the larger branch known as Memetic Computing. In the literature, several population-based methods were proposed for solving the ETTP: ant colony [11], particle swarm optimisation [8], fish swarm optimisation algorithm [25, 26], and honeybee mating optimisation [23].

In previous research undertaken [16], the ETTP was approached by an adaptation of the Shuffled Frog-Leaping Algorithm (SFLA) [14]. The SFLA is, by its turn, based on the Shuffled Complex Evolution (SCE) approach [12]. Both are evolutionary algorithms (EA) containing structured populations and an efficient exploitation phase (local search). They were successfully applied to Global optimisation problems. An important feature that must be observed when designing an EA-based approach is the diversity management [20], as a poor population diversity leads the algorithm to stagnate prematurely. Another emergent approach that uses a structured population is the Cellular Genetic Algorithm (cGA) [3], which promotes a smooth actualization of solutions through the population, therefore maintaining the diversity.

In [17], the authors proposed a memetic algorithm for solving the ETTP. The method, coined SCEA—Shuffled Complex Evolution Algorithm, inherits features from the SCE and SFLA approaches, namely the population is organized into sub-populations called *complexes* (*memeplexes* in SFLA). Population diversity is maintained in the SCEA using both a crossover operator and a special solution update mechanism. The method is hybridised with the single-solution method Great Deluge Algorithm (GDA) [13]. The GDA is a simulated annealing variant which comprises a deterministic acceptance function of neighbouring solutions.

The SCEA is applied to the *uncapacitated Toronto benchmarks* [7], comprising 13 ETTP real instances (single examination epoch), and also to the uncapacitated ETTP instance of the Department of Electronics, Telecommunications and Computer Engineering at the Lisbon Polytechnic Institute (ISEL–DEETC), which comprises two examination epochs. The second problem is a new problem emerged from practice, where two examination epochs are considered, with different number of time slots allotted for each epoch.

In the present work, the two-epoch problem is tackled using a new model with the objective of spreading away exams in the first epoch that conflict with exams in the second epoch.

The paper is organized as follows. Section 2 presents the ETTP formulations of the considered ETTP instances. In Sect. 3 we describe the proposed memetic algorithm for solving the ETTP. Section 4 presents simulation results and analysis on the algorithm performance. Finally, conclusions and future work are presented in Sect. 5.

2 Problem Description

In this section the two studied examination timetabling problems, namely the single-epoch and the two-epoch problems, are described.

2.1 The Single-epoch Problem

In the single-epoch problem there exists a single examination epoch comprising a fixed number of time slots. The problem formulation given next was adapted from [9]. Given a set of examinations, $\mathcal{E} = \{e_1, e_2, \ldots, e_{|\mathcal{E}|}\}$, and a set of time slots, $\mathcal{T} = \{1, 2, \ldots, |\mathcal{T}|\}$, the optimisation goal is to find the optimal timetable represented by the set h of ordered pairs (t, e) where $t \in \mathcal{T}$ and $e \in \mathcal{E}$. The obtained timetable is called *feasible* if it satisfies all hard constraints. Otherwise, the timetable is said to be *unfeasible*. Table 1 describe the hard and soft constraints of the ETTP instances analysed in this work.

The following additional symbols were defined:

Table 1 Hard and soft constraints of the uncapacitated ETTP

Constraint and Type	Explanation
H_1 (Hard)	There cannot exist students sitting for more than one exam simultaneously
H_2 (Hard)	A minimum distance between the two exams of a course must be observed
S_1 (Soft)	Exams should be spread out evenly through the timetable

- $C = (c_{ij})_{|\mathcal{E}| \times |\mathcal{E}|}$ (*Conflict matrix*), is a symmetric matrix of size $|\mathcal{E}|$ where each element, denoted by c_{ij} ($i, j \in \{1, \ldots, |\mathcal{E}|\}$), represents the number of students attending exams e_i and e_j. The diagonal elements c_{ii} denote the total of students enrolled in exam e_i;
- N_s, is the total number of students;
- t_k ($t_k \in \mathcal{T}$) denotes the assigned time slot for exam e_k ($e_k \in \mathcal{E}$).

The model presented next represents an Uncapacitated Exam Proximity problem (UEPP) [9], where the classroom seating capacity is not considered.

$$\text{minimise} \quad f = \frac{1}{N_s} \sum_{i=0}^{4} w_{i+1} \, u_2|_{q=|\mathcal{T}|, N=i} \tag{1}$$

where

$$u_2|_{q=|\mathcal{T}|, N=x} = \frac{1}{2} \sum_{k=1}^{|\mathcal{T}|-(x+1)} \sum_{i=1}^{|\mathcal{E}|} \sum_{j=1}^{|\mathcal{E}|} c_{ij} \varepsilon_{ik} \varepsilon_{jk+(x+1)}, \tag{2}$$

$$\forall k \text{ where } k \bmod |\mathcal{T}| \neq 0,$$

$$\text{subject to} \quad \sum_{k=1}^{|\mathcal{T}|} \sum_{i=1}^{|\mathcal{E}|} \sum_{j=1}^{|\mathcal{E}|} c_{ij} \varepsilon_{ik} \varepsilon_{jk} = 0. \tag{3}$$

Equation (1) represents the standard soft constraint S_1 as proposed by Carter et al. [7]. Function f counts the number of students having 0–4 free timeslots between exams, penalising more the conflicting exams that are close to each other. The weighting factors are $w_1 = 16$, $w_2 = 8$, $w_3 = 4$, $w_4 = 2$, and $w_5 = 1$.

In (2), $\varepsilon_{jk} \in \{0, 1\}$ is a binary quantity with $\varepsilon_{jk} = 1$ if exam e_j is assigned to time slot k. Otherwise, $\varepsilon_{jk} = 0$. Equation (3) represents the hard constraint H_1, that is, a timetable is feasible if (3) is satisfied.

2.1.1 Toronto Datasets.

For evaluation purposes, the single-epoch Uncapacitated Toronto benchmarks, Version I [22], were used.

2.2   The Two-epoch Problem

The two-epoch problem is an extension of the standard problem where two exam-ination epochs of different lengths are considered instead of just a single epoch. A snapshot of timetables having two epochs is given in Tables 5 and 6. A new hard constraint (constraint H_2 in Table 1) is specified in the two-epoch problem for guar-anteeing a minimum number of time slots between the two examinations of the same course, in order to give the necessary study time, exam correction and proofing.

In the two-epoch problem formulation, the following terms were added to the single-epoch problem formulation:

- set of time slots of the second examination epoch, $\mathcal{L} = \{|\mathcal{T}| + 1, |\mathcal{T}| + 2, \ldots, |\mathcal{T}| + |\mathcal{L}|\}$;
- $C_{relax} = (c_{relax_{ij}})_{|\mathcal{E}| \times |\mathcal{E}|}$ (*Relaxed conflict matrix*), is a relaxed version of the conflict matrix C. This matrix is used in the generation of the second examination epoch. Further details are given below;
- u_k ($u_k \in \mathcal{L}$) denotes the assigned time slot for exam e_k ($e_k \in \mathcal{E}$) in the second examination epoch.

The hard constraint H_2 is specified by:

$$u_k - t_k \geq L_{min}, \tag{4}$$

where L_{min} is the minimum time slot distance between the first and second epoch exams of a given course.

The devised algorithm for solving the two-epoch problem executes the following steps:

1. Divide the two-epoch problem into two distinct single-epoch problems. Solve the second epoch problem by considering the standard single-epoch formulation presented in Sect. 2.1, but using the relaxed conflict matrix, C_{relax}.
2. Then, solve the first epoch problem using a variation of the standard formulation given in Sect. 2.1, in order to be able to spread out first epoch exams from con-flicting exams in the second epoch. In addition, the constraint H_2, defined by (4), is also included in the extended formulation. More details are given next.
3. At the end of the optimisation step, join the two timetables forming the two-epoch timetable that respects both constraints H_1 and H_2. The cost of the two-epoch timetable is the sum of the costs of the individual timetables.

Some final details are now given. In the ISEL–DEETC dataset, the second epoch is more constrained than the first epoch, having less time slots (18 and 12 time slots, respectively, for the first and second epochs). In order to be possible to generate feasible initial solutions for the second epoch, some entries (with very few students enrolled) were set to zero, resulting in a relaxed conflict matrix. However, this pre-processing is problem dependent and for that reason not obligatory.

2.2.1 The Extended Exam Proximity Problem

The UEPP model presented earlier by Eqs. (1)–(3) is now extended in order to relate first epoch exams with second epoch exams. As mentioned in the introduction, the goal of using this extended model is to devise a two-epoch optimisation algorithm that could spread away first epoch exams from conflicting exams in the second epoch, in a neighbourhood of 0–4 time slots. The extended UEPP is presented next.

$$\text{minimise} \quad f_{epoch1} = \frac{1}{N_s} \sum_{i=0}^{4} w_{i+1} \, u_{2_{extended}}|_{q=|\mathcal{T}|+1,N=i,K=i+1} \tag{5}$$

where

$$u_{2_{extended}}|_{q=|\mathcal{T}|+1,N=x,K=l} = \frac{1}{2} \sum_{k=1}^{|\mathcal{T}|-(x+1)+l} \sum_{i=1}^{|\mathcal{E}|} \sum_{j=1}^{|\mathcal{E}|} c_{ij} \varepsilon_{ik} \varepsilon_{jk+(x+1)} , \tag{6}$$

$$\forall k \text{ where } k \bmod |\mathcal{T}| + 1 \neq 0 . \tag{7}$$

Figure 1 illustrates how the proximity cost function represented by (5) is computed.

We mention that while the timetables were optimised using the fitness function specified in (5) and (1), respectively, for the first and second epochs, in the collected results the fitness function published in [18] is used instead, that only considers conflicts from two consecutive time slots, and do not consider conflicts on Saturdays. This later measure is used in order to be able to compare results with the method published in [18].

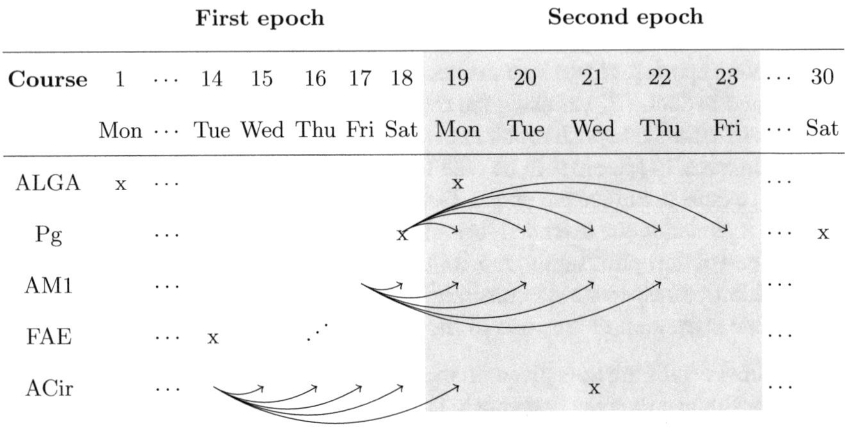

		First epoch						Second epoch						
Course	1	$\cdots$	14	15	16	17	18	19	20	21	22	23	$\cdots$	30
	Mon	$\cdots$	Tue	Wed	Thu	Fri	Sat	Mon	Tue	Wed	Thu	Fri	$\cdots$	Sat
ALGA	x	$\cdots$						x					$\cdots$	
Pg		$\cdots$					x						$\cdots$	x
AM1		$\cdots$											$\cdots$	
FAE		$\cdots$	x										$\cdots$	
ACir		$\cdots$									x		$\cdots$	

Fig. 1 Exam proximity cost computation for the two-epoch problem. When computing the first epoch proximity cost, an extended version of the chromosome is considered, having more five periods from the second epoch, marked in *greyscale*. The goal of using this extended model is to spread away first epoch exams that conflict with exams in the second epoch in a neighbourhood of zero to four time slots away

A final detail relates to the necessary modification of the algorithm in order to include the hard constraint H_2 in the first epoch optimisation. Modified versions of the initialisation procedure, the crossover and the neighbourhood operators, that work with feasible timetables satisfying both hard constraints, H_1 and H_2, were applied.

2.2.2 ISEL–DEETC Dataset

The problem instance considered in this work is the DEETC timetable of the winter semester of the 2009/2010 academic year. This benchmark data is detailed in [17, 18].

3 Shuffled Complex Evolution Algorithm for Examination Timetabling

In this section we describe the SCEA algorithm for solving the ETTP. The SCEA is based on ideas from the SCE [12] and the SFLA [14] approaches. In the SCE, the population is organized into *complexes* whereas in the SFLA it is organized into *memeplexes*. In the text, we use these two terms interchangeably.

In the SCE and SFLA approaches, global search is managed as a process of natural evolution. The sample points form a population that is partitioned into distinct groups called complexes (memeplexes). Each of these evolve independently, by searching the space into different directions. After completing a certain number of generations, the complexes are combined, and new complexes are formed through the process of shuffling. These procedures enhance survivability by a sharing of information about the search space, constructed independently by each complex [12].

The SCEA main steps are illustrated in Fig. 2, whereas the SCEA local search step is depicted in Fig. 2. The main loop of SCEA is identical to the SCE and SFLA main loop, where complexes are formed by creating random initial solutions that span the search space. Here, instead of points, the solutions correspond to complete and feasible timetables.

The local search step (Fig. 2b) was fully redesigned from the SCE and SFLA methods, in order to operate with ETTP solutions. Like the SFLA, we maintain the best and worst solutions of the memeplex, denoted respectively as P_b and P_w, and elitism is achieved by maintaining the best global solution, denoted as P_g. The local search step starts by selecting, randomly, and according with crossover probability cp, two parent solutions, P_1 and P_2, for recombination in order to produce a new offspring. P_1 must be different from the complex's best solution. The solution P_1 is recombined with solution P_2 (see the crossover operation depicted in Fig. 3) and the resulting offspring replaces the parent P_1. After this, the complex is sorted in order of increasing objective function value. The crossover operator was adapted from the crossover operator of [2, 23].

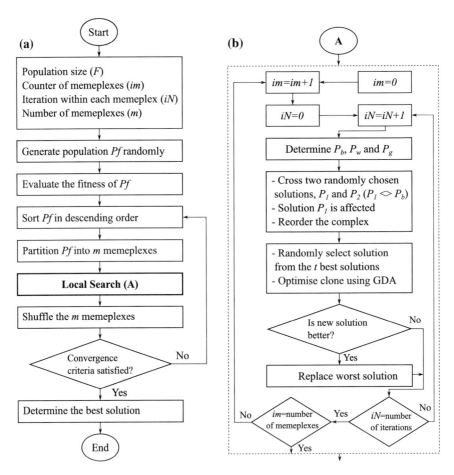

Fig. 2 Shuffled complex evolution algorithm: **a** Main algorithm steps; **b** SCEA local search step

After the crossover, a solution in the complex is selected for improvement according to an improvement probability, *ip*. The solution is improved by employing the local search meta-heuristic GDA (*Great Deluge Algorithm*) [13]. The template of GDA is presented in Algorithm 3.1.

The selection of the solution to improve is made on the group of the top *t* best solutions. The exploitation using GDA is done on a clone of the original solution, selected from this group. If the optimised solution is better than the original, then it will replace the complex's worst solution. This updating step in conjunction with the crossover operator guarantees a reasonable diversity, in an implicit fashion.

The GDA was integrated in the SCEA in the following fashion. We use as the initial solution s_0 the chosen solution to improve. The level *LEVEL* is set to the fitness value of this initial solution s_0. The search stops when the water level is equal to the solution fitness.

(a) Solution P_1

t_1	e_2	e_{14}	e_{10}	e_3	e_{16}
t_2	e_1	e_{11}	e_4		
t_3	e_9	e_{20}	e_5	e_{18}	
t_4	e_6	e_{13}	e_7		
t_5	e_8	e_{12}	e_{15}	e_{17}	e_{19}

(b) Solution P_2

t_1	e_{15}	e_{20}			
t_2	e_9	e_2	e_{12}	e_{10}	e_7
t_3	e_6	e_1	e_{17}	e_{13}	
t_4	e_5	e_{18}	e_4	e_{16}	
t_5	e_8	e_{14}	e_{11}	e_3	e_{19}

(c) New solution P_1

t_1	e_2	~~e_{14}~~	e_{10}	~~e_3~~	e_{16}			
t_2	e_1	e_{11}	e_4	~~e_8~~	e_{14}	~~e_{11}~~	e_3	e_{19}
t_3	e_9	e_{20}	e_5	e_{18}				
t_4	e_6	e_{13}	e_7					
t_5	e_8	e_{12}	e_{15}	e_{17}	~~e_{19}~~			

Fig. 3 Crossover between P_1 and P_2. The resulting solution P_1 in (**c**) is the result of combining the solution P_1 (**a**) with other solution P_2 (**b**). The operator inserts into P_1, at a random time slot (shown *dark shaded* in (**a**) and (**c**)), exams chosen from a random time slot from solution P_2 (shown *dark shaded* in (**b**)). When inserting these exams (shown *light gray* in (**c**)), some could be infeasible or already existing in that time slot (respectively, the case of e_8 and e_{11} in (**c**)). These exams are not inserted. The duplicated exams in the other time slots are removed

3.1 Solution Construction

The construction of the initial feasible solutions is done by a heuristic algorithm which uses the Saturation Degree graph colouring heuristic [7].

3.2 GDA's Neighbourhood

In the local search with GDA we used the Kempe chain neighbourhood [10].

3.3 Two-epoch Feasibility

In order to be able to execute the SCEA on the two-epoch problem, we have to implement a different version of the initialisation procedure, and the crossover and neighbourhood operators, that could manage the hard constraint H_2 mentioned in Sect. 2. This different version is executed in the first epoch generation, while the original version is executed in the second epoch generation.

Algorithm 1: Template of the Great Deluge Algorithm.

1: **Input:**	
2: $- s_0$	// Initial solution
3: $-$ Initial water level *LEVEL*	
4: $-$ Rain speed *UP*	// $UP > 0$
5: $s = s_0$;	// Generation of the initial solution
6: **repeat**	
7: Generate a random neighbour s'	
8: If $f(s') < LEVEL$ Then $s = s'$	// Accept the neighbour solution
9: $LEVEL = LEVEL - UP$	// update the water level
10: **until** Stopping criteria satisfied	
11: **Output:** Best solution found.	

4 Experiments

4.1 Settings

The algorithm is programmed in the C++ language using the ParadisEO framework [24]. The hardware and software specifications are: Intel Core i7-2630QM, CPU @ 2.00 GHz × 8, with 8 GB RAM; OS: Ubuntu 14.04, 64 bit; Compiler used: GCC v. 4.8.2. The parameters of SCEA are: Population size $F = 24$, Memeplex count $m = 3$, Memeplex size $n = 8$ (no sub-memeplexes were defined), and Number of time loops (convergence criterion) $L = 100000000$. The number of best solutions to consider for selection on a complex is given by $t = n/4 = 2$. The Great Deluge parameter, *UP*, was set to: $UP = 1e - 7$. The crossover and improvement probabilities, respectively, *cp* and *ip*, were set equal to 0.2 and 1.0. The parameter values were chosen empirically. To obtain our simulation results, the SCEA was run five times on each instance with different random seeds. The running time of the algorithm was limited to 24 h to all datasets except for the Toronto's `pur93` benchmark. For this larger dataset, the running time was limited to 48 h.

4.2 Single-epoch Problem

Tables 2 and 3 show the best results of the SCEA on the Toronto datasets as well as a selection of the best results available in the literature. The listed methods, dated until 2008, include only results validated by [22]. In the last two rows of each table, the *TP* and *TP (11)* indicate, respectively, the total penalty for the 13 instances and the total penalty except the `pur93` and `rye92` instances. For the SCEA we present the lowest penalty value f_{min}, the average penalty value f_{ave}, and the standard deviation σ over five independent runs. For the reference algorithms we present the best and average (where available) results and the number of runs. From the analysed works

Table 2 Simulation results of SCEA and comparison with selection of best algorithms from literature

Instance	Carter et al. (1996) [7]	Burke and Bykov (2006) [4]	Abdullah et al. (2009) [1] (five runs)		Burke et al. (2010) [6]
	f_{min}	f_{min}	f_{min}	f_{ave}	f_{min}
car91	7.10	4.42	4.42	4.81	4.90
car92	6.20	**3.74**	3.76	3.95	4.10
ear83	36.40	32.76	**32.12**	33.69	33.20
hec92	10.80	10.15	**9.73**	10.10	10.30
kfu93	14.00	12.96	**12.62**	**12.97**	13.20
lse91	10.50	**9.83**	10.03	10.34	10.40
pur93	**3.90**	–	–	–	–
rye92	**7.30**	–	–	–	–
sta83	161.50	157.03	156.94	157.30	**156.90**
tre92	9.60	7.75	7.86	8.20	8.30
uta92	3.50	3.06	**2.99**	3.32	3.30
ute92	25.80	24.82	24.90	25.41	24.90
yor83	41.70	34.84	34.95	36.27	36.30
TP (11)	327.10	301.36	**300.32**	306.36	305.80
TP	338.30	–	–	–	–

Values in bold represent the best results reported. "–" indicates that the corresponding instance is not tested or a feasible solution cannot be obtained

Table 3 Simulation results of SCEA (continuation)

Instance	Abdullah et al. (2010) [2]	Demeester et al. (2012) [10] (ten runs)		**SCEA** (five runs)		
	f_{min}	f_{min}	f_{ave}	f_{min}	f_{ave}	σ
car91	**4.35**	4.52	4.64	4.41	**4.45**	0.03
car92	3.82	3.78	3.86	3.75	**3.77**	0.01
ear83	33.76	32.49	**32.69**	32.62	**32.69**	0.07
hec92	10.29	10.03	**10.06**	10.03	**10.06**	0.03
kfu93	12.86	12.90	13.24	12.88	13.00	0.13
lse91	10.23	10.04	10.21	9.85	**9.93**	0.12
pur93	–	5.67	5.75	4.10	**4.17**	0.05
rye92	–	8.05	8.20	7.98	**8.06**	0.06
sta83	**156.90**	157.03	157.05	157.03	**157.03**	0.00
tre92	8.21	**7.69**	**7.79**	7.75	7.80	0.05
uta92	3.22	3.13	3.17	3.08	**3.15**	0.05
ute92	25.41	**24.77**	24.88	24.78	**24.81**	0.02
yor83	36.35	34.64	34.83	**34.44**	**34.73**	0.17
TP (11)	305.40	301.02	302.42	300.62	**301.42**	
TP	–	314.74	316.37	**312.70**	**313.65**	

that mention running times, the registered computation ranges in the interval [several minutes—1 h, several hours (12 h maximum)].

The best results obtained by SCEA are competitive with the ones produced by state-of-the-art algorithms. It attains a new lower bound on the `yor83` dataset. We also observe that the SCEA obtains the lowest sum of average cost on the *TP* and *TP (11)* quantities, and the lowest sum of best costs on the *TP* quantity, for the Toronto datasets. This demonstrates that SCEA can optimise very different datasets with good efficiency. A negative aspect of SCEA is the time taken compared with other algorithms. The time taken is a reflex of the high diversity of the method mixed with the low decreasing rate *UP*. A low *UP* value is needed in order for the GDA to find the best exam movements. If the *UP* value is higher, the optimisation is faster but with worse results, because the initial, larger conflict, exams are scheduled into sub optimal time slots, and thus the remainder exams, as the GDA's level decreases, could not be scheduled in the optimal fashion.

4.3 Two-epoch Problem

For the ISEL–DEETC, we compare the automatic solution with a manual solution available from the ISEL academic services. The L_{min} parameter was set to $L_{min} = 10$ (first and second epoch examinations of a given course are 10 time slots apart). Table 4 presents the costs for the manual solution and the automatic solution produced by the SCEA. The SCEA could generate timetables with lower cost comparing with the manual solution. We note that SCEA optimised the merged timetable comprising the five timetables and not the individual timetables, so in some cases, some programs timetables have worse cost (e.g., MEIC, 2nd epoch). The results produced in the first epoch, are comparable with the results published in [18]. Tables 5 and 6 illustrate, respectively, the manual and automatic solutions for the most difficult timetable, the LEETC timetable.

Table 4 Number of clashes for the manual and automatic solutions in DEETC dataset

Timetable	Manual sol.		Automatic sol.	
	1st ep.	2nd ep.	1st ep.	2nd ep.
LEETC	287	647	238	550
LEIC	197	442	171	418
LERCM	114	208	125	195
MEIC	33	63	23	66
MEET	50	144	23	124
Combined	**549**	**1163**	**447**	**1060**
Sum	**1712**		**1507**	

For the automatic solution, the best cost out of five runs is presented

Table 5 Manual solution for the LEETC examination timetable

Course	First epoch																		Second epoch											
	1	2	3	4	5	6	7	8	9	10	11	12	13	14	15	16	17	18	19	20	21	22	23	24	25	26	27	28	29	30
	Mo	Tu	Wd	Tr	Fr	Sa	Mo	Tu	Wd	Tr	Fr	Sa	Mo	Tu	Wd	Tr	Fr	Sa	Mo	Tu	Wd	Tr	Fr	Sa	Mo	Tu	Wd	Tr	Fr	Sa
ALGA			x																		x									
Pg							x																x							
AM1												x																x		
FAE																	x								x					
ACir														x											x					
POO								x																x						
AM2											x											x								
LSD																		x											x	
E1															x												x			
MAT					x															x										
PE													x													x				
ACp																	x													x
EA									x										x											
E2	x																						x							
SS						x																								x
RCp											x										x									
PICC/CPg															x													x		
PR							x													x										
FT							x																	x						
SEAD1																		x									x			
ST														x															x	
RCom										x									x											
RI													x									x								
SE1					x																					x				
AVE								x																x						

(continued)

Table 5 (continued)

Course	First epoch						Second epoch					
SCDig				x						x		
SOt					x				x			x
PI			x	x				x				
SCDist					x						x	
EGP	x			x			x					
OGE	x			x			x					
SG	x			x			x					

The courses marked in bold face are shared with other programs. The number of clashes of this timetable is 287 and 647, respectively, for the first and second epochs

Table 6 Automatic solution for the LEETC examination timetable

Course	First epoch																		Second epoch											
	1	2	3	4	5	6	7	8	9	10	11	12	13	14	15	16	17	18	19	20	21	22	23	24	25	26	27	28	29	30
	Mo	Tu	Wd	Tr	Fr	Sa	Mo	Tu	Wd	Tr	Fr	Sa	Mo	Tu	Wd	Tr	Fr	Sa	Mo	Tu	Wd	Tr	Fr	Sa	Mo	Tu	Wd	Tr	Fr	Sa
ALGA															x															
Pg	x																		x											x
AM1					x																						x			
FAE								x																x						
ACir											x										x									
POO		x																								x				
AM2				x																x										
LSD							x																		x					
EI													x										x							
MAT																	x												x	
PE										x												x								
ACp								x																x						
EA						x													x									x		
E2												x																		
SS																x														
RCp				x													x					x								x
PICC/CPg																											x			
PR									x					x															x	
FT	x										x																			
SEAD1											x										x									
ST			x																						x					
RCom													x													x				
RI																												x		
SE1										x												x								
AVE		x																		x										

(continued)

Table 6 (continued)

Course	First epoch											Second epoch										
SCDig		x																x				
SOt			x																x			
PI			x							x												
SCDist					x												x					
EGP				x		x																x
OGE				x																		x
SG				x																		x

The courses marked in bold face are shared with other programs. The number of clashes of this timetable is 238 and 550, respectively, for the first and second epochs. As can be observed, all first epoch examinations respect the minimum distance (L_{min} = 10) to the corresponding exam time slot in the second epoch

5 Conclusions

We presented a memetic algorithm that combines features from the SCE and the GDA meta-heuristics. The experimental evaluation of the SCEA shows that it is competitive with state-of-the-art methods. In the set of the 13 instances of the Toronto benchmark data it attains the lowest cost on one dataset, and the lowest sum of best and average cost with a low standard deviation. The algorithm main disadvantage is the time taken on the larger instances. Further studies should address the diversity management in order to accelerate the algorithm while maintaining a satisfactory diversity. As future research, we intend to apply our solution method to the instances of the 1st Track (Examination Timetabling) of the 2nd International Timetabling Competition (ITC2007), which contain more hard and soft constraints.

Acknowledgments This work was partially supported by national funds through Fundação para a Ciência e a Tecnologia (FCT) under project PEst-OE/EEI/ LA0009/2013, and by the PROTEC Program funds under the research grant SFRH/ PROTEC/67953/2010.

References

1. Abdullah, S., Turabieh, H., McCollum, B.: A hybridization of electromagnetic-like mechanism and great deluge for examination timetabling problems. In: Blesa, M.J., Blum, C., Gaspero, L.D., Roli, A., Sampels, M., Schaerf, A. (eds.) HM. LNCS, vol. 5818, pp. 60–72. Springer (2009)
2. Abdullah, S., Turabieh, H., McCollum, B., McMullan, P.: A tabu-based memetic approach for examination timetabling problems. In: Yu, J., Greco, S., Lingras, P., Wang, G., Skowron, A. (eds.) RSKT. LNCS, vol. 6401, pp. 574–581. Springer (2010)
3. Alba, E., Dorronsoro, B.: Cellular Genetic Algorithms, 1st edn. Springer Publishing Company, Incorporated (2008)
4. Burke, E.K., Bykov, Y.: Solving exam timetabling problems with the flex-deluge algorithm. In: Sixth International Conference on the Practice and Theory of Automated Timetabling. pp. 370–372 (2006). ISBN: 80-210-3726-1
5. Burke, E.K., McCollum, B., McMullan, P., Parkes, A.J.: Multi-objective aspects of the examination timetabling competition track. In: 7th International Conference on the Practice and Theory of Automated Timetabling. Montreal, Canada (Aug 2008)
6. Burke, E., Eckersley, A., McCollum, B., Petrovic, S., Qu, R.: Hybrid variable neighbourhood approaches to university exam timetabling. Eur. J. Oper. Res. **206**(1), 46–53 (2010). http://www.sciencedirect.com/science/article/pii/S0377221710000780
7. Carter, M., Laporte, G., Lee, S.Y.: Examination timetabling: algorithmic strategies and applications. J. Oper. Res. Soc. **47**(3), 373–383 (1996). http://www.jstor.org/stable/3010580?origin=crossref
8. Chu, S.C., Chen, Y.T., Ho, J.H.: Timetable scheduling using particle swarm optimization. In: First International Conference on Innovative Computing, Information and Control - vol. 3. pp. 324–327. ICICIC '06, IEEE Computer Society, Washington, DC, USA (2006). http://dx.doi.org/10.1109/ICICIC.2006.541
9. Côté, P., Wong, T., Sabourin, R.: Application of a hybrid multi-objective evolutionary algorithm to the uncapacitated exam proximity problem. In: Proceedings of the 5th International Conference on Practice and Theory of Automated Timetabling (PATAT 2004), pp. 151–167 (2004)

10. Demeester, P., Bilgin, B., Causmaecker, P.D., Berghe, G.V.: A hyperheuristic approach to examination timetabling problems: benchmarks and a new problem from practice. J. Sched. **15**(1), 83–103 (2012)
11. Dowsland, A., Thompson, M.: Ant colony optimization for the examination scheduling problem. J. Oper. Res. Soc. **56**(4), 426–438 (Apr 2005). http://dx.doi.org/10.1057/palgrave.jors.2601830
12. Duan, Q., Gupta, V., Sorooshian, S.: Shuffled complex evolution approach for effective and efficient global minimization. J. Optim. Theory Appl. **76**(3), 501–521 (1993). http://dx.doi.org/10.1007/BF00939380
13. Dueck, G.: New optimization heuristics: the great deluge algorithm and the record-to-record travel. J. Comput. Phys. **104**(1), 86–92 (1993). http://www.sciencedirect.com/science/article/pii/S0021999183710107
14. Eusuff, M., Lansey, K., Pasha, F.: Shuffled frog-leaping algorithm: a memetic meta-heuristic for discrete optimization. Eng. Optim. **38**(2), 129–154 (Mar 2006). http://dx.doi.org/10.1080/03052150500384759
15. Kamil, A., Krebs, J., Pulliam, H.: Foraging Behavior. Plenum Press (1987)
16. Leite, N., Melício, F., Rosa, A.: Solving the examination timetabling problem with the shuffled frog-leaping algorithm. In: 5th International Joint Conference on Computational Intelligence, pp. 175–180 (2013)
17. Leite, N., Melício, F., Rosa, A. C.: A shuffled complex evolution based algorithm for examination timetabling - benchmarks and a new problem focusing two epochs. In: Proceedings of the International Conference on Evolutionary Computation Theory and Applications (ECTA), pp. 112–124. SciTePress (2014). http://dx.doi.org/10.5220/0005164801120124
18. Leite, N., Neves, R.F., Horta, N., Melício, F., Rosa, A.C.: Solving an uncapacitated exam timetabling problem instance using a hybrid NSGA-II. In: Rosa, A.C., Correia, A.D., Madani, K., Filipe, J., Kacprzyk, J. (eds.) IJCCI, pp. 106–115. SciTePress (2012)
19. McCollum, B., McMullan, P., Parkes, A.J., Burke, E.K., Qu, R.: A new model for automated examination timetabling. Ann. Oper. Res. **194**, 291–315 (2012). http://dx.doi.org/10.1007/s10479-011-0997-x
20. Neri, F.: Diversity Management in Memetic Algorithms. In: Neri et al. [21], pp. 153–165
21. Neri, F., Cotta, C., Moscato, P. (eds.): Handbook of Memetic Algorithms, Studies in Computational Intelligence, vol. 379. Springer (2012)
22. Qu, R., Burke, E., McCollum, B., Merlot, L.T.G., Lee, S.Y.: A survey of search methodologies and automated system development for examination timetabling. J. Sched. **12**, 55–89 (2009)
23. Sabar, N.R., Ayob, M., Kendall, G.: Solving examination timetabling problems using honey-bee mating optimization (ETP-HBMO). In: Blazewicz, J., Drozdowski, M., Kendall, G., McCollum, B. (eds.) 4th Multidisciplinary International Scheduling Conference: Theory and Applications, pp. 399–408 (Aug 2009)
24. Talbi, E.G.: Metaheuristics - From Design to Implementation. Wiley (2009)
25. Turabieh, H., Abdullah, S.: A hybrid fish swarm optimisation algorithm for solving examination timetabling problems. In: Coello, C.A.C. (ed.) LION. LNCS, vol. 6683, pp. 539–551. Springer (2011)
26. Turabieh, H., Abdullah, S.: An integrated hybrid approach to the examination timetabling problem. Omega **39**(6), 598–607 (2011)

Part II
Fuzzy Computation Theory
and Applications

Static and Dynamic Methods for Fuzzy Signal Processing of Sound and Electromagnetic Environment Based on Fuzzy Observations

Akira Ikuta and Hisako Orimoto

Abstract The real observed data in sound and electromagnetic waves often contain fuzziness due to confidence limitations in sensing devices, permissible errors in the experimental data, and quantizing errors in digital observations. In this study, by paying attention to the specific signal in the real sound and electromagnetic environment, which exhibits complex probability distribution forms, a signal processing method is considered for estimating the probability distribution and the fluctuation wave form of the specific signal based on the observation with fuzziness. First, a static signal processing method is considered for predicting the probability distribution of electromagnetic wave leaked from several kinds of electronic information equipment in the real working environment based on the observed fuzzy data of the sound. Next, a dynamic state estimation method is proposed for estimating only the specific signal by removing background noise based on the fuzzy observation data in the sound environment under the existence of background noise. The effectiveness of the theoretically proposed static and dynamic signal processing methods is experimentally confirmed by applying those to real data in the sound and electromagnetic environment.

Keywords Probability measure of fuzzy events · Fuzzy signal processing · Sound and electromagnetic environment

A. Ikuta (✉) · H. Orimoto
Department of Management Information Systems, Prefectural University of Hiroshima,
1-1-71 Ujina-higashi, Minaki-ku, Hiroshima 734-8558, Japan
e-mail: ikuta@pu-hiroshima.ac.jp

H. Orimoto
e-mail: orimoto@pu-hiroshima.ac.jp

© Springer International Publishing Switzerland 2016
J.J. Merelo et al. (eds.), *Computational Intelligence*,
Studies in Computational Intelligence 620, DOI 10.1007/978-3-319-26393-9_11

1 Introduction

The Probability distribution of a specific signal in the real sound and electromagnetic environment can take various forms, not necessarily characterized by a standard Gaussian distribution. This is due to the diverse nature of factors affecting the properties of the signal [1]. Therefore, it is necessary for the estimation of the evaluation quantities such as the peak value, the amplitude probability distribution, the average crossing rate, the pulse spacing distribution, and the frequent distribution of occurrence etc. of the specific signal, to consider the lower order statistical properties of the signal such as mean and variance as well as the higher order statistics associated with non-Gaussian properties.

On the other hand, the observed data often contain fuzziness due to confidence limitations in sensing devices, permissible errors in the experimental data, and quantizing errors in digital observations [2]. For reasons of simplicity, many previously proposed estimation methods have not considered fuzziness in the observed data under the restriction of Gaussian type fluctuations [3–7]. Although several state estimation methods for a stochastic environment system with non-Gaussian fluctuations and many analyses based on Gaussian Mixture Models have previously been proposed [8–12], the fuzziness contained in the observed data has not been considered in these studies. Therefore, it is desirable to develop a method that is flexible and is applicable to ill-conditioned fuzzy observations.

In this study, a new estimation theory is proposed for a signal based on observations with non-Gaussian properties, from both static and dynamic viewpoints by regarding the observation data with fuzziness as fuzzy observations.

First, a static signal processing method considering not only linear correlation but also the higher order nonlinear correlation information is proposed on the basis of fuzzy observation data, in order to find the mutual relationship between sound and electromagnetic waves leaked from electronic information equipment. More specifically, a conditional probability expression for fuzzy variables is derived by applying probability measure of fuzzy events [13] to a joint probability function in a series type expression reflecting various correlation relationships between the variables. By use of the derived probability expression, a method for estimating precisely the correlation information based on the observed fuzzy data is theoretically proposed. On the basis of the estimated correlation information, the probability distribution for a specific variable (e.g. electromagnetic wave) based on the observed fuzzy data of the other variable (e.g. sound) can be predicted.

Next, a dynamic state estimation method for estimating a specific signal based on fuzzy observations with the existence of background noise is proposed in a recursive form suitable for use with a digital computer. More specifically, by paying attention to the power state variable for a specific signal in the sound environment, a new type of signal processing method for estimating a specific signal on a power scale is proposed. In the case of considering the power state variable, a physical mechanism of contamination by background noise can be reflected in the state estimation algorithm by using the additive property between the specific signal and the background noise. There is a restriction for power state variables fluctuating only in the non-negative region (i.e., any fluctuation width around the mean value has necessary to tend zero when the mean value tends zero), and it is obvious that the Gaussian distribution and Gaussian Mixture Models regarding the mean and variance as independent parameters are not adequate for power state variables. The proposed method positively utilizes Gamma distribution and Laguerre polynomial suitable to represent the power state variable, which fluctuates only within the positive region [14].

The effectiveness of the theoretically proposed static and dynamic fuzzy signal processing methods for estimating the specific signal is experimentally confirmed by applying those to real data in the sound and electromagnetic environment.

2 Static Signal Processing Based on Fuzzy Observations in Sound and Electromagnetic Environment

2.1 Prediction for Probability Distribution of Specific Signal from Fuzzy Fluctuation Factor

The observed data in the real sound and electromagnetic environment often contain fuzziness due to several factors such as limitations in the measuring instruments, permissible error tolerances in the measurement, and quantization errors in digitizing the observed data. In this study, the observation data with fuzziness are regarded as fuzzy observations.

In order to evaluate quantitatively the complicated relationship between sound and electromagnetic waves leaked from an identical electronic information equipment, let two kinds of variables (i.e. sound and electromagnetic waves) be x and y, and the observed data based on fuzzy observations be X and Y respectively. There exist the mutual relationships between x and y, and also between X and Y. Therefore, by finding the relations between x and X, and also between y and Y, based on probability measure of fuzzy events [13], it is possible to predict the true value y (or x) from the observed fuzzy data X (or Y). For example, for the prediction of the probability density function $P_s(y)$ of y from X, averaging the conditional probability density function $P(y|X)$ on the basis of the observed fuzzy data X, $P_s(y)$ can be obtained as: $P_s(y) = \langle P(y|X) \rangle_X$. The conditional probability density function $P(y|X)$ can be expressed under the employment of the well-known Bayes' theorem:

$$P(y|X) = \frac{P(X, y)}{P(X)} \qquad (1)$$

The joint probability distribution $P(X, y)$ is expanded into an orthonormal polynomial series on the basis of the fundamental probability distribution $P_0(X)$

and $P_0(y)$, which can be artificially chosen as the probability function describing approximately the dominant parts of the actual fluctuation pattern, as follows:

$$P(X, y) = P_0(X) P_0(y) \sum_{m=0}^{\infty} \sum_{n=0}^{\infty} A_{mn} \varphi_m(X) \phi_n(y),$$

$$A_{mn} \equiv \langle \varphi_m(X) \phi_n(y) \rangle,$$

(2)

where $\langle \cdot \rangle$ denotes the averaging operation with respect to the random variables. The information on the various types of linear and nonlinear correlations between X and y is reflected in each expansion coefficient A_{mn}. When X is a fuzzy number expressing an approximated value, it can be treated as a discrete variable with a certain level difference. Therefore, as $P_0(X)$, the generalized binomial distribution with a level difference interval h_X can be chosen [1]:

$$P_0(X) = \frac{\left(\frac{N_X - M_X}{h_X}\right)!}{\left(\frac{X - M_X}{h_X}\right)! \left(\frac{N_X - X}{h_X}\right)!} p_X^{\frac{X - M_X}{h_X}} (1 - p_X)^{\frac{N_X - X}{h_X}},$$

$$p_X \equiv \frac{\mu_X - M_X}{N_X - M_X}, \; \mu_X \equiv \langle X \rangle,$$

(3)

where M_X and N_X are the maximum and minimum values of X. Furthermore, as the fundamental probability density function $P_0(y)$ of y, the standard Gaussian distribution is adopted:

$$P_0(y) = \frac{1}{\sqrt{2\pi\sigma_y^2}} e^{-\frac{(y - \mu_y)^2}{2\sigma_y^2}}, \qquad \mu_y \equiv \langle y \rangle, \; \sigma_y^2 \equiv \langle (y - \mu_y)^2 \rangle.$$

(4)

The orthonormal polynomials $\varphi_m(X)$ and $\phi_n(y)$ with the weighting functions $P_0(X)$ and $P_0(y)$ can be determined as [1]

$$\phi_m(X) = \left\{ \left(\frac{N_X - M_X}{h_X}\right)^{(m)} m! \right\}^{-\frac{1}{2}} \left(\frac{1 - p_X}{p_X}\right)^{\frac{m}{2}} \frac{1}{h_X^m}$$

$$\sum_{j=0}^{m} \frac{m!}{(m-j)! j!} (-1)^{m-j} \left(\frac{p_X}{1 - p_X}\right)^{m-j} (N_X - X)^{(m-j)} (X - M_X),$$

$$\left(X^{(n)} \equiv X(X - h_X) \cdots (X - (n-1)h_X), \; X^{(0)} \equiv 1 \right),$$

(5)

$$\varphi_n(y) = \frac{1}{\sqrt{n!}} H_n\left(\frac{y - \mu_y}{\sigma_y}\right); \text{ Hermite polynomial.}$$

(6)

Thus, the predicted probability density function $P_s(y)$ can be expressed in an expansion series form:

$$P_s(y) = P_0(y) \sum_{n=0}^{\infty} \langle \frac{\sum_{m=0}^{\infty} A_{mn} \phi_m(X)}{\sum_{m=0}^{\infty} A_{m0} \phi_m(X)} \rangle_X \varphi_n(y). \tag{7}$$

2.2 Estimation of Correlation Information Based on Fuzzy Observation Data

The expansion coefficient A_{mn} in (2) has to be estimated on the basis of the fuzzy observation data X and Y, when the true value y is unknown. Let the joint probability distribution of X and Y be $P(X, Y)$. By applying probability measure of fuzzy events [13], $P(X, Y)$ can be expressed as:

$$P(X, Y) = \frac{1}{K} \int \mu_Y(y) P(X, y) \, dy, \tag{8}$$

where K is a constant satisfying the normalized condition: $\sum_X \sum_Y P(X, Y) = 1$. The fuzziness of Y can be characterized by the membership function $\mu_Y(y) (= \exp\{-\alpha(y - Y)^2\}$, α; a parameter).

Substituting (2) in (8), the following relationship is derived.

$$P(X, Y) = \frac{1}{K} P_0(X) \sum_{m=0}^{\infty} \sum_{n=0}^{\infty} A_{mn} a_n \varphi_m(X),$$

$$a_n = \int e^{-\alpha(y - Y)^2} P_0(y) \phi_n(y) \, dy. \tag{9}$$

The conditional N-th order moment of the fuzzy variable X is given from (9) as

$$\langle X^N \mid Y \rangle = \sum_X P(X \mid Y) = \frac{\sum_X X^N P(X, Y)}{P(Y)}$$

$$= \frac{\sum_X P_0(X) X^N \sum_{m=0}^{\infty} \sum_{n=0}^{\infty} A_{mn} a_n \phi_m(X)}{\sum_{n=0}^{\infty} A_{0n} a_n}. \tag{10}$$

After expanding X^N in an orthogonal series expression, by considering the orthonormal relationship of $\varphi_m(X)$, (10) is expressed explicitly as

$$\langle X^N \mid Y \rangle = \frac{\sum\limits_{m=0}^{\infty} \sum\limits_{n=0}^{\infty} d_m^N A_{mn} a_n}{\sum\limits_{n=0}^{\infty} A_{0n} a_n},$$

$$\left(X^N \equiv \sum_{m=0}^{N} d_m^N \phi_m(X), \ d_m^N; \text{ appropriate constant} \right). \tag{11}$$

The right side of the above equation can be evaluated numerically from the fuzzy observation data. Accordingly, by regarding the expansion coefficients A_{mn} as unknown parameters, a set of simultaneous equations in the same form as in (11) can be obtained by selecting a set of N and/or Y values equal to the number of unknown parameters. By solving the simultaneous equations, the expansion coefficients A_{mn} can be estimated. Furthermore, using these estimates, the probability density function $P_s(y)$ can be predicted from (7).

3   Dynamic Signal Processing Based on Fuzzy Observations in Sound Environment

3.1   Formulation of Fuzzy Observation Under Existence of Background Noise

Consider a sound environmental system with background noise having a non-Gaussian distribution. Let the specific signal power of interest in the environment at a discrete time k be x_k, and the dynamical model of the specific signal be:

$$x_{k+1} = F x_k + G u_k, \tag{12}$$

where u_k denotes the random input power with known statistics, and F, G are known system parameters and can be estimated by use of the system identification method [15] when these parameters cannot be determined on the basis of the physical mechanism of system.

The observed data in the real sound environment often contain fuzziness due to several factors, as indicated earlier. Therefore, in addition to the inevitable background noise, the effects of the fuzziness contained in the observed data have to be considered in developing a state estimation method for the specific signal of interest. From a functional viewpoint, the observation equation can be considered as involving two types of operation:

1. The additive property of power state variable with the background noise can be expressed as:
$$y_k = x_k + v_k,$$
where it is assume that the statistics of the background noise power v_k are known in advance.

2. The fuzzy observation z_k is obtained from y_k. The fuzziness of z_k is characterized by the membership function $\mu_{z_k}(y_k)$.

3.2 State Estimation Based on Fuzzy Observation Data

To obtain an estimation algorithm for the signal power x_k based on the fuzzy observation z_k, the Bayes' theorem for the conditional probability density function can be considered [9].

$$P(x_k \mid Z_k) = \frac{P(x_k, z_k \mid Z_{k-1})}{P(z_k \mid Z_{k-1})}, \tag{13}$$

where $Z_k (\equiv (z_1, z_2, \dots, z_k))$ is a set of observation data up to a time k. By applying probability measure of fuzzy events [13] to the right side of (13), the following relationship is derived.

$$P(x_k \mid Z_k) = \frac{\int_0^\infty \mu_{z_k}(y_k) P(x_k, y_k \mid Z_{k-1}) dy_k}{\int_0^\infty \mu_{z_k}(y_k) P(y_k \mid Z_{k-1}) dy_k}. \tag{14}$$

The conditional probability density function of x_k and y_k can be generally expanded in a statistical orthogonal expansion series.

$$P(x_k, y_k \mid Z_{k-1}) = P_0(x_k \mid Z_{k-1}) P_0(y_k \mid Z_{k-1})$$
$$\sum_{m=0}^\infty \sum_{n=0}^\infty B_{mn} \theta_m^{(1)}(x_k) \theta_n^{(2)}(y_k), \tag{15}$$

$$B_{mn} \equiv \langle \theta_m^{(1)}(x_k) \theta_n^{(2)}(y_k) \mid Z_{k-1} \rangle, \tag{16}$$

where the functions $\theta_m^{(1)}(x_k)$ and $\theta_n^{(2)}(y_k)$ are the orthogonal polynomials of degrees m and n with weighting functions $P_0(x_k \mid Z_{k-1})$ and $P_0(y_k \mid Z_{k-1})$, which can be artificially chosen as the probability density functions describing the dominant parts of $P(x_k \mid Z_{k-1})$ and $P(y_k \mid Z_{k-1})$. These two functions must satisfy the following orthonormal relationships:

$$\int_0^\infty \theta_m^{(1)}(x_k) \theta_{m'}^{(1)}(x_k) P_0(x_k \mid Z_{k-1}) dx_k = \delta_{mm'}, \tag{17}$$

$$\int_0^\infty \theta_n^{(2)}(y_k)\theta_{n'}^{(2)}(y_k)P_0(y_k\,|\,Z_{k-1})dy_k = \delta_{nn'}. \tag{18}$$

By substituting (15) into (16), the conditional probability density function $P(x_k\,|\,Z_k)$ can be expressed as:

$$P(x_k\,|\,Z_k) = \frac{\sum_{m=0}^{\infty}\sum_{n=0}^{\infty} B_{mn}P_0(x_k\,|\,Z_{k-1})\theta_m^{(1)}(x_k)I_n(z_k)}{\sum_{n=0}^{\infty} B_{0n}I_n(z_k)} \tag{19}$$

with

$$I_n(z_k) \equiv \int_0^\infty \mu_{z_k}(y_k)P_0(y_k\,|\,Z_{k-1})\theta_n^{(2)}(y_k)dy_k. \tag{20}$$

Based on (19), and using the orthonormal relationship of (17), the recurrence algorithm for estimating an arbitrary N-th order polynomial type function $f_N(x_k)$ of the specific signal can be derived as follows:

$$\hat{f}_N(x_k) \equiv \langle f_N(x_k)\,|\,Z_k\rangle$$
$$= \frac{\sum_{m=0}^{N}\sum_{n=0}^{\infty} B_{mn}C_{Nm}I_n(z_k)}{\sum_{n=0}^{\infty} B_{0n}I_n(z_k)}, \tag{21}$$

where C_{Nm} is the expansion coefficient determined by the equality:

$$f_N(x_k) = \sum_{m=0}^{N} C_{Nm}\theta_m^{(1)}(x_k). \tag{22}$$

In order to make the general theory for estimation algorithm more concrete, the well-known Gamma distribution is adopted as $P_0(x_k\,|\,Z_{k-1})$ and $P_0(y_k\,|\,Z_{k-1})$, because this probability density function is defined within positive region and is suitable to the power state variables.

$$P_0(x_k\,|\,Z_{k-1}) = P_\Gamma(x_k; m_{x_k}^*, s_{x_k}^*),$$
$$P_0(y_k\,|\,Z_{k-1}) = P_\Gamma(y_k; m_{y_k}^*, s_{y_k}^*) \tag{23}$$

with

$$P_\Gamma(x; m, s) \equiv \frac{x^{m-1}}{\Gamma(m)s^m} e^{-\frac{x}{s}},$$

$$m_{x_k}^* \equiv (x_k^*)^2/\Gamma_k, \quad s_{x_k}^* \equiv x_k^*/m_{x_k}^*,$$

$$x_k^* = \langle x_k \mid Z_{k-1} \rangle, \quad \Gamma_k \equiv \langle (x_k - x_k^*)^2 \mid Z_{k-1} \rangle,$$

$$m_{y_k}^* \equiv (y_k^*)^2/\Omega_k, \quad s_{y_k}^* \equiv y_k^*/m_{y_k}^*,$$

$$y_k^* = \langle y_k \mid Z_{k-1} \rangle = x_k^* + \langle v_k \rangle,$$

$$\Omega_k \equiv \langle (y_k - y_k^*)^2 \mid Z_{k-1} \rangle = \Gamma_k + \langle (v_k - <v_k>)^2 \rangle. \tag{24}$$

Then, the orthonormal functions with two weighting probability density functions in (23) can be given in the Laguerre polynomial [14]:

$$\theta_m^{(1)}(x_k) = \sqrt{\frac{\Gamma(m_{x_k}^*)m!}{\Gamma(m_{x_k}^* + m)}} L_m^{(m_{x_k}^* - 1)}\left(\frac{x_k}{s_{x_k}^*}\right),$$

$$\theta_n^{(2)}(y_k) = \sqrt{\frac{\Gamma(m_{y_k}^*)n!}{\Gamma(m_{y_k}^* + n)}} L_n^{(m_{y_k}^* - 1)}\left(\frac{y_k}{s_{y_k}^*}\right). \tag{25}$$

As the membership function $\mu_{z_k}(y_k)$, the following function suitable for the Gamma distribution is newly introduced.

$$\mu_{z_k}(y_k) = (z_k^{-\beta} e^\beta) y_k^\beta \exp\{-\frac{\beta}{z_k} y_k\}, \tag{26}$$

where $\beta(> 0)$ is a parameter. Accordingly, by considering the orthonormal condition of Laguerre polynomial [14], (20) can be given by

$$I_n(z_k) = \frac{z_k^{-\beta} e^\beta}{\Gamma(m_{y_k}^*)(s_{y_k}^*)^{m_{y_k}^*}} \Gamma(M_k) D_k^{M_k}$$

$$\int_0^\infty \frac{y_k^{M_k - 1}}{\Gamma(M_k) D_k^{M_k}} e^{-\frac{1}{D_k}} \sqrt{\frac{\Gamma(m_{y_k}^*)n!}{\Gamma(m_{y_k}^* + n)}} \sum_{r=0}^n d_{nr} L_r^{(M_k - 1)}\left(\frac{y_k}{D_k}\right) dy_k \tag{27}$$

$$= \frac{z_k^{-\beta} e^\beta}{\Gamma(m_{y_k}^*)(s_{y_k}^*)^{m_{y_k}^*}} \Gamma(M_k) D_k^{M_k} \sqrt{\frac{\Gamma(m_{y_k}^*)n!}{\Gamma(m_{y_k}^* + n)}} d_{n0}$$

with

$$M_k \equiv m_{y_k}^* + \beta, \quad D_k \equiv \frac{s_{y_k}^* z_k}{\beta s_{y_k}^* + z_k}, \tag{28}$$

where d_{nr} $(r = 0, 1, 2, \ldots, n)$ are the expansion coefficients in the equality:

$$L_n^{(m_{y_k}^* - 1)}\left(\frac{y_k}{s_{y_k}^*}\right) = \sum_{r=0}^{n} d_{nr} L_r^{(M_k - 1)}\left(\frac{y_k}{D_k}\right). \tag{29}$$

Especially, the estimates for mean and variance can be obtained as follows:

$$\hat{x}_k \equiv \langle x_k \mid Z_k \rangle$$
$$= \frac{\sum\limits_{n=0}^{\infty} \{B_{0n} C_{10} + B_{1n} C_{11}\} I_n(z_k)}{\sum\limits_{n=0}^{\infty} B_{0n} I_n(z_k)}, \tag{30}$$

$$P_k \equiv \langle (x_k - \hat{x}_k)^2 \mid Z_k \rangle$$
$$= \frac{\sum\limits_{n=0}^{\infty} \{B_{0n} C_{20} + B_{1n} C_{21} + B_{2n} C_{22}\} I_n(z_k)}{\sum\limits_{n=0}^{\infty} B_{0n} I_n(z_k)} \tag{31}$$

with

$$\begin{aligned}
C_{10} &= m_{x_k}^* s_{x_k}^*, \quad C_{11} = -\sqrt{m_{x_k}^*} s_{x_k}^*, \\
C_{20} &= \hat{x}_k^2 - 2m_{x_k}^* s_{x_k}^* \{\hat{x}_k - (m_{x_k}^* + 1) s_{x_k}^*\} - m_{x_k}^* (m_{x_k}^* + 1) s_{x_k}^{*\,2}, \\
C_{21} &= 2\sqrt{m_{x_k}^*} s_{x_k}^* \{\hat{x}_k - (m_{x_k}^* + 1) s_{x_k}^*\}, \\
C_{22} &= \sqrt{2 m_{x_k}^* (m_{x_k}^* + 1)} s_{x_k}^*.
\end{aligned} \tag{32}$$

Finally, by considering (12), the prediction step which is essential to perform the recurrence estimation can be given by

$$\begin{aligned}
x_{k+1}^* &\equiv \langle x_{k+1} \mid Z_k \rangle = F\hat{x}_k + G\langle u_k \rangle, \\
\Gamma_{k+1} &\equiv \langle (x_{k+1} - x_{k+1}^*)^2 \mid Z_k \rangle \\
&= F^2 \Gamma_k + G^2 \langle (u_k - \langle u_k \rangle)^2 \rangle.
\end{aligned} \tag{33}$$

By replacing k with $k + 1$, the recurrence estimation can be achieved.

4 Application to Sound and Electromagnetic Environment

4.1 Prediction of Sound and Electric Field in PC Environment

By adopting a personal computer (PC) in the real working environment as specific information equipment, the proposed static method was applied to investigate the mutual relationship between sound and electromagnetic waves leaked from the PC under the situation of playing a computer game. In order to eliminate the effects of sound from outside, the PC was located in an anechoic room (cf. Fig. 1). The RMS value (V/m) of the electric field radiated from the PC and the sound intensity level (dB) emitted from a speaker of the PC were simultaneously measured. The data of electric field strength and sound intensity level were measured by use of an electromagnetic field survey mater and a sound level meter respectively. The slowly changing non-stationary 600 data for each variable were sampled with a sampling interval of 1 (s). Two kinds of fuzzy data with the quantized level widths of 0.1 (v/m) for electric field strength and 5.0 (dB) for sound intensity level were obtained.

Based on the 400 data points, the expansion coefficients A_{mn} were first estimated by use of (11). Furthermore, the parameters of the membership functions in (8) for sound level and electric field strength with rough quantized levels were decided so as to express the distribution of data as precisely as possible, as shown in Figs. 2 and 3. Next, the 200 sampled data within the different time interval which were non-stationary different from data used for the estimation of the expansion coefficients were adopted for predicting the probability distributions of (i) the electric field based on sound and (ii) the sound based on electric field.

The experimental results for the prediction of electric field strength and sound level are shown in Figs. 4 and 5 respectively in a form of cumulative distribution. From these figures, it can be found that the theoretically predicted curves show good agreement with experimental sample points by considering the expansion coefficients with several higher orders.

For comparison, the generalized regression analysis method [1] without using fuzzy theory was applied to the fuzzy data X and Y. The prediction results are shown in Figs. 6 and 7. As compared with Figs. 4 and 5, it is obvious that the proposed method considering fuzzy theory is more effective than the previous method.

Fig. 1 A schematic drawing of the experiment

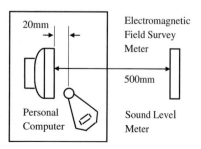

Fig. 2 Membership function
of sound level

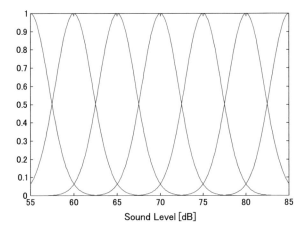

Fig. 3 Membership function
of electric field

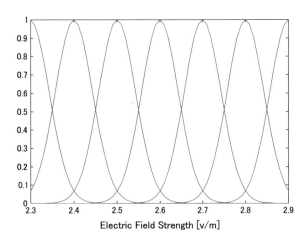

Fig. 4 Prediction of the
cumulative distribution for the
electric field strength based on
the fuzzy observation of
sound

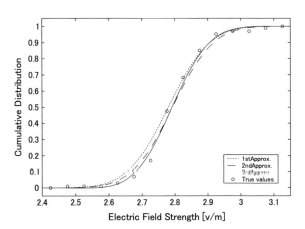

Fig. 5 Prediction of the cumulative distribution for the sound level based on the fuzzy observation of electric field

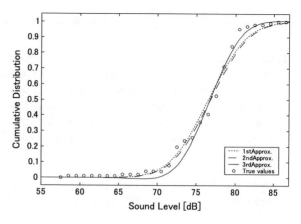

Fig. 6 Prediction of the cumulative distribution for the electric field strength by use of the extended regression analysis method

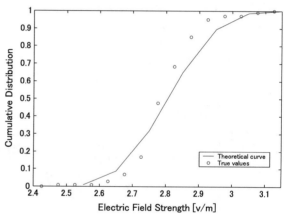

Fig. 7 Prediction of the cumulative distribution for the sound level by use of the extended regression analysis method

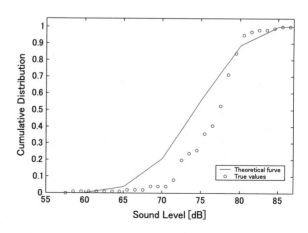

4.2 Estimation of Specific Signal in Sound Environment

In order to examine the practical usefulness of the proposed dynamic signal processing based on the fuzzy observation, the proposed method was applied to the real sound environmental data. The road traffic noise was adopted as an example of a specific signal with a complex fluctuation form. Applying the proposed estimation method to actually observed data contaminated by background noise and quantized with 1 dB width, the fluctuation wave form of the specific signal was estimated. The statistics of the specific signal and the background noise used in the experiment are shown in Table 1.

Figures 8, 9 and 10 show the estimation results of the fluctuation wave form of the specific signal. In this estimation, the finite number of expansion coefficients

Table 1 Statistics of the specific signal and the background noise

Statistics of specific signal		Statistics of background noise	
Mean (watt/m^2)	Standard deviation (watt/m^2)	Mean (watt/m^2)	Standard deviation (watt/m^2)
2.9×10^{-5}	2.8×10^{-5}	2.9×10^{-5}	1.4×10^{-6}

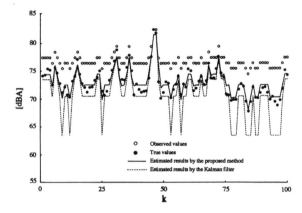

Fig. 8 State estimation results for the road traffic noise during a discrete time interval of [1, 100] s, based on the quantized data with 1 dB width

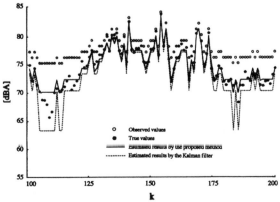

Fig. 9 State estimation results for the road traffic noise during a discrete time interval of [101, 200] s, based on the quantized data with 1 dB width

Fig. 10 State estimation results for the road traffic noise during a discrete time interval of [201, 300] s, based on the quantized data with 1 dB width

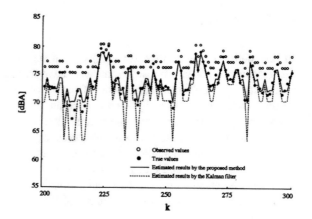

$B_{mn}(m, n \leq 2)$ was used for the simplification of the estimation algorithm. In these figures, the horizontal axis shows the discrete time k, of the estimation process, and the vertical axis expresses the sound level taking a logarithmic transformation of power-scaled variables, because the real sound environment usually is evaluated on dB scale connected with human effects. For comparison, the estimation results calculated using the usual method without considering any membership function are also shown in these figures. Since Kalman's filtering theory is widely used in the field of stochastic system [4–6], this method was also applied to the fuzzy observation data as a trail.

The results estimated by the proposed method considering the membership function show good agreement with the true values. On the other hand, there are great discrepancies between the estimates based on the standard type dynamical estimation method (i.e., Kalman filter) without consideration of the membership function and the true values, particularly in the estimation of the lower level values of the fluctuation.

The squared sum of the estimation error is shown in Table 2. These results clearly show the effectiveness of the proposed method for application to the observation of fuzzy data.

Table 2 Root-mean squared error of the estimation (in dBA)

$\sqrt{\frac{1}{N} \sum_{k=1}^{N} (x_k - \hat{x}_k)^2}$	Proposed method	Kalman filter
	0.90	2.52

5 Conclusions

In this study, a new method for estimating a specific signal embedded in fuzzy observations has been proposed from two viewpoints of static and dynamic signal processing. The nonlinear correlation of higher order as well as the linear correlation between the specific signal and observations has been considered in order to derive the estimation method. More specifically, in order to treat fuzzy observation data on the sound and electromagnetic waves, by applying probability measure of fuzzy events to the probability expression, a prediction method of the probability distribution of sound and electromagnetic waves based on the fuzzy observations has been theoretically derived. Next, an estimation method of the fluctuation wave of the specific signal based on the fuzzy observation data under the existence of background noise has been derived.

The proposed approach is still at the early stage of study, and there are left a number of practical problems to be continued in the future. For example, the proposed method has to be applied to many other actual data of sound and electromagnetic environment. Furthermore, the proposed theory has to be extended to more complicated situations involving multi-signal sources, and an optimal number of expansion terms in the proposed estimation and prediction algorithms of expansion type has to be found.

References

1. Ikuta, A., Ohta, M., Ogawa, H.: Various regression characteristics with higher order among light, sound and electromagnetic waves leaked from VDT. J. Int. Measur. Confederation **21**, 25–33 (1997)
2. Ikuta, A., Ohta, M., Siddique, M.N.H.: Prediction of probability distribution for the psychological evaluation of noise in the environment based on fuzzy theory. Int. J. Acoust. Vibr. **10**, 107–114 (2005)
3. Bell, B.M., Cathey, F.W.: The iterated Kalman filter update as a Gaussian-Newton methods. IEEE Trans. Autom. Control **38**, 294–297 (1993)
4. Kalman, R.E.: A new approach to linear filtering and prediction problems. Trans. ASME Series D J. Basic Eng. **82**, 35–45 (1960)
5. Kalman, R.E., Buch, R.S.: New results in linear filtering and prediction theory. Trans. ASME Series D J. Basic Eng. **83**, 95–108 (1961)
6. Kushner, H.J.: Approximations to optimal nonlinear filter. IEEE Trans. Autom. Control **12**, 546–556 (1967)
7. Julier, S.J.: The scaled unscented transformation. Proc. Am. Control Conf. **6**, 4555–4559 (2002)
8. Kitagawa, G.: Monte carlo filter and smoother for non-Gaussian nonlinear state space models. J. Comput. Graph. Stat. **5**, 1–25 (1996)
9. Ohta, M., Yamada, H.: New methodological trials of dynamical state estimation for the noise and vibration environmental system. Acustica **55**, 199–212 (1984)
10. Ikuta, A., Tokhi, M.O., Ohta, M.: A cancellation method of background noise for a sound environment system with unknown structure, IEICE Trans. Fundam. Electron. Commun. Comput. Sci. **E84-A**, 457–466 (2001)

11. H. Orimoto, A. Ikuta: State estimation method of sound environment system with multiplicative and additive noise. Int. J. Circ. Syst. Sig. Process. **8**, 307–312 (2014)
12. Guoshen, Y.: Solving inverse problems with piecewise linear estimators: from gaussian mixture models to structured sparsity. IEEE Trans. Image Process. **21**, 2481–2499 (2012)
13. Zadeh, L.A.: Probability measures of fuzzy events. J. Math. Anal. Appl. **23**, 421–427 (1968)
14. Ohta, M., Koizumi, T.: General statistical treatment of response of a non-linear rectifying device to a stationary random input. IEEE Trans. Inf. Theory **14**, 595–598 (1968)
15. Eykhoff, P.: System Identification: Parameter and State Estimation. Wiley, New York (1984)

The Ordinal Controversy and the Fuzzy Inference System Through an Application and Simulation to Teaching Activity Evaluation

Michele Lalla and Tommaso Pirotti

Abstract The handling of ordinal variables presents many difficulties in both the measurements phase and the statistical data analysis. Many efforts have been made to overcome them. An alternative approach to traditional methods used to process ordinal data has been developed over the last two decades. It is based on a fuzzy inference system and is presented, here, applied to the student evaluations of teaching data collected via Internet in Modena, during the academic year 2009/10, by a questionnaire containing items with a four-point Likert scale. The scores emerging from the proposed fuzzy inference system proved to be approximately comparable to scores obtained through the practical, but questionable, procedure based on the average of the item value labels. The fuzzification using a number of membership functions smaller than the number of modalities of input variables yielded outputs that were closer to the average of the item value labels. The Center-of-Area defuzzification method showed good performances and lower dispersion around the mean of the value labels.

Keywords Ordinal scales · Likert scale · Student evaluation · Fuzzification · Defuzzification

1 Introduction

The analysis of an ordinal variable presents some difficulties deriving from the nature of the collected data, which is not quantitative. Even some simple statistical indicators, such as the sample average and standard deviation of a variable, are not meaningful quantities to represent data [1, 2]. In the context of regression analysis, the dependent ordinal variable requires specific models, such as ordinal logit models

M. Lalla · T. Pirotti(✉)
Department of Economics Marco Biagi, University of Modena and Reggio Emilia,
Viale Berengario 51, Modena, MO, Italy
e-mail: tommaso.pirotti@unimore.it

M. Lalla
e-mail: michele.lalla@unimore.it

© Springer International Publishing Switzerland 2016   189
J.J. Merelo et al. (eds.), *Computational Intelligence*,
Studies in Computational Intelligence 620, DOI 10.1007/978-3-319-26393-9_12

[3–5], while the independent ordinal variables could be introduced in the models both directly as it is and transformed into dummy variables, one for each modality. Many procedural efforts have been made to improve measurement tools aimed at obtaining more accurate quantities in order to move beyond the few techniques offered by nonparametric statistics, such as the thermometer and Juster scales. However, none of the traditional methods allow for transformation of the nature of the qualitative variables into quantitative variables.

Over the last two decades, a new class of models, termed the fuzzy inference system (FIS), has been developed based on the mathematical theory of fuzzy sets, which was originally proposed to represent indeterminacy and to formalize qualitative concepts that generally have no precise boundaries [6]. Therefore, FIS may be an ideal tool to process ordinal data because it allows for handling these data with fewer restrictions than traditional statistical techniques, even though some ontological and intrinsic issues remain unsolved, such as the unidimensionality of the measured concept or the equidistance (or the fixed distance) between the modalities of the scale.

The purpose of the present paper is twofold. Firstly, it briefly discusses the limits of some statistical indices in representing synthetically ordinal variables, which constitute the current or traditional procedures. Secondly, the fuzzy approach, based on the FIS, is applied to data concerning student evaluations of teaching activity (SETA). In fact, our data set contains prevailingly ordinal information from an online survey conducted by the University of Modena and Reggio Emilia, for the Academic Year 2009/2010. The data analysis was restricted to the evaluated courses in the "Economics and International Management" degree program of the Faculty of Economics. The fuzzy approach through FIS offers a clear advantage over traditional methods because it is highly flexible in handling data and avoids the usual complications related to measurement methodology. For example, the FIS permits handling of both the four- or five-point Likert scale and any other type of scale without theoretical difficulties and great flexibility with a large variety of solutions. A comparison between the results obtained from current procedures and the proposed FIS will be analyzed and described, illustrating the strengths and weaknesses of both. A fuzzy inference model may well be a new and different way to analyze ordinal variables and, specifically, student evaluations of teaching activity.

2   Background of Ordinal Scale

The objective of the measurements process is to obtain information that is valid (i.e., it succeeds in evaluating what it is intended to evaluate), reliable (i.e., the results can be reproduced upon replication of the procedure, yielding identical or very similar values), and precise (i.e., the multiples or submultiples of the unit of measurement are contained by the available device). Given that it is not always possible to establish or find the unit of measurement of social concepts, preciseness remains a real difficulty of each intensity concept's evaluation, which is generally classified on the basis of its nature and preciseness [7], where the lowest level is based on discrimination (nominal) and the subsequent level is based on an order relation (ordinal). The basic assumptions of almost all ordinal scales are (1) the unidimensionality of the surveyed concept, (2) the location of the concept on a *continuum*, (3) the non-equidistance between the modalities constituting the observable level of the intensity of the concept. However, in practice, this latter assumption is often violated to obtain the sum or the mean of the scores for two or more items. Therefore, the validity of these operations makes ordinal scales equivalent to interval scales, which constitutes a real difficulty of the traditional approach.

Many techniques of scales have been developed since the 1920 s to study attitudes and, to a lesser extent, psychophysical and psychometric behavior [8–11]. The ordinal scales most used in practice are 'summated' scales and one of the first successful procedures to obtain an ordinal variable, whose values denote the intensity level of its denoted concept, was proposed by Likert [12] to measure attitudes and opinions through statements. The intensity of each statement was rated with graduated response keys (modalities), originally seven: strongly agree, mildly agree, agree, uncertain, disagree, mildly disagree, and strongly disagree (seven-point Likert scale). Subsequently, the alternatives containing "mildly" were dropped, obtaining a five-point scale. The neutral point presents a theoretical and empirical, unsolved issue because many empirical results do not give strong indications about the advisability of its presence/absence. However, it is often eliminated, assuming that (1) it attracts people who are careless, lazy or have no opinion, (2) respondents tend toward one of the two nearest alternatives, (3) respondents who really are neutral, randomly choose a polar alternative [13].

Let i be an index denoting the interviewed subject. Let j be an index denoting a concept, and k, a statement or item about the jth concept. The corresponding score, y_{ijk}, belongs to $\{1, \dots, M\} \subset \mathbb{N}$ for any statement favorable to the concept and it belongs to $\{M, \dots, 1\} \subset \mathbb{N}$ for any statement not favoring the concept, where M is the number of points of the scale (5 or 7) and $\mathbb{N}$ is the set of natural numbers. The jth concept is often measured through K_j items (variables), forming a battery and semantically connected to it. Each item, k, has a Likert scale with M_k modalities, in general, but often M_k is the same for all items. The answer of the ith respondent gives an outcome x_{ijk} in $(1, 2, 3, 4 [, 5, 6, 7])$. The sum (x_{ij}) or the mean $(\bar{x}_{ij})$ of the K_j natural numbers yields a measure of the intensity of the jth concept

$$\text{(a)} \quad x_{ij} = \sum\nolimits_{k=1}^{K_j} x_{ijk} \quad \text{or}$$

$$\text{(b)} \quad \bar{x}_{ij} = (1/K_j) \sum\nolimits_{k=1}^{K_j} x_{ijk} \tag{1}$$

The sum is sometimes rescaled to one (or ten), y_{ij}, through the expression $y_{ij} = (x_{ij} - x_{\min, j})/(x_{\max, j} - x_{\min, j})$, for the ith individual and the jth concept, where the $x_{\min, j}$ and $x_{\max, j}$ are, respectively, the maximum and the minimum of x_{ij} in the data set. However, this calculus is not admissible as the average and the sum because the device generates only ordinal data.

The semantic differential scale is another ordinal scale [14, 15] and in its usual or standard format, it consists of a set of seven categories, but they may vary in number, associated with bipolar adjectives or phrases. For each bipolar item, the respondent indicates the extent to which one descriptor represents the concept under examination. The semantic differential scale is aimed at measuring direction (with the choice of one of two terms, such as 'useful' or 'useless') and extent/amount (by selection of one of the provided categories expressing the intensity of the choice). The volume of measurements is generally high and the interpretation of the results of word scales is theoretically based on three factors ('evaluation', 'potency', and 'activity'), which involves fairly complex analyses requiring expensive data-processing procedures. Therefore, the objectives of these theoretical scales may necessarily involve long-term research, limiting their applicability or often subjecting them to simplified analysis and thus reducing some of their potential [16].

The Stapel scale is a ten-point non-verbal rating scale, ranging from +5 to −5 without a zero point and measuring direction and intensity simultaneously. It has been clearly stated that *"it cannot be assumed that the intervals are equal or that ratings are additive"* [17], but the Stapel scale is often used under the same assumptions as the Likert scale. With respect to semantic differential, the Stapel scale presents (measures) each adjective or phrase separately and the points are identified by number (but frequently the scale positions of the semantic differential are numbered too). The use of a ten-point scale is more intuitive and common than the seven-point scale.

The self-anchoring scale is another type of ordinal scale and, in its usual or standard format, it consists of a graphic, non-verbal scale, such as the ten-point ladder scale [18, 19], where respondents are asked to define their own end points (anchors). The best is at the top, if the ladder is in vertical position (case 1), or at the right, if the ladder is in horizontal position (case 2). The worst is at bottom in the first case and on the left in the second case. It is a direct outgrowth of the transactional theory of human behavior in which the 'reality world' of each of us is always to some extent unique, the outcomes of our perceptions being conceived as ongoing extrapolations of the past related to sensory stimulation. The scale may solve some problems and biases typical of category scales, but it is often used as fixed anchoring rating scale, where the anchor of the scale is already defined, assuming, implicitly, the existence of an objective reality. However, the two strategies seem psychometrically equivalent, although there are some differences between them (see, among others, [20]).

The feeling thermometer scale was developed by Aage R. Clausen for social groups and was first used in the American National Election Survey (ANES) [21]. It was later modified by Weisberg and Rusk [22]. Basically, its format is like a segment of a 0-to-100-degree temperature scale, which reports some specific values. In the evaluation of political candidates, it was *"a card listing nine temperatures throughout the scale range and their corresponding verbal meanings as to intensity of 'hot' or 'cold' feelings was handed to the respondent"* [22].

Roughly speaking, the Stapel, self-anchoring, and feeling thermometer scales are structurally similar to thermometer scales that have a long history, although they are often ascribed to Crespi [23, 24] as cited, for example, by Bernberg [25]. However, the thermometer scales used in social sciences do not provide values on interval scales, as does a thermometer used to measure temperatures.

The Juster scale is a technique for predicting the purchase of consumer durables consisting of an 11-point scale, like a decimal scale. It is used for each question asking people to assign probabilities to the likelihood of their adopting the described behavior on that question [26, 27].

Among other ordinal scales, there is the Guttman scale, which is a method of discovering and using the empirical intensity structure among a set of given indicators of a concept. The Bogardus social distance scale measures the degree to which a person would be willing to associate with a given class of people - such as an ethnic minority [28].

There is no rationale in the practice of handling the figures assigned to the modalities of an ordinal scale as real numbers, also under the assumption of equidistance between the categories, as that distance is not necessarily equal to 1. It is possible to envisage the selection of a modality as an output of a normal variable underlying a random discriminatory process, which could justify the use of equations (1) exploiting the properties of the normal random variables. However, if the modalities are subordinate only to a relation order, the use of the sum and the mean remains problematic.

3 Student Evaluation of Teaching Activity

The students' opinions about teaching activity rose to the attention of the academic administrations in the 1920 s and some US universities such as Harvard, the University of Washington and Texas, Purdue University, and other institutions, introduced student evaluations as a standard practice [29]). Since then, many aspects have been largely investigated, such as the reliability, validity, unbiasedness, efficiency, and efficacy of SETA. Moreover, many more universities in the US and other countries have also introduced the practice of evaluating teachers and course organization.

3.1 The Course-Evaluation Questionnaire

In Italy, the evaluation of university teaching activities and research is regulated by Law no. 370 (of 19/10/1999, Official Gazette, General Series, no. 252 of 26/10/1999), which also does not allow administrations failing to comply with it to apply for certain grants. The same law established the National Committee for University System Evaluation (*Comitato Nazionale di Valutazione del Sistema Universitario*, CNVSU), replacing the Observatory for University System Evaluation. A research group of the CNVSU [30] proposed a standard course evaluation questionnaire with a minimum set (battery) of fifteen items for all universities. Each item includes the following four-point Likert scale: ① Definitely no, ② No rather than yes, ③ Yes rather than no,

Table 1 Questionnaire items with median (md), mean ($\bar{x}$), standard deviation (σ), and number of valid cases (n) for the "Economics and International Management" degree program: Academic Year 2009/2010

Questionnaire items—Total $n = 4537$	Acronym	md	$\bar{x}$	σ	n
S. II—Organization of this course					
I01: Adequacy of the Work Load required by the course	AWL	7	7	2	4500
I02: Adequacy of the Teaching Materials	ATM	7	7	2	4472
I03: Usefulness of Supplementary Teaching Activity (STA)	USTA	7	7	2	2503
I04: Clarity of the Forms and rules of the Exams	CFE	7	7	2	4429
S. III—Elements concerning the teacher					
I05: Reliability of the Official Schedule of Lectures	ROSL	7	8	2	4460
I06: Teacher Availability for Explanations	TAE	7	8	2	4442
I07: Motivation and Interest generated by Teacher	MIT	7	7	2	4449
I08: Clarity and Preciseness of the Teacher's Presentations	CPTP	7	7	2	4427
S. IV—Lecture room and resource room					
I09: Adequacy of the Lecture Room	ALR	7	7	2	4444
I10: Adequacy of the Room and Equipment for STA	ARESTA	7	7	2	2475
S. V—Background-interest-satisfaction					
I11: Sufficiency of Background Knowledge	SBK	7	7	2	4442
I12: Level of Interest in the Subject matter	LIS	7	7	2	4444
I13: Level of Overall Satisfaction with the course	LOS	7	7	2	4418
S. VI—Organization of all courses in the degree program					
I14: Adequacy of the Total Work Load of current courses	ATWL	7	6	2	4452
I15: Feasibility of the Total Organization (lect. & exams)	FTO	7	6	2	4445

④ Definitely yes (CNVSU scale). A traditional item-by-item analysis was generally carried out, using means and variances of numerical values obtained by translating the categories (or labels) into a ten-point scale as follows:① = 2, ② = 5, ③ = 7, ④ = 10, hereinafter referred to as numeric values of labels [31]. One could argue that the absence of a mid value on this ordinal scale could violate the linearity assumption and the mean and variance analysis cannot validly be used. Moreover, the meaning of the labels might not be clear to all students. Consequently, the intensities, or degree of certainty, associated with these labels often correspond to a high level of vagueness; investigations of these topics are reported elsewhere [32, 33].

In the Academic Year 2004/2005, the Committee for Technical Evaluation of the University of Modena and Reggio Emilia adopted the questionnaire proposed by CNVSU [30] and in the Academic Year 2005/2006, it introduced the online survey for SETA [34]. Some minor changes involved a slight modification of the wording of some items (to make their meaning clearer) and the order of the items (to reduce the halo effect). The overall questionnaire contained seven sections, but Sections I (*personnel data*, containing information about the course, teacher, and some student characteristics) and VII (*remarks and suggestions*, listing nine items with dichotomous choices) are not presented here.

Sections from II to VI represent the core of the questionnaire and contain a 15-item battery with the four-point Likert scale to achieve the standard evaluation (Table 1).

The current procedure generates the evaluation of a single item or domain, which is performed using the traditional procedure of averaging (over the sample) the numerical labels or the assigned values corresponding to a ten-point scale

$$\bar{x}_{jk} = (1/n) \sum_{k=1}^{n} x_{ijk} \qquad (2)$$

Although the items included in a domain are assumed to have the same importance in the arithmetic mean, one could argue that certain variables indicating the efficiency of the course are more important than others. To take into account these differences, one approach is to use a weighted average. Still, the choice of the weights is a controversial point due its arbitrariness. Moreover, it could be noted that the median, which is the correct statistical index for ordinal variables, is less informative of the mean in summarizing the distribution of the answers, as is evident, but questionable, from the data reported in Table 1 for the academic year 2009/2010 in the Economics and International Management degree program of the Faculty of Economics.

4 Background of Fuzzy Inference System

The structure and functioning of a FIS follow a sequence of hierarchical steps [35, 36] involving: *(i)* issue identification, *(ii)* fuzzification of input variables, *(iii)* block rules construction, *(iv)* block rules aggregation, *(v)* defuzzification of output variable(s), and *(vi)* model tuning. They will be illustrated through an empirical application to the data described above.

4.1 The Fuzzy Inference System for Teaching Evaluation

Issue identification (i) involves a stepwise procedure that could be carried out through a top-down or bottom-up strategy. In the first case, it is similar to a scientific inquiry. It starts from the output variables and attempts to identify macro-indicators—possibly including multiple input variables—that can adequately explain the output. The macro-indicators are subsequently broken down into smaller indicators that include fewer variables. The stepwise process continues until the single input variables are isolated. The final product is a modular tree-patterned system, where several fuzzy modules are interlinked. In the bottom-up strategy, the input variables are already available, as is the case examined here, and the goal consists in subsequently aggregating them. Each aggregation generates a fuzzy module. The various modules are interlinked with each other. The final emerging arrangement is like a tree-patterned structure generating a single (or multiple) output(s). An example of the latter is the model considered for SETA, which includes the 15-item battery and item aggregation, from input to final output (Fig. 1). The variables enter the system at different levels of importance, which heavily affects the final output. Roughly speaking, the approach corresponds to a weighted average, where the weights are unknown and higher for variables entering the last steps. The aggregating function is generally unknown and not necessarily linear, as in a weighted average. In the FIS, each aggregation of variables gives an intermediate solution, which is a fuzzy set variable and it does not have necessarily a particular meaning attached to it. Sometimes, however, the intermediate variables do have a useful meaning. For example, the aggregation of AWL and ATM generates the new intermediate fuzzy variable OMW (Organization of Materials and Workload), while the aggregation of USTA and CFE generates the new intermediate fuzzy variable OEE (Organization of Exercises and Exams). In a subsequent step, the aggregation of OMW and OEE generates OC (Organization of the Course). The merging of fuzzy modules will continue up to the aggregation of OC with TTC (Total Teaching Capability), obtaining SET (Student Evaluation of Teacher). To account for student satisfaction, SET is aggregated with LOS, the level of overall satisfaction of students, obtaining SETS (Student Evaluation of Teacher plus Satisfaction). The adopted level of inclusion involves a strong influence of satisfaction on SETS, while a mitigation of its effect could be obtained, firstly, by combining TTC or OC with LOS and, secondly, by merging the result with OC or TTC, as reaffirmed in the comments on the results reported below.

The fuzzification of input, (ii), involves the specifications concerning the shapes and the number of the membership functions (mf) for the input variables. A membership function defines the extent to which each value of a numerical variable belongs to some specified categorical labels. In the following, we briefly describe the most popular approaches used to determine their shapes [37, 38]. (1) The survey approach determines the shapes of the membership functions based on information from a specifically designed sample. For this purpose, a common choice is to use empirical sampling distributions from the particular collected data concerning the intensity of the value labels. (2) The comparative judgment approach defines the functions

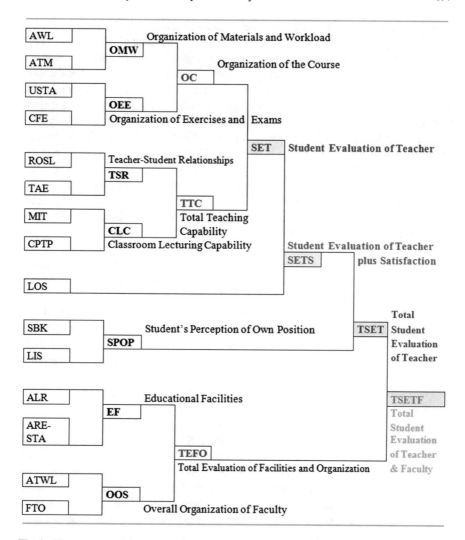

Fig. 1 The structure of the Fuzzy Inference System for student evaluation of teaching activity

through a comparison of stimuli and some given features or prototypes. (3) The expert scaling approach characterizes the function in accordance with the subjective experience of an expert. (4) The formalistic approach selects functions with specific mathematical properties. (5) Machine-learning builds up the functions from a set of past data (training and testing data set) and transfers the same structure to the present and future.

The aim of SETA is to collect student opinions about the teachers and the course organization. Therefore, approach (1) is preferable because it allows for direct measurement of the meaning that students attribute to the linguistic options/categories on the 15-item battery. In particular, each sampled student should assign scores for each category of the CNVSU scale denoted by a label, for each of the 15-items. This permits construction of frequency polygons, which give approximate representations of the vagueness level of the category choices. The polygons translate the decimal value of each category into a corresponding membership level for the population. This strategy could be costly, as all surveys are costly, and it presents some degree of difficulty due to the nuisance, laboriousness, and repetitiveness of the task. Actually, each student should assign scores for 15×4 elements. In fact, the evaluation of the same four response options is repeated fifteen times, the number of items on the CNVSU questionnaire. As a consequence, the procedure requires a large sample of students and a well-designed strategy for data collection. Moreover, given that the empirical frequency polygons are somewhat irregular, their final shapes could be determined taking into account method (2) in order to smooth and simplify the forms of the frequency polygons with respect to both the theoretical constraints and the aims of the FIS. In general, the frequency polygons are well fitted by probability distribution, such as normal, gamma, and beta [39], but they are also well approximated by triangular (a, α, β) or trapezoidal (a, b, α, β) shapes centered about the means of the score distributions for each modality. With respect to other strategies, approach (3) is inadequate for teaching evaluation because the experts' opinions might not match those of students. Approach (4) does not fit our purposes simply because a priori mathematical properties do not necessarily fit the reality of our data (although in many situations they are recommended as they produce smoothed and tractable functions). Finally, approach (5) also appears to be inappropriate because past data on the numerical values for the scale categories are not available and are probably not time-invariant.

The membership functions of some input variables for the FIS, in Fig. 1, could be deduced from the survey carried out in October 2000, where the modalities of eight items were evaluated by students using a decimal scale [32, 33]. Only eight items (seven of them about the teacher) were available out of fifteen, but the scores refer to different formats and ten years have already passed. Therefore, a simple fuzzification was adopted, assuming as membership functions trapezoidal or triangular shapes and considering the symmetry about some specific values in the decimal scale range or the value labels of the modalities (*inter alia* [40–43]). In fact, the relative frequency polygons could be well approximated by triangular shapes centered about the means of the score distributions for each category. The triangularization of the membership functions of the input variable is a common practice, but it involves a right-angled triangle for the first and the last modality, i.e., the first and the last triangle. Hence, the following types of fuzzification were considered. The first type used three membership functions (mf3); this number is lower then the number of modalities (4) to allow the activation of two membership functions for the internal modalities. The triangular fuzzy numbers (a, α, β) had peaks $a \in \{2, 6, 10\}$ and the left width α was equal to the right width β, i.e., $\alpha = \beta = 4$, as in Fig. 2 generated by "fuzzyTECH" [44]. The domain of the membership functions ranged from 2 to 10 and the response of the FIS was restricted to the interval [2, 10] because the traditional means of the numerical values attributed to the CNVSU scale categories, $D = \{2, 5, 7, 10\}$,

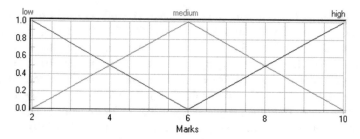

Fig. 2 Fuzzification of an input variable using three triangles as membership functions (mf)

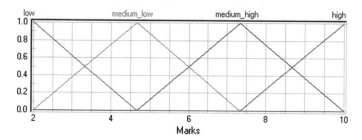

Fig. 3 Fuzzification of an input variable using four triangles as membership functions (mf)

clearly ranged from 2 to 10. Note that there are also many other possible choices and modifications of the shapes for improvement of the performances of the FIS.

There is also the possibility of using trapezoidal fuzzy numbers.

For example, the first membership function is a trapezoidal fuzzy number, (a, b, α, β), with peak $(a = 2, b = 3)$, left width $\alpha = 0$ and right width $\beta = 3$. The second function is triangular shaped (a, α, β), with peaks $a = 6$, left width $\alpha = 3$, and right width $\beta = 3$. The third, which is also the last, would be a trapezoidal fuzzy number (a, b, α, β) again, with peak $(a = 9, b = 10)$, left width $\alpha = 3$ and right width $\beta = 0$. The structure with two trapezoids, at the extremes of the decimal scale, could emphasize the FIS scores towards the upper and lower bounds of the support in some defuzzification methods [33]. However, it was not used here.

The second type used four membership functions (mf4). This number was equal to the number of modalities, implying that, in the absence of any other information out of the symmetry and the range of numeric values of the CNVSU scale categories, the triangular fuzzy numbers (a, α, β) had peaks coinciding with the numeric values $a = (2, 4.\bar{6}, 7.\bar{3}, 10)$ and the left width α was equal to the right width β, i.e., $\alpha = \beta = 2.\bar{6}$, as generated by "fuzzyTECH" [44] in Fig. 3. In this case, the most natural pattern may appear to be a fuzzification with peaks in $a = (2, 5, 7, 10)$ and different values of the left width α and the right width β. In other terms, the membership function associated with a fixed modality is represented by a triangle with the peak centered on its value on the scale and the amplitude ranging from the first lower to the first upper modality. However, it may tend to confine the results strictly to the selected modalities and the FIS does not work completely, but only through the rule-blocks, thus partially losing its nature.

In formal terms, the following could be stated. The indices j and k of x_{ijk} are summarized in a single index l to simplify the formalism. Therefore, let x_{il}, $l = 1, \ldots , L$, be the input variables ($L = 15$ in the examined case) provided by the i–th student with range U_l and let y be the output variable with range V. Let $M(l)$ be the number of categories of x_l. Generally, such a number could change from one variable to another, but in this case, $M(l) = 4$ (the number of CNVSU scale options/categories) for all $l = 1, \ldots , L$. Therefore, in general, an effective fuzzification of input requires a number of membership functions greater than one and less than $M(l)$. Moreover, each category of x_l is described by a fuzzy number, $A^l_{j(l)}$, $\forall j(l) \in [1, \ldots , M(l)]$, and the set $A^l = \{ A^l_1, \ldots , A^l_{M(l)} \}$ denotes the fuzzy input x_l, while the fuzzy output of y is defined by $B = \{ B_1, \ldots , B_{M(y)} \}$ where $M(y)$ denotes the number of membership functions (or categories or modalities) for y. Each set has a membership function:

$$\mu_{A^l_{j(l)}} (x) : \; U_l \; \rightarrow \; [0, 1] \qquad \mu_{B_m}(x) : \; V \; \rightarrow \; [0, 1] \qquad (3)$$

The construction of rule-blocks, (iii), concerns the relationships between the input linguistic variables and output linguistic variables. It involves a multicriteria situation, described by a number of rules like:

$$R_s : \; \text{IF } [x_1 \text{ is } A^1_{j(1)} \otimes \ldots \otimes x_L \text{ is } A^L_{j(L)}] \quad \text{THEN } (y \text{ is } B_m) \qquad (4)$$

for all combinations of $j(l) \in [1, \ldots , M(l)]$ and $m \in [1, \ldots , M(y)]$. The left-hand side of *THEN* is the antecedent (protasis or premise) and the right-hand side is the consequent (apodosis or conclusion). The symbol $\otimes$ (otimes) denotes an aggregation operator, one of several t-norms (if the aggregation is an *AND* operation) or t-conorms (if the aggregation is an *OR* operation). The aggregation operator, *AND*, that was chosen, produces a numerical value $\alpha_{s,m} \in [0, 1]$ and the latter represents the execution of the antecedent in rule R_s. The $\alpha_{s,m}$ number should be applied to the consequent membership function of B_m, in order to calculate the output of each rule. The *AND* aggregation operator was used once again, but in a slightly different context: $\otimes$ works on a number and the membership function of a fuzzy set B_m, whereas in the case of the R_s rule, it is applied on two numbers [45, 46]. An example of the rule-block is presented for the fuzzy module OMW (organization of material and workload) aggregating AWL and ATM (Fig. 1) and using the numeric values instead of labels, for the sake of brevity (three membership functions):

- IF AWL is mf1 and ATM is mf1, THEN OMW is mf1
- IF AWL is mf1 and ATM is mf2, THEN OMW is mf2
- IF AWL is mf1 and ATM is mf3, THEN OMW is mf3
- IF AWL is mf2 and ATM is mf1, THEN OMW is mf2
- IF AWL is mf2 and ATM is mf2, THEN OMW is mf3
- IF AWL is mf2 and ATM is mf3, THEN OMW is mf4
- IF AWL is mf3 and ATM is mf1, THEN OMW is mf3
- IF AWL is mf3 and ATM is mf2, THEN OMW is mf4
- IF AWL is mf3 and ATM is mf3, THEN OMW is mf5.

It is possible to generate these rules automatically through an algorithm, but an expert may express them in a form that more closely fits the reality.

The aggregation of rule-blocks, (iv), is the step of the evidential reasoning incorporating the process of unification of the outputs of all the rules in a single output Y. For every rule (R_s) involved in the numerical inputs, $\mu(\alpha_{s,m} \otimes B_m)$, a different output is obtained. These membership functions of fuzzy sets have to be aggregated by an *OR* operation using a t-conorm. The most frequently used are the max conorm, the probabilistic conorm, and the Lukasiewicz conorm usually known as the bounded sum. Now, the response of a module is ready, but it is still in a fuzzy form. If the module needs to be aggregated with other modules, the aggregation process continues. Otherwise, it is an output module, even if it is not the last output module, implying that it needs to be changed back into a number to provide an easy understanding of the system response.

Defuzzification of output, (v), is the process that maps the output fuzzy set $\mu_B(y)$ into a crisp value, y_{ij}, for the ith student and jth fuzzy module or concept; i.e., it concentrates the vagueness expressed by the polygon resulting from the activated output membership functions into a single summary figure that best describes the central location of an entire polygon. There is no universal technique to perform defuzzification, i.e., to summarize this output polygon by a number, as each algorithm exhibits suitable properties for particular classes of applications [47]. The selection of a proper method requires an understanding of the process that underlies the mechanism generating the output and the meaning of the different possible responses on the basis of two criteria: the "best compromise" and the "most plausible result". Moreover, in an ordinal output, with modalities described by linguistic expressions, their corresponding real values are always given by the membership definitions, where the understanding of their meaning plays a key role.

For the first criterion, one of the most popular methods is the Center-of-Maxima (CoM), which yields the best compromise between the activated rules [45]. Given that more than one output membership function could be activated or evaluated as a possible response for the ith student and jth fuzzy module or concept, let $M_{F;ij}$ be the number of output-activated membership functions. Let y_{ijm} be the abscissa of the maximum in the mth activated output membership function. If the latter has a maximizing interval, y_{ijm} will be the median of this interval. The final output crisp value, $y_{CoM;ij}$, is given by an average of membership maxima weighted by their corresponding level of activation, $\mu_{out;k}$,

$$y_{CoM;ij} = \sum_{m=1}^{M_{F;ij}} \mu_{out;m}\, y_{ijm} \Big/ \sum_{m=1}^{M_{F;ij}} \mu_{out;m} \tag{5}$$

The method of the center of area/gravity (CoA/G) was excluded because it cannot reach the extremes of the range [2, 10] without fuzzification of input and output on an interval wider than [2, 10], which may seem unnatural. In fact, it would have been possible to pick a more suitable fuzzification of the input, but, in general, FIS might produce an output greater than the maximum or lower than the minimum of the scale. However, with singleton membership functions, CoA and CoM methods provide the same results.

For the second criterion, the Mean-of-Maxima (MoM) method yields the most plausible result, determining the system output only for the membership function with the highest resulting degree of the support. If the maximum is not unique, i.e., it is a maximizing interval, the mean of the latter will be given as the response

$$y_{MmM;ij} = \max_{l <= m <= M_{F;ij}} (y_{ijm}) \tag{6}$$

This approach selects the typical value of the terms that is most valid, instead of balancing out the different inference results [45]. Therefore, it is often used in pattern recognition and classification applications, as in the case of an ordinal output whose modalities are described by linguistic expressions, because the most plausible solution is more appropriate instead of the mean. There are many other sophisticated defuzzification strategies, but they are rarely used in practice.

In addition, the sensitivity analysis, (vi), is a possible sixth step that may be carried out to adapt the FIS to the real situations that it would represent. The FIS is handled as a parametric model, relating input variables to membership functions, to fuzzy rules, to hedges operations, to aggregations, and so on. The tuning of the performance of the FIS is carried out by means of four steps: (1) definition of the objective functions for the output fuzzy variables; (2) changes of parameters for input data, membership functions, and fuzzy rules; (3) validation of results by comparing the objective functions and output functions; (4) repetition of steps (2) and (3) until the difference between the objective and output functions is smaller than the chosen error criterion.

5 Empirical Results

The academic year 2009/2010 was fixed as the reference date and the degree program in Economics and International Management was selected out of three undergraduate degree programs of the Faculty of Economics. Some restrictions were imposed to the overall dataset of 4537 (evaluating) students (Table 1), even if some analyses suggesting such restrictions are not reported here for the sake of brevity. The course-teacher is a unique combination because the same teacher may teach more than one course and in the same course there may be more than one teacher. Five courses had less than twenty evaluating students and these courses were thus eliminated, leading to a reduction of 55 cases. In practice, the score provided by just one student is also useful for the teacher, but a sufficiently large sample is preferable for each single course-teacher to compare the mean of the value labels with the output of a FIS.

The complete elimination of nonresponses is not the ideal strategy as it is too costly, it implies a loss of cases and biases the estimates because nonresponses do not randomly occur. However, considering the nature of SETA data, if all items concerning the teacher (I01-I08, I13), or all items concerning the organization (I09, I10, I14, I15), were missing in a case, that case was dropped. Fifty-four cases were lost through this control, prevailingly owing to missing values for all items referring to the teacher. Background knowledge (I11) was used to replace the level of interest in the subject matter (I12) when the latter was missing and vice versa; if both were missing, they were replaced with the mean of teacher items. Therefore, they did not involve a loss of cases. If in the 15-item battery there were more than 8 (threshold) missing values in a single case (student), that case was dropped: 17 cases crossed the threshold. In the end, a total of 4411 cases were used.

The remaining missing values were replaced on the basis of the available data for each single case and considering that an evaluating student expressed an opinion about three main areas (Fig. 1): Student evaluation of teacher plus satisfaction (SETS), student's perception of own position (SPOP), and total evaluation of facilities and organization (TEFO). For each student, i, the kth item belonging to a certain area with a missing value was replaced with the mean of the values for the non-missing items of the same area provided by the same student and not by the mean of the kth item for the total sample, as is usual. For example, let $I02(i)$, which belongs to SETS, be missing; it was then replaced by the mean of $[I01(i), I03(i), I04(i), I05(i), I06(i), I07(i), I08(i), I13(i)]$. Let $I02(i)$ and $I13(i)$ be missing; they were then replaced by the mean of $[I01(i), I03(i), I04(i), I05(i), I06(i), I07(i), I08(i)]$. The rationale of this procedure relies on the core of the evaluation process, which is the evaluator. Therefore, the value used in the substitution is anchored to his/her average level of judgment and not to the average level of the total sample. The number of replaced values varied from one item to another, ranging from 0.1 % to 1 %, except for supplementary teaching activity (I03) and adequacy of the room and equipment for the supplementary teaching activity (I10), as those activities were not always present in a course, implying an obvious high rate of absence of evaluations for that particular item.

5.1 Student Evaluations of Teachers

The analysis has been prevailingly restricted to the subsystems concerning the student evaluations of teachers (SET) and SET plus satisfaction (SETS), as indicated in Fig. 1, for the sake of brevity and owing to the possibility to simulate the input data, as indicated below. The traditional evaluation of teachers currently in use, from the ith student, is given by the mean of the value labels assigned

Table 2 Output data: mean ($\bar{x}_{set;i}$) and fuzzy outputs for different conditions (3 or 4 membership functions in the fuzzification of input, CoM and MoM methods in the defuzzification of output)

$\bar{x}_{set;i}$	$x_{CoM3;i}$	$\bar{x}_{set;i} - x_{CoM4;i}$		$\bar{x}_{set;i} - x_{MoM3;i}$		$\bar{x}_{set;i} - x_{MoM4;i}$		$\bar{x}_{set;i} - x_{MoM4;i}$
		$x_{CoM3;i}$		$x_{CoM4;i}$		$x_{MoM3;i}$		
10.00	10.00	0.00	10.00	0.00	10.00	0.00	10.00	0.00
9.63	10.00	0.37	10.00	0.37	10.00	0.37	10.00	0.37
9.00	10.0	1.00	10.00	1.00	10.00	1.00	10.00	1.00
9.00	8.48	0.52	8.67	0.33	7.71	1.29	8.86	0.14
8.63	9.77	1.14	10.00	1.37	10.00	1.37	10.00	1.37
8.00	8.37	0.37	7.46	0.54	7.71	0.29	7.71	0.29
7.63	8.09	0.46	7.49	0.14	7.71	0.08	7.71	0.08
7.00	7.52	0.52	7.19	0.19	7.71	0.71	7.71	0.71
6.50	5.89	0.61	5.43	1.07	6.57	0.07	5.43	1.07
6.00	6.57	0.57	6.36	0.36	6.57	0.57	7.71	1.71
5.33	4.51	0.82	4.81	0.52	5.43	0.10	4.29	1.04
4.11	3.57	0.54	3.52	0.59	4.29	0.18	3.14	0.97
3.00	2.86	0.14	3.29	0.29	3.14	0.14	3.14	0.14
3.00	2.00	1.00	2.00	1.00	2.00	1.00	2.00	1.00
2.00	2.00	0.00	2.00	0.00	2.00	0.00	2.00	0.00

to the four modalities of each item:

$$(a) \quad \bar{x}_{set;i} = (x_{01;i} + \cdots + x_{08;i})/8$$
$$(b) \quad \bar{x}_{sets;i} = (x_{01;i} + \cdots + x_{08;i} + x_{13;i})/9 \tag{7}$$

Note that $\bar{x}_{set;i}$ is examined and reported in the tables below, while $\bar{x}_{sets;i}$ is referred to only in the comments.

The output of a FIS, $x_{FIS;i}$, depends on the decisions taken at each step. Specifically, the CoM method used in the defuzzification step, with 3 or 4 membership functions in the fuzzification of input, CoM3 or CoM4, generated $x_{CoM3;i}$ and $x_{CoM4;i}$, respectively. Analogously, the MoM method used in the defuzzification step, with 3 or 4 membership functions in the fuzzification of input, MoM3 or MoM4, generated $x_{MoM3;i}$ and $x_{MoM4;i}$, respectively. The fuzzy data process was carried out using the support of "fuzzyTECH" [44]. An example of the output for different values of the output range [2, 10] is given in Table 2, which also reports the differences between the mean of the value labels, $\bar{x}_{set;i}$, and the four fuzzy evaluations.

The rank of course-teachers may be a useful tool to identify critical situations, where to offer suggestions to the teacher or to urge him/her to improve his/her behavior, the scope of the program, the organization of teaching materials, exercises or exams, and so on. For this purpose, an item-by item analysis could help persons in charge of academic organization and/or teachers, but here the results are limited only to the overall evaluation of the teacher. The first and last positions of the rank are reported in Table 3. The mean of the value labels, $\bar{x}_{set}$, was lower than the fuzzy outputs ($\bar{x}_{CoM3}, \bar{x}_{MoM3}, \bar{x}_{CoM4}, \bar{x}_{MoM4}$). The CoM3 fuzzy evaluations were higher than those obtained by the mean of the value labels and the mean of differences was 0.55, with the lowest standard deviation (sd) being 0.45. The MoM3 provided crisp values that were close to CoM3 evaluations and closer to $\bar{x}_{set;i}$. In fact, the mean difference was 0.41 (sd = 0.63). Assuming the mean of the value

Table 3 First and last five teachers in the rank obtained through the mean of the value labels ($\bar{x}_{set}$) with fuzzy outputs for different conditions (3 or 4 membership functions in the fuzzification of input, CoM and MoM methods in the defuzzification of output)

Order	Teacher	n	$\bar{x}_{set}$	$\bar{x}_{CoM3}$	$\bar{x}_{MoM3}$	$\bar{x}_{CoM4}$	$\bar{x}_{MoM4}$
1	Xy01	109	8.38	8.89	8.70	8.89	9.23
2	Xy02	21	8.33	8.71	8.60	8.72	9.05
3	Xy03	168	8.30	8.87	8.81	8.93	9.19
4	Xy04	43	8.30	8.81	8.70	8.79	9.07
5	Xy05	69	8.28	8.83	8.65	8.78	9.15
...	...	...	...	...	...	...	...
37	Xy37	44	6.38	6.82	6.94	6.55	6.67
38	Xy38	114	6.36	6.82	6.91	6.53	6.85
39	Xy39	90	6.33	6.78	6.68	6.56	6.92
40	Xy40	90	6.22	6.59	6.77	6.38	6.53
41	Xy41	121	6.16	6.61	6.56	6.32	6.73
	Total	4411	7.32	7.87	7.73	7.81	8.18

labels as a benchmark, $\bar{x}_{set;i}$, the differences proved to be slightly higher than 5 %, on the average. Moreover, the fuzzification with 4 membership functions did not work as well as the fuzzification with 3 membership functions because the outputs generally showed an increase in the differences with respect to $\bar{x}_{set;i}$. Opposite results were obtained by CoM4 and MoM4, i.e., the CoM4 fuzzy evaluations yielded crisp values closer to $\bar{x}_{set;i}$ than those yielded by MoM4. In fact, the means of the differences with respect to $\bar{x}_{set;i}$ were 0.49 (sd = 0.64) and 0.86 (sd = 0.99), respectively. Note that the fuzzification "centered" on the modalities tended to reduce the vagueness and to restrict the output to a single modality, as the effective fuzzy mechanism was not activated and worked prevailingly only with the rule-blocks.

The fuzzy outputs, $x_{FIS;i}$, and the mean of the value labels, $x_{set;i}$, for the ith student measure the same concept, i.e., the performance of a teacher. Therefore, they should be correlated and an analysis of the relationships between the different fuzzy outputs and $\bar{x}_{set;i}$ clarifies the structure of some differences. The scatter-plots of fuzzy evaluations of teachers (SET) against the mean of the corresponding value labels are reported in Fig. 4. The estimates of the linear regression parameters between the four dependent variables ($x_{CoM3;i}, x_{MoM3;i}, x_{CoM4;i}, x_{MoM4;i}$) on $\bar{x}_{set;i}$, as the independent variable, are reported in Table 4 and the corresponding residuals are plotted in Fig. 5. If $x_{FIS;i}$ and $\bar{x}_{set;i}$ are the same, one can expect a slope (β_1) of the regression line equal to 1 and an intercept (β_0) equal to 0: in the tables their estimates are denoted b_1 and b_0, respectively. The results of the t-tests on estimated slopes and intercepts showed that these hypotheses were always rejected, but the relationships were always approximately linear. The assumption of constant variance was refused in all models and the coefficients of determination were sufficiently high.

The result closer to the hypotheses, notwithstanding their rejection, was given by CoM3, which showed a distribution of residuals with an undesirable shape, but more concentrated than other fuzzy outputs having the lowest standard deviation: sd(res)=0.45. MoM3 provided a distribution of residuals that was more acceptable than that of CoM3, but it was also slightly more dispersed: sd(res)=0.63. Correspondence between the mean of the value labels and the fuzzy outputs by CoM4 and MoM4 was poorer in terms of the slope and the shapes or dispersion of residuals: sd(res-CoM4)=0.58 and sd(res-MoM4)=0.95. Therefore, as expected, FIS works better when the number of membership functions for each x_{il} input variable is lower than its number of modalities, $M(l)$. Reasonably, the number of input membership functions for the x_{il} input variable should range from

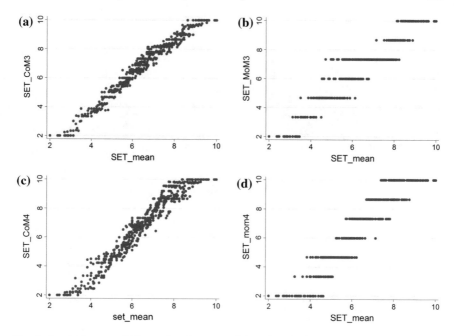

Fig. 4 Output fuzzy variables against the mean of the value labels. **a** SET_{CoM3}. **b** SET_{MoM3}. **c** (c) SET_{CoM4}. **d** SET_{MoM4}

Table 4 Parameter estimates for the regression of fuzzy outputs on the mean of the value labels (real data)

Dependent	b_0	$SE(b_0)$	$t(b_0 = 0)$	b_1	$SE(b_1)$	$t(b_1 = 1)$	R^2	Het^a
SET CoM3	0.367	0.031	11.72	1.026	0.004	6.12	0.932	0
SET MoM3	0.848	0.044	19.34	0.940	0.006	−10.28	0.854	0
SET CoM4	−0.711	0.041	−17.53	1.164	0.005	30.38	0.913	0
SET MoM4	−0.426	0.066	−6.41	1.176	0.009	19.83	0.799	0

[a]Breusch-Pagan test for heteroskedasticity, where H_0 is constant variance

2 to $[M(l) - 1]$. For fuzzy output ordinal variables, MoM is more suitable than CoM because it chooses the most plausible result among the possible $M_{F;ij}$ results, while, in general, a weighted average of the possible M_F; ij results have no meaning for an ordinal variable.

5.2 Simulated Data: All Possible Input Patterns

The mean of the value labels and of the FIS outputs yielded measurements that often did not coincide, as noted regarding the results in Tables 3–4. The differences between the mean of the FIS outputs and the mean of the value labels were statistically different from zero for both the total sample and the single course teacher. However, the surveyed data did not present all the possible

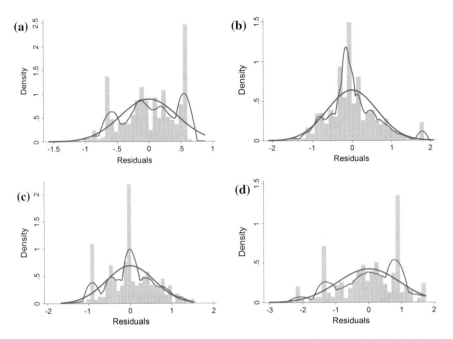

Fig. 5 Distributions of residuals for the four regression models. **a** 3-input membership functions/ CoM. **b** 3-input membership functions/MoM. **c** 4-input membership functions/CoM. **d** 4-input membership functions/CoM

combinations of input values because many evaluation patterns were frequently repeated and others were never expressed by students. Therefore, the previous analysis was repeated using a simulated dataset, which contained all the possible combinations of the values of the input variables.

The generation of the dataset considered the output termed SETS in Fig. 1, although the more attention was focused on SET. Given that for SETS there were nine input variables and there were four modalities for each variable, the various possible combinations were given by four raised to nine (or 4 to the 9th power) equal to 262144. Each combination corresponded to an evaluation of a potential student, which was different from the other 262143.

In the simulated dataset, the input variables are perfectly uncorrelated to each other, as each pattern appears once, while in the surveyed datasets there are often a correlation because the input variables are like paired variables or repeated measurements. The differences between the fuzzy outputs and the mean of the value labels have been plotted in Fig. 6. Differing from the above results, CoM3 showed a distribution of residuals with an acceptable shape and it was more concentrated than other fuzzy outputs. The resulting differences were less marked than those observed in the surveyed data: $x_{CoM3} - \bar{x}_{set} = 0.22$ (sd=0.44), $x_{MoM3} - \bar{x}_{set} = 0.28$ (sd = 0.65), $x_{CoM4} - \bar{x}_{set} = 0.20$ (sd=0.78), $x_{MoM4} - \bar{x}_{set} = 0.20$ (sd = 0.93).

Analogously, the parameters of the regression between the fuzzy outputs ($x_{CoM3;i}$, $x_{MoM3;i}$, $x_{CoM4;i}$, $x_{MoM4;i}$), as dependent variables, on the mean of the value labels, $\bar{x}_{set;i}$, as the independent variable, were estimated (Table 5) and the corresponding residuals are shown in Fig. 7. The usual hypotheses about the parameters were tested, with the slope equal to 1 and the intercept equal to 0, and rejected. Again, the assumption of constant variance was refused in all models and the coefficients of determination were sufficiently high. In a different direction, the result closest to the hypotheses, notwithstanding their rejection, was given by MoM3, which showed a distribution of residuals with an acceptable shape, though less concentrated [sd(res − MoM3) = 0.61] than in the case of CoM3 [sd(res − CoM3) = 0.32]. Moreover, CoM3 showed a bimodal

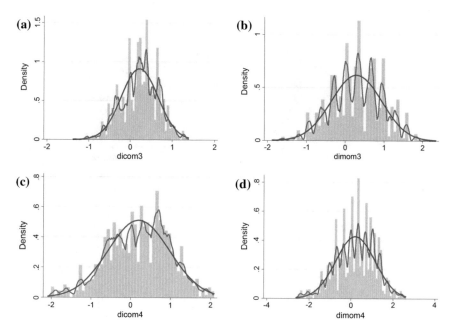

Fig. 6 Distributions of differences between the fuzzy outputs and the mean of the value labels ($\bar{x}_{set;i}$). **a** 3-input membership functions/ CoM. **b** 3-input membership functions/MoM. **c** 4-input membership functions/CoM. **d** 4-input membership functions/CoM

Table 5 Parameter estimates for the regression of fuzzy outputs on the means of the value labels (simulated data)

Dependent	b_0	$SE(b_0)$	$t(b_0 = 0)$	b_1	$SE(b_1)$	$t(b_1 = 1)$	R^2	Het^a
SET CoM3	−1.540	0.004	−420.2	1.293	0.001	487.1	0.946	0
SET MoM3	−0.947	0.007	−134.6	1.205	0.001	177.7	0.806	0
SET CoM4	−3.622	0.006	−562.6	1.577	0.001	605.5	0.913	0
SET MoM4	−3.623	0.008	−473.1	1.637	0.001	506.7	0.866	0

[a]Breusch-Pagan test for heteroskedasticity, where H_0 is constant variance

distribution. As for the surveyed data, the correspondence between the mean of the value labels and the fuzzy outputs by CoM4 and MoM4 was poorer in the slope, but CoM4 showed a better coefficient of determination and shape of the histogram of residuals than MoM4: sd(res − CoM4) = 0.50, sd(res − MoM4) = 0.66. However, this was not the same for SETS, in which the coefficients of determination decreased by about 50 %. These substantial differences mainly depended on the structure of the tree reported in Fig. 1, where satisfaction (LOS) is combined directly with SET involving a high weight of LOS on the fuzzy output for SETS and the absence of correlation between LOS and $\bar{x}_{set;i}$ increased the reduction of the determination coefficients. In fact, with the surveyed data, this reduction was not observed because it was negligible, for in that case, LOS was correlated with other input variables and the output of various fuzzy modules. A possible solution to remedy this discrepancy could consist in the combination of LOS with a fuzzy module in a previous node of the tree illustrated in Fig. 1; for example, LOS could be grouped with CLC, TSR or TTC.

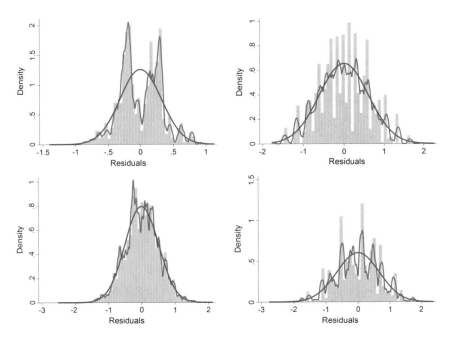

Fig. 7 Distributions of residuals for the four regression models estimated using the simulated dataset

Many other decisions should be made in steps (*iii*) and (*iv*), the construction and aggregation of rule-blocks, which could change the output of a FIS, but those examined here are structurally issue-/ situation-/ dependent and represent the more important decisions for any FISs.

6 Comments and Remarks

FIS offers the possibility of handling verbal terms via approximately quantitative values and avoids some methodological issues inherent in traditional procedures concerning the measurement of concepts and the consequent limitation of statistical data analysis of ordinal, but also nominal variables. For example, the use of the mean (sample average) becomes irrelevant because the response of the FIS could be maintained as ordinal. However, if a numerical output is desired, as in the case-study presented here, then many problems still hold conceptually, but some of them are operatively irrelevant as the vagueness weakens the sharpness. In other words, by construction, the nature of the fuzzy inputs mitigates the certainty that we would normally have about the distances of the numbers on a given scale. Therefore, the issue concerning the value attributed to a modality (e.g. 7 assigned to "⑤ Yes rather than no" leading to questions like "Why 7 and not 7.5 or 6.5 or 8 or 6") is less restrictive because the fuzzification spreads the choice over the support, even if all choices affect the output. In any case, the FIS-based approach could represent a bridge between qualitative and quantitative analysis for a consistent treatment of concepts that are measured differently. This potentiality would be useful in many fields of application.

In the social sciences an attribute or concept is often measured by a scale, i.e., a battery of several items, which are generally statements based on a logical or empirical structure and semantically linked to the attribute to be measured [28]. Each item provides a set of ordered options (Likert or thermometer type) for the responses of individuals and the intension of the attribute is evaluated through the sum or the average of the numeric values of labels corresponding to the selected options from the set of options provided for each item of the battery, as indicated in Eq. 1. However, these latter equations violate the assumption of nonlinearity defining an ordinal variable [48], implying a transition of an ordinal scale to an interval scale. In fact, the operations involved in Eq. 1 are not admissible for ordinal variables, as the items constituting the battery are measured at the ordinal level. The opposite point of view asserts that "the numbers do not know where they come from" [49], suggesting the possibility for a wide use of any mathematical and statistical technique for ordinal data analysis. There are scientists who opt for a compromise solution and are more open to use of the parametric approach as an exploratory strategy for analysis of the ordinal variables to discover and to describe some structural relationship for the phenomena under observation, serving at least as orientation and not as an exact impact evaluation [4, 50]. Even the original propounder of this theory of measurement wrote: "for this 'illegal' statisticizing there can be invoked a kind of pragmatic sanction: in numerous instances it leads to fruitful results" (p. 26) [51]. FIS does not present any problems in the application of the analogous operations of Eq. 1 and the final results belong to the same level of measurement, i.e., it provides an ordinal response, which is a true and valuable advantage, although this response does not bring out the ordinal level of measurement as it depends on the starting domain. If the starting domain is defined through meaningless numbers, the domain of the FIS outcome will also have meaningless numbers. This topic still requires much work and calibration as concerns the isomorphism between the numbers and the intensions of an attribute.

FIS, however, also presents difficulties at each construction step. In the identification of the issue (step *i*), the order in which the input variables are aggregated in the system affects the output. Particularly, input variables in the first nodes of the tree affect the output less than those forming the subsequent nodes. Moreover, the exponential explosion of the number of rules limits the input of fuzzy modules to two or three variables. The fuzzification of input (step *ii*) is not a straightforward step and leaves a kind of indeterminacy. The construction of block rules (step *iii*) is a subjective process open to criticism by all, as there is no rule to make rules. In fact, the heuristic fuzzy rules constitute a controversial issue. Certainly, the flexibility derived from these rules allows for adequately representing the actual phenomenon, but for this same reason, the choices of the decision-maker play a key role in the pattern of combinations involving the wording of the items. There are many methods and possibilities for the aggregation of block rules (step *iv*), but they must be selected on the bases of the knowledge of their functioning. Defuzzification (step *v*) also offers a large variety of techniques that might puzzle final users, although it can be seen as not being a part of the core of a FIS or of the fuzzy set theory [47].

Overall, the FIS generates reasonable and reliable results, showing remarkable flexibility and more manageability than the official evaluation systems, in spite of discrepancies with respect to the means of the value labels, which are the official results used by the persons in charge of academic organization. However, part of this manageability could originate from the arbitrary choices required by the construction steps, especially from the heuristic fuzzy rules (*if − then* rules) and from the fuzzy inference method (selection of aggregation's operators for precondition and conclusion). Despite some unavoidable degree of arbitrariness in some modeling choices, the results were satisfactory. The final outcomes resembled those of the traditional procedure, but the values were slightly higher than those of the official evaluations. The procedure illustrated above represents the initial steps to construct an FIS of teaching activity that is adequate and accounts for the complexity and the multiform aspects of the observed phenomenon. The results point out that even a raw and approximate system, like the four-point Likert-like scale, can provide reasonable evaluations of the courses. This implies an advantage of this procedure because public statistics tend to prefer an objective and transparent strategy, since the method of calculation is known in advance and does not involve subjective intervention. On the other hand, the FIS could also be transparent if the subjective decisions are made known to the final users and everyone accepts them. However, the procedure is more complicated and it is difficult to comprehend all the details of its functioning. Therefore, a public institution may prefer a procedure characterized by a transparency and simplicity of the strategy to be applied.

References

1. Coombs, C.H.: Theory and methods of social measurement. In: Festinger, L., Katz, D. (eds.) Research Methods in the Behavioral Sciences, pp. 471–535. New York, Dryden (1953)
2. Linneman, T.J.: Social Statistics: The Basics and Beyond. Routledge, New York (2011)
3. Agresti, A.: Categorical Data Analysis. Wiley, New York (2002)
4. Amemiya, T.: Qualitative response models: a survey. J. Econ. Lit. **XIX**, 1483–1538 (1981)
5. Greene, W.H.: Econometric analysis. Pearson Education India, New Delhi (2003)
6. Zadeh, L.A.: Fuzzy sets. Inf. Control **8**, 338–353 (1965)
7. Stevens, S.S.: On the theory of scales of measurement. Science **103**, 677–680 (1946)
8. Thurstone, L.L.: A law of comparative judgment. Psychol. Rev. **34**, 273–286 (1927a)
9. Thurstone, L.L.: The method of paired comparison for social values. J. Abnorm. Soc. Psychol. **21**, 384–397 (1927b)
10. Thurstone, L.L.: Attitudes can be measured. Am. J. Sociol. **33**, 529–554 (1928)
11. White, M.: Psychological technique and social problems. Southwest. Polit. Soc. Sci. Q. **2**, 58–73 (1926)
12. Likert, R.: A technique for the measurement of attitudes. Arch. Psychol. Monogr. **140**, 1–50 (1932)
13. Schuman, H., Presser, S.: Questions and Answers in Attitude Surveys: Experiments on Question Form, Wording, and Context. Sage, Thousand Oaks (1996)
14. Osgood, C.E.: The nature and measurement of meaning. Psychol. Bull. **49**, 197–237 (1952)
15. Osgood, C.E., Suci, G.J., Tannenbaum, P.H.: The Measurement of Meaning. University of Illinois Press, Urbana (1957)
16. Yu, J.H., Albaum, G., Swenson, M.: Is a central tendency error inherent in the use of semantic differential scales in different cultures? Int. J. Mark. Res. **45**, 213–228 (2003)
17. Crespi, I.: Use of a scaling technique in surveys. J. Mark. **25**, 69–72 (1961)
18. Cantril, H., Free, L.A.: Hopes and fears for self and country: the self-anchoring striving scale in cross-cultural research. Am. Behav. Sci. **6**, 4–30 (1962)
19. Kilpatrick, F.P., Cantril, H.: Self-anchoring scaling: a measure of individuals' unique reality worlds. J. Individ. Psychol. **16**, 158–173 (1960)
20. Hofmans, J., Theuns, P., Van Acker, F.: Combining quality and quantity. a psychometric evaluation of the self-anchoring scale. Qual. Quant. **43**, 703–716 (2009)
21. ANES: American National Election Studies 1964: Pre-post election study. Survey Research Center (S473). http://www.electionstudies.org/studypages/1964prepost/int1964.txt (1964). Accessed 12 Feb 2015
22. Weisberg, H.F., Rusk, J.G.: Dimensions of candidate evaluation. Am Polit. Sci. Rev. **64**, 1167–1185 (1970)
23. Crespi, L.P.: Public opinion toward conscientious objectors: Ii. measurement of national approval-disapproval. J. Psychol. **19**, 209–250 (1945a)
24. Crespi, L.P.: Public opinion toward conscientious objectors: Iii. intensity of social rejection in stereotype and attitude. J. Psychol. **19**, 251–276 (1945b)
25. Bernberg, R. E.: Socio-psychological factors in industrial morale; I, the prediction of specific indicators. J. Soc. Psychol. **36**, 73–82 (1952)
26. Juster, T.F.: Prediction and consumer buying intentions. Am. Econ. Rev. **50**, 604–617 (1960)
27. Juster, T.F.: Consumer buying intentions and purchase probability: an experiment in survey design. J. Am. Stat. Assoc. **61**, 658–696 (1966)

28. Babbie, E.R.: Introduction to Social Research. Cengage Learning, Wadsworth (2010)
29. Marsh, H.W.: Students' evaluations of university teaching: research findings, methodological issues, and directions for future research. Int. J. Educ. Res. **11**, 253–388 (1987)
30. CNVSU: Proposta di un insieme minimo di domande per la valutazione della didattica da parte degli studenti frequentanti. CNVSU, Doc 09/02, Rome, Retrieved from http://www.cnvsu.it (2002). Accessed 28 July 2011
31. Chiandotto, B., Gola, M.M.: Questionario di base da utilizzare per l'attuazione di un programma per la valutazione della didattica da parte degli studenti. Rapporto finale del gruppo di ricerca (RdR 01/00), CNVSU, Rome, Retrieved from http://www.cnvsu.it (2000). Accessed 28 July 2011
32. Lalla, M., Facchinetti, G.: Measurement and fuzzy scales. In: Atti della XLII Riunione Scientifica: Sessioni Plenarie e Specializzate, pp. 351–362. SIS—University of Bari, Bari, 9–11 June 2004
33. Lalla, M., Facchinetti, G., Mastroleo, G.: Ordinal scales and fuzzy set systems to measure agreement: an application to the evaluation of teaching activity. Qual. Quant. **38**, 577–601 (2004b)
34. Lalla, M., Ferrari, D.: Web-based versus paper-based data collection for the evaluation of teaching activity: empirical evidence from a case study. Assess. Eval. High. Educ. **36**, 347–365 (2011)
35. Dubois, D., Prade, H.: Fundamentals of Fuzzy Sets. Kluwer Academic Publishers, Boston (2000)
36. Kasabov, N.K.: Foundations of Neural Networks, Fuzzy Systems, and Knowledge Engineering. MIT Press, Cambridge (1996)
37. Smithson, M.: Fuzzy Set Analysis for Behavioral and Social Sciences. Springer, Heidelberg (1987)
38. Smithson, M.: Fuzzy set theory and the social sciences: the scope for applications. Fuzzy Sets Syst. **26**, 1–21 (1988)
39. DasGupta, A.: Fundamentals of Probability: a First Course. Springer, Heidelberg (2010)
40. Grzegorzewski, P., Mrówka, E.: Trapezoidal approximations of fuzzy numbers. Fuzzy Sets Syst. **153**, 115–135 (2005)
41. Grzegorzewski, P., Mrówka, E.: Trapezoidal approximations of fuzzy numbers—revisited. Fuzzy Sets Syst. **158**, 757–768 (2007)
42. Grzegorzewski, P.: Trapezoidal approximations of fuzzy numbers preserving the expected interval—algorithms and properties. Fuzzy Sets Syst. **159**, 1354–1364 (2008)
43. Yeh, C.T.: Weighted trapezoidal and triangular approximations of fuzzy numbers. Fuzzy Sets Syst. **160**, 3059–3079 (2009)
44. INFORM-GmbH: Manual, FuzzyTech Users. Inform Software Corporation (2007)
45. Von Altrock, C.: Fuzzy Logic and NeuroFuzzy Applications in Business and Finance. Prentice Hall PTR, Upper Saddle River (1997)
46. Zimmermann, H.J.: Fuzzy Set Theory Appl. Kluwer Academic Publishers, Boston (1996)
47. Van Leekwijck, W., Kerre, E.E.: Defuzzification: criteria and classification. Fuzzy Sets Syst. **108**, 159–178 (1999)
48. Kampen, J., Swyngedouw, M.: The ordinal controversy revisited. Qual. Quant. **34**, 87–102 (2000)
49. Lord, F.M.: On the statistical treatment of football members. Am. Psychol. **8**, 750–751 (1953)
50. Velleman, P.F., Wilkinson, L.: Ordinal, interval, and ratio typologies are misleading. Am. Stat. **47**, 65–72 (1993)
51. Stevens, S.S.: Mathematics, measurement, and psychophysics. In: Stevens, S.S. (ed.) Handbook of Experimental Psychology, pp. 1–49. Wiley, New York (1951)

Unsatisfiable Formulae of Gödel Logic with Truth Constants and Δ Are Recursively Enumerable

Dušan Guller

Abstract This paper brings a solution to the open problem of recursive enumerability of unsatisfiable formulae in the first-order Gödel logic. The answer is affirmative even for a useful expansion by intermediate truth constants and the projection operator Δ. The affirmative result for unsatisfiable prenex formulae of G_∞^Δ has been stated in [1]. In [2], we have generalised the well-known hyperresolution principle to the first-order Gödel logic for the general case. We now propose a modification of the hyperresolution calculus suitable for automated deduction with explicit partial truth.

Keywords Gödel logic · Resolution · Many-valued logics · Automated deduction

1 Introduction

Current research in many-valued logics is mainly concerned with left-continuous t-norm based logics including the three fundamental fuzzy logics: Gödel, Łukasiewicz, and Product ones. From a syntactical point of view, classical many-valued deduction calculi are widely studied, especially Hilbert-style ones. In addition, a perspective from automated deduction has received attractivity during the last two decades. A considerable effort has been made in development of *SAT* solvers for the problem of Boolean satisfiability. *SAT* solvers may exploit either complete solution methods (called complete or systematic *SAT* solvers) or incomplete or hybrid ones. Complete *SAT* solvers are mostly based on the Davis-Putnam-Logemann-Loveland procedure (*DPLL*) [3, 4] or resolution proof methods [5–7], improved by various features, [8]. t-norm based logics are logics of comparative truth: the residuum of a t-norm satisfies, for all $x, y \in [0, 1]$, $x \to y = 1$ if and only if $x \leq y$. Since implication is interpreted by a residuum, in the propositional case, a formula of

This work is partially supported by VEGA Grant 1/0592/14.

D. Guller (✉)
Department of Applied Informatics, Comenius University,
Mlynská Dolina, 842 48 Bratislava, Slovakia
e-mail: guller@fmph.uniba.sk

© Springer International Publishing Switzerland 2016
J.J. Merelo et al. (eds.), *Computational Intelligence*,
Studies in Computational Intelligence 620, DOI 10.1007/978-3-319-26393-9_13

the form $\phi \to \psi$ is a consequence of a theory if $\|\phi\|^{\mathfrak{A}} \leq \|\psi\|^{\mathfrak{A}}$ for every model $\mathfrak{A}$ of the theory. Most explorations of t-norm based logics are focused on tautologies and deduction calculi with the only distinguished truth degree 1, [9]. However, in many real-world applications, one may be interested in representation and inference with explicit partial truth; besides the truth constants 0, 1, intermediate truth constants are involved in. In the literature, two main approaches to expansions with truth constants, are described. Historically, the first one has been introduced in [10], where the propositional Łukasiewicz logic is augmented by truth constants $\bar{r}$, $r \in [0,1]$, Pavelka's logic (*PL*). A formula of the form $\bar{r} \to \phi$ evaluated to 1 expresses that the truth value of ϕ is greater than or equal to r. In [11], further development of evaluated formulae, and in [9], Rational Pavelka's logic (*RPL*)—a simplification of *PL*, are described. Another approach relies on traditional algebraic semantics. Various completeness results for expansions of t-norm based logics with countably many truth constants are investigated, among others, in [12–18].

Concerning the three fundamental first-order fuzzy logics, the set of logically valid formulae is Π_2-complete for Łukasiewicz logic, Π_2-hard for Product logic, and Σ_1-complete for Gödel logic, as with classical first-order logic. Among these fuzzy logics, only Gödel logic is recursively axiomatisable. Hence, it was necessary to provide a proof method suitable for automated deduction, as one has done for classical logic. In contrast to classical logic, we cannot make shifts of quantifiers arbitrarily and translate a formula to an equivalent (satisfiable) prenex form. In [2, 19], we have generalised the well-known hyperresolution principle to the first-order Gödel logic for the general case. Our approach is based on translation of a formula of Gödel logic to an equivalent satisfiable finite order clausal theory, consisting of order clauses. We have introduced a notion of quantified atom: a formula a is a quantified atom if $a = Qx\,p(t_0, \ldots, t_\tau)$ where Q is a quantifier ($\forall$, $\exists$); $p(t_0, \ldots, t_\tau)$ is an atom; x is a variable occurring in $p(t_0, \ldots, t_\tau)$; for all $i \leq \tau$, either $t_i = x$ or x does not occur in t_i (t_i is a free term in the quantified atom). The notion of quantified atom is all important. It permits us to extend classical unification to quantified atoms without any additional computational cost. Two quantified atoms $Qx\,p(t_0, \ldots, t_\tau)$ and $Q'x'\,p'(t_0', \ldots, t_\tau')$ are unifiable if $Q = Q'$, $x = x'$, $p = p'$, and the left-right sequence of free terms of $Qx\,p(t_0, \ldots, t_\tau)$ is unifiable with the left-right sequence of free terms of $Q'x'\,p'(t_0', \ldots, t_\tau')$ in the standard manner. An order clause is a finite set of order literals of the form $\varepsilon_1 \diamond \varepsilon_2$ where ε_i is an atom or a quantified atom, and $\diamond$ is the connective $\preceq$ or $\prec$. $\preceq$ and $\prec$ are interpreted by the equality and standard strict linear order on $[0,1]$, respectively. On the basis of the hyperresolution principle, a calculus operating over order clausal theories, has been devised. The calculus is proved to be refutation sound and complete for the countable case with respect to the standard $\boldsymbol{G}$-algebra $\boldsymbol{G} = ([0,1], \leq, \vee, \wedge, \Rightarrow, \overline{}, \preceq, \prec, 0, 1)$ augmented by binary operators $\preceq$ and $\prec$ for $\preceq$ and $\prec$, respectively. As another step, one may incorporate a countable set of intermediate truth constants $\bar{c}$, $c \in (0,1)$, to get a modification of the hyperresolution calculus suitable for automated deduction with explicit partial truth [20]. We shall investigate the so-called canonical standard completeness, where the semantics of Gödel logic is given by the standard $\boldsymbol{G}$-algebra $\boldsymbol{G}$ and truth constants are interpreted by 'themselves'. We say that a set $\{0, 1\} \subseteq X$ of truth constants is admissible with respect to suprema and infima if, for all $\emptyset \neq Y_1, Y_2 \subseteq X$ and $\bigvee Y_1 = \bigwedge Y_2, \bigvee Y_1 \in Y_1, \bigwedge Y_2 \in Y_2$. Then the hyperresolution calculus is refutation sound and complete for a countable order clausal theory if the set of all truth constants occurring in the theory, is admissible with respect to suprema and infima. This condition obviously covers the case of finite order clausal theories. As an interesting consequence, we get an affirmative solution to the open problem of recursive enumerability of unsatisfiable formulae in Gödel logic with intermediate truth constants and the projection operator $\Delta : [0, 1] \longrightarrow [0, 1]$,

$$\Delta a = \begin{cases} 1 & \text{if } a = 1, \\ 0 & \text{else}; \end{cases}$$

which strengthens a similar result for prenex formulae of G_∞^Δ stated in Conclusion of [1].

The paper is organised as follows. Section 2 gives the basic notions and notation concerning the first-order Gödel logic. Section 3 deals with clause form translation. In Sect. 4, we propose a hyperresolution calculus with truth constants and prove its refutational soundness, completeness. Section 5 brings conclusions.

2 First-Order Gödel Logic

Throughout the paper, we shall use the common notions and notation of first-order logic.[1] By $\mathscr{L}$ we denote a first-order language. We assume truth constants—nullary predicate symbols $0, 1 \in Pred_{\mathscr{L}}$, $ar_{\mathscr{L}}(0) = ar_{\mathscr{L}}(1) = 0$; 0 denotes the false and 1 the true in $\mathscr{L}$. Let $\mathbb{C}_{\mathscr{L}} \subseteq (0, 1)$ be countable. In addition, we assume a countable set of nullary predicate symbols $\overline{C}_{\mathscr{L}} = \{\bar{c} \mid \bar{c} \in Pred_{\mathscr{L}}, ar_{\mathscr{L}}(\bar{c}) = 0, c \in \mathbb{C}_{\mathscr{L}}\} \subseteq Pred_{\mathscr{L}}$. $0, 1, \bar{c} \in \overline{C}_{\mathscr{L}}$ are called truth constants. We denote $Tcons_{\mathscr{L}} = \{0, 1\} \cup \overline{C}_{\mathscr{L}} \subseteq Pred_{\mathscr{L}}$. Let $X \subseteq Tcons_{\mathscr{L}}$. We denote $\overline{X} = \{0 \mid 0 \in X\} \cup \{1 \mid 1 \in X\} \cup \{c \mid \bar{c} \in X \cap \overline{C}_{\mathscr{L}}\} \subseteq [0, 1]$. By $Form_{\mathscr{L}}$ we designate the set of all formulae of $\mathscr{L}$ built up from $Atom_{\mathscr{L}}$ and $Var_{\mathscr{L}}$ using the connectives: $\neg$, negation, Δ, Delta, $\wedge$, conjunction, $\vee$, disjunction, $\rightarrow$, implication, and the quantifiers: $\forall$, the universal quantifier, $\exists$, the existential one. In addition, we introduce new binary connectives $\approx$, equality, and $\prec$, strict order. We denote $Con = \{\neg, \Delta, \wedge, \vee, \rightarrow, \approx, \prec\}$. By $OrdForm_{\mathscr{L}}$ we designate the set of all so-called order formulae of $\mathscr{L}$ built up from $Atom_{\mathscr{L}}$ and $Var_{\mathscr{L}}$ using the connectives in Con and the quantifiers: $\forall, \exists.$[2] In the paper, we shall assume that $\mathscr{L}$ is a countable first-order language; hence, all the above mentioned sets of symbols and expressions are countable. By $varseq(\phi)$, $vars(varseq(\phi)) \subseteq Var_{\mathscr{L}}$, we denote the sequence of all variables of $\mathscr{L}$ occurring in ϕ which is built up via the left-

---

[1]Cf. http://ii.fmph.uniba.sk/~guller/sci15.pdf, Sect. 2.

[2]We assume a decreasing connective and quantifier precedence: $\forall, \exists, \neg, \Delta, \wedge, \rightarrow, \approx, \prec, \vee$.

right preorder traversal of ϕ. For example, $varseq(\exists w \,(\forall x\, p(x, x, z) \vee \exists y\, q(x, y, z))) = w, x, x, x, z, y, x, y, z, \,|w, x, x, x, z, y, x, y, z| = 9.$

Gödel logic is interpreted by the standard $\boldsymbol{G}$-algebra augmented by binary operators $\boldsymbol{\varkappa}$ and $\prec$ for $\varkappa$ and $\prec$, respectively.

$$\boldsymbol{G} = ([0, 1], \leq, \vee, \wedge, \Rightarrow, \overline{}, \boldsymbol{\varDelta}, \boldsymbol{\varkappa}, \prec, 0, 1)$$

where $\vee \mid \wedge$ denotes the supremum $\mid$ infimum operator on $[0, 1]$;

$$a \Rightarrow b = \begin{cases} 1 & \text{if } a \leq b, \\ b & \text{else}; \end{cases} \qquad \overline{a} = \begin{cases} 1 & \text{if } a = 0, \\ 0 & \text{else}; \end{cases}$$

$$a \boldsymbol{\varkappa} b = \begin{cases} 1 & \text{if } a = b, \\ 0 & \text{else}; \end{cases} \qquad a \prec b = \begin{cases} 1 & \text{if } a < b, \\ 0 & \text{else}. \end{cases}$$

Recall that $\boldsymbol{G}$ is a complete linearly ordered lattice algebra; $\vee \mid \wedge$ is commutative, associative, idempotent, monotone; $0 \mid 1$ is its neutral element; the residuum operator $\Rightarrow$ of $\wedge$ satisfies the condition of residuation:

$$\text{for all } a, b, c \in \boldsymbol{G}, \ a \wedge b \leq c \Longleftrightarrow a \leq b \Rightarrow c; \tag{1}$$

Gödel negation $\overline{}$ satisfies the condition:

$$\text{for all } a \in \boldsymbol{G}, \ \overline{a} = a \Rightarrow 0; \tag{2}$$

the following properties, which will be exploited later, hold[3]:

for all $a, b, c \in \boldsymbol{G}$,

$$a \vee b \wedge c = (a \vee b) \wedge (a \vee c), \quad \text{(distributivity of } \vee \text{ over } \wedge) \tag{3}$$
$$a \wedge (b \vee c) = a \wedge b \vee a \wedge c, \quad \text{(distributivity of } \wedge \text{ over } \vee) \tag{4}$$
$$a \Rightarrow (b \vee c) = a \Rightarrow b \vee a \Rightarrow c, \tag{5}$$
$$a \Rightarrow b \wedge c = (a \Rightarrow b) \wedge (a \Rightarrow c), \tag{6}$$
$$(a \vee b) \Rightarrow c = (a \Rightarrow c) \wedge (b \Rightarrow c), \tag{7}$$
$$a \wedge b \Rightarrow c = a \Rightarrow c \vee b \Rightarrow c, \tag{8}$$
$$a \Rightarrow (b \Rightarrow c) = a \wedge b \Rightarrow c, \tag{9}$$
$$((a \Rightarrow b) \Rightarrow b) \Rightarrow b = a \Rightarrow b, \tag{10}$$
$$(a \Rightarrow b) \Rightarrow c = ((a \Rightarrow b) \Rightarrow b) \wedge (b \Rightarrow c) \vee c, \tag{11}$$
$$(a \Rightarrow b) \Rightarrow 0 = ((a \Rightarrow 0) \Rightarrow 0) \wedge (b \Rightarrow 0). \tag{12}$$

An interpretation $\mathscr{I}$ for $\mathscr{L}$ is a triple $\left(\mathscr{U}_{\mathscr{I}}, \{f^{\mathscr{I}} \mid f \in Func_{\mathscr{L}}\}, \{p^{\mathscr{I}} \mid p \in Pred_{\mathscr{L}}\} \right)$ defined as follows: $\mathscr{U}_{\mathscr{I}} \neq \emptyset$ is the universum of $\mathscr{I}$; every $f \in Func_{\mathscr{L}}$

---

[3]We assume a decreasing operator precedence: $\overline{}, \boldsymbol{\varDelta}, \wedge, \Rightarrow, \boldsymbol{\varkappa}, \prec, \vee$.

is interpreted as a function $f^{\mathscr{I}} : \mathscr{U}_{\mathscr{G}}^{ar_{\mathscr{L}}(f)} \longrightarrow \mathscr{U}_{\mathscr{G}}$; every $p \in Pred_{\mathscr{L}}$ is interpreted as a $[0,1]$-relation $p^{\mathscr{I}} : \mathscr{U}_{\mathscr{G}}^{ar_{\mathscr{L}}(p)} \longrightarrow [0,1]$; particularly, $0^{\mathscr{I}} = 0$, $1^{\mathscr{I}} = 1$, for all $\bar{c} \in \overline{C}_{\mathscr{L}}$, $\bar{c}^{\mathscr{I}} = c$. We denote $tcons(\phi) = \{0,1\} \cup (preds(\phi) \cap \overline{C}_{\mathscr{L}}) \subseteq Tcons_{\mathscr{L}}$ and $tcons(T) = \{0,1\} \cup (preds(T) \cap \overline{C}_{\mathscr{L}}) \subseteq Tcons_{\mathscr{L}}$.

3 Translation to Clausal Form

We firstly define a notion of quantified atom. Let $a \in Form_{\mathscr{L}}$. a is a quantified atom of $\mathscr{L}$ iff $a = Qxp(t_0, \dots, t_\tau)$ where $p(t_0, \dots, t_\tau) \in Atom_{\mathscr{L}}$, $x \in vars(p(t_0, \dots, t_\tau))$, either $t_i = x$ or $x \notin vars(t_i)$. $QAtom_{\mathscr{L}} \subseteq Form_{\mathscr{L}}$ denotes the set of all quantified atoms of $\mathscr{L}$. Let $Qxp(t_0, \dots, t_\tau) \in QAtom_{\mathscr{L}}$ and $p(t'_0, \dots, t'_\tau) \in Atom_{\mathscr{L}}$. We denote

$$boundindset(Qxp(t_0, \dots, t_\tau)) = \{i \mid i \leq \tau, t_i = x\} \neq \emptyset.$$

Let $I = \{i \mid i \leq \tau, x \notin vars(t_i)\}$ and $r_1, \dots, r_k, r_i \leq \tau, k \leq \tau$, for all $1 \leq i < i' \leq k, r_i < r_{i'}$, be a sequence such that $\{r_i \mid 1 \leq i \leq k\} = I$. We denote

$$freetermseq(Qxp(t_0, \dots, t_\tau)) = t_{r_1}, \dots, t_{r_k},$$
$$freetermseq(p(t'_0, \dots, t'_\tau)) = t'_0, \dots, t'_\tau.$$

We further introduce conjunctive normal form (*CNF*) in Gödel logic. In contrast to two-valued logic, we have to consider an augmented set of literals appearing in *CNF* formulae. Let $l, \phi \in Form_{\mathscr{L}}$. l is a literal of $\mathscr{L}$ iff either $l = a$ or $l = b \rightarrow c$ or $l = (a \rightarrow d) \rightarrow d$ or $l = a \rightarrow e$ or $l = e \rightarrow a$ or $l = a \rightarrow \Delta f$ or $l = \Delta f \rightarrow a$, $a, f \in Atom_{\mathscr{L}} - Tcons_{\mathscr{L}}$, $b \in Atom_{\mathscr{L}} - \{0,1\}$, $c \in Atom_{\mathscr{L}} - \{1\}$, $d \in (Atom_{\mathscr{L}} - Tcons_{\mathscr{L}}) \cup \{0\}$, $e \in QAtom_{\mathscr{L}}$, $\{b,c\} \not\subseteq Tcons_{\mathscr{L}}$. The set of all literals of $\mathscr{L}$ is designated as $Lit_{\mathscr{L}} \subseteq Form_{\mathscr{L}}$. ϕ is a conjunctive | disjunctive normal form of $\mathscr{L}$, in symbols *CNF* | *DNF*, iff either $\phi \in Tcons_{\mathscr{L}}$ or $\phi = \bigwedge_{i \leq n} \bigvee_{j \leq m_i} l_j^i \mid \phi = \bigvee_{i \leq n} \bigwedge_{j \leq m_i} l_j^i$, $l_j^i \in Lit_{\mathscr{L}}$.

We finally introduce order clauses in Gödel logic. Let $l \in OrdForm_{\mathscr{L}}$. l is an order literal of $\mathscr{L}$ iff $l = \varepsilon_1 \diamond \varepsilon_2, \varepsilon_i \in Atom_{\mathscr{L}} \cup QAtom_{\mathscr{L}}, \diamond \in \{\preceq, \prec\}$. The set of all order literals of $\mathscr{L}$ is designated as $OrdLit_{\mathscr{L}} \subseteq OrdForm_{\mathscr{L}}$. An order clause of $\mathscr{L}$ is a finite set of order literals of $\mathscr{L}$. An order clause $\{l_1, \dots, l_n\}$ is written in the form $l_1 \vee \cdots \vee l_n$. The order clause $\emptyset$ is called the empty order clause and denoted as $\square$. An order clause $\{l\}$ is called a unit order clause and denoted as l. We designate the set of all order clauses of $\mathscr{L}$ as $OrdCl_{\mathscr{L}}$. Let $l, l_0, \dots, l_n \in OrdLit_{\mathscr{L}}$ and $C, C' \in OrdCl_{\mathscr{L}}$. We define the size of C as $|C| = \sum_{l \in C} |l|$. By $l \vee C$ we denote $\{l\} \cup C$ where $l \notin C$. Analogously, by $l_0 \vee \cdots \vee l_n \vee C$ we denote $\{l_0\} \cup \cdots \cup \{l_n\} \cup C$ where, for all $i, i' \leq n, i \neq i', l_i \notin C$ and $l_i \neq l_{i'}$. By $C \vee C'$ we denote $C \cup C'$. C is a subclause of C', in symbols $C \sqsubseteq C'$, iff $C \subseteq C'$. An order clausal theory of $\mathscr{L}$ is a set of order clauses of $\mathscr{L}$. A unit order clausal theory is a set of unit order clauses.

Let $\phi, \phi' \in OrdForm_{\mathscr{L}}$, $T, T' \subseteq OrdForm_{\mathscr{L}}$, $S, S' \subseteq OrdCl_{\mathscr{L}}$, $\mathscr{I}$ be an interpretation for $\mathscr{L}$, $e \in \mathscr{S}_{\mathscr{J}}$. C is true in $\mathscr{I}$ with respect to e, written as $\mathscr{I} \vDash_e C$, iff there exists $l^* \in C$ such that $\mathscr{I} \vDash_e l^*$. $\mathscr{I}$ is a model of C, in symbols $\mathscr{I} \vDash C$, iff, for all $e \in \mathscr{S}_{\mathscr{J}}$, $\mathscr{I} \vDash_e C$. $\mathscr{I}$ is a model of S, in symbols $\mathscr{I} \vDash S$, iff, for all $C \in S$, $\mathscr{I} \vDash C$. $\phi' \mid T' \mid C' \mid S'$ is a logical consequence of $\phi \mid T \mid C \mid S$, in symbols $\phi \mid T \mid C \mid S \vDash \phi' \mid T' \mid C' \mid S'$, iff, for every model $\mathscr{I}$ of $\phi \mid T \mid C \mid S$ for $\mathscr{L}$, $\mathscr{I} \vDash \phi' \mid T' \mid C' \mid S'$. $\phi \mid T \mid C \mid S$ is satisfiable iff there exists a model of $\phi \mid T \mid C \mid S$ for $\mathscr{L}$. $\phi \mid T \mid C \mid S$ is equisatisfiable to $\phi' \mid T' \mid C' \mid S'$ iff $\phi \mid T \mid C \mid S$ is satisfiable if and only if $\phi' \mid T' \mid C' \mid S'$ is satisfiable. We denote $tcons(S) = \{0, 1\} \cup (preds(S) \cap \overline{C}_{\mathscr{L}}) \subseteq Tcons_{\mathscr{L}}$. Let $S \subseteq_{\mathscr{F}} OrdCl_{\mathscr{L}}$. We define the size of S as $|S| = \sum_{C \in S} |C|$. l is a simplified order literal of $\mathscr{L}$ iff $l = \varepsilon_1 \diamond \varepsilon_2$, $\{\varepsilon_1, \varepsilon_2\} \nsubseteq Tcons_{\mathscr{L}}$, $\{\varepsilon_1, \varepsilon_2\} \nsubseteq QAtom_{\mathscr{L}}$. The set of all simplified order literals of $\mathscr{L}$ is designated as $SimOrdLit_{\mathscr{L}} \subseteq OrdLit_{\mathscr{L}}$. We denote $SimOrdCl_{\mathscr{L}} = \{C \mid C \in OrdCl_{\mathscr{L}}, C \subseteq SimOrdLit_{\mathscr{L}}\} \subseteq OrdCl_{\mathscr{L}}$. Let $\tilde{f}_0 \notin Func_{\mathscr{L}}$; $\tilde{f}_0$ is a new function symbol. Let $\mathbb{I} = \mathbb{N} \times \mathbb{N}$; $\mathbb{I}$ is an infinite countable set of indices. Let $\tilde{P} = \{\tilde{p}_i \mid i \in \mathbb{I}\}$ such that $\tilde{P} \cap Pred_{\mathscr{L}} = \emptyset$; $\tilde{P}$ is an infinite countable set of new predicate symbols.

From a computational point of view, the worst case time and space complexity will be estimated using the logarithmic cost measurement. Let $\mathscr{A}$ be an algorithm. $\#\mathcal{O}_{\mathscr{A}}(In) \geq 1$ denotes the number of all elementary operations executed by $\mathscr{A}$ on an input In.

3.1 Substitutions

We assume the reader to be familiar with the standard notions and notation of substitutions.[4] Let $X = \{x_i \mid 1 \leq i \leq n\} \subseteq Var_{\mathscr{L}}$. A substitution ϑ of $\mathscr{L}$ is a mapping $\vartheta : X \longrightarrow Term_{\mathscr{L}}$. ϑ may be written in the form $x_1/\vartheta(x_1), \ldots, x_n/\vartheta(x_n)$. We denote $dom(\vartheta) = X \subseteq_{\mathscr{F}} Var_{\mathscr{L}}$ and $range(\vartheta) = \bigcup_{x \in X} vars(\vartheta(x)) \subseteq_{\mathscr{F}} Var_{\mathscr{L}}$. The set of all substitutions of $\mathscr{L}$ is designated as $Subst_{\mathscr{L}}$. Let $Qx\,a \in QAtom_{\mathscr{L}}$. ϑ is applicable to $Qx\,a$ iff $dom(\vartheta) \supseteq freevars(Qx\,a)$ and $x \notin range(\vartheta|_{freevars(Qx\,a)})$. We define the application of ϑ to $Qx\,a$ as $(Qx\,a)\vartheta = Qx\,a(\vartheta|_{freevars(Qx\,a)} \cup x/x) \in QAtom_{\mathscr{L}}$. Let ε and ε' be expressions. ε' is an instance of ε of $\mathscr{L}$ iff there exists $\vartheta^* \in Subst_{\mathscr{L}}$ such that $\varepsilon' = \varepsilon\vartheta^*$. ε' is a variant of ε of $\mathscr{L}$ iff there exists a variable renaming $\rho^* \in Subst_{\mathscr{L}}$ such that $\varepsilon' = \varepsilon\rho^*$. Let $C \in OrdCl_{\mathscr{L}}$ and $S \subseteq OrdCl_{\mathscr{L}}$. C is an instance $\mid$ a variant of S of $\mathscr{L}$ iff there exists $C^* \in S$ such that C is an instance $\mid$ a variant of C^* of $\mathscr{L}$. We denote $Inst_{\mathscr{L}}(S) = \{C \mid C \text{ is an instance of } S \text{ of } \mathscr{L}\} \subseteq OrdCl_{\mathscr{L}}$ and $Vrnt_{\mathscr{L}}(S) = \{C \mid C \text{ is a variant of } S \text{ of } \mathscr{L}\} \subseteq OrdCl_{\mathscr{L}}$.

Let E be a set of expressions. ϑ is a unifier of $\mathscr{L}$ for E iff $E\vartheta$ is a singleton set. Let $\theta \in Subst_{\mathscr{L}}$. θ is a most general unifier of $\mathscr{L}$ for E iff θ is a unifier of $\mathscr{L}$ for E, and for every unifier ϑ of $\mathscr{L}$ for E, there exists $\gamma^* \in Subst_{\mathscr{L}}$ such that $\vartheta|_{freevars(E)} = \theta|_{freevars(E)} \circ \gamma^*$. By $mgu_{\mathscr{L}}(E) \subseteq Subst_{\mathscr{L}}$ we denote the set of all most general unifiers of $\mathscr{L}$ for E. Let $\overline{E} = E_0, \ldots, E_n$, $E_i \subseteq A_i$, either $A_i = Term_{\mathscr{L}}$ or

---

[4]Cf. http://ii.fmph.uniba.sk/~guller/sci15.pdf, Appendix, Sect. 5.1.

$A_i = Atom_{\mathscr{L}}$ or $A_i = QAtom_{\mathscr{L}}$ or $A_i = OrdLit_{\mathscr{L}}$. ϑ is a unifier of $\mathscr{L}$ for $\overline{E}$ iff, for all $i \leq n$, ϑ is a unifier of $\mathscr{L}$ for E_i. θ is a most general unifier of $\mathscr{L}$ for $\overline{E}$ iff θ is a unifier of $\mathscr{L}$ for $\overline{E}$, and for every unifier ϑ of $\mathscr{L}$ for $\overline{E}$, there exists $\gamma^* \in Subst_{\mathscr{L}}$ such that $\vartheta|_{freevars(\overline{E})} = \theta|_{freevars(\overline{E})} \circ \gamma^*$. By $mgu_{\mathscr{L}}(\overline{E}) \subseteq Subst_{\mathscr{L}}$ we denote the set of all most general unifiers of $\mathscr{L}$ for $\overline{E}$.

Theorem 1 (Unification Theorem) *Let* $\overline{E} = E_0, \ldots, E_n$, *either* $E_i \subseteq_{\mathscr{F}} Term_{\mathscr{L}}$ *or* $E_i \subseteq_{\mathscr{F}} Atom_{\mathscr{L}}$. *If there exists a unifier of* $\mathscr{L}$ *for* $\overline{E}$, *then there exists* $\theta^* \in mgu_{\mathscr{L}}(\overline{E})$ *such that* $range(\theta^*|_{vars(\overline{E})}) \subseteq vars(\overline{E})$.

Proof By induction on $\|vars(\overline{E})\|$; a modification of the proof of Theorem 2.3 (Unification Theorem) in [21], Sect. 2.4, pp. 5–6. □

Theorem 2 (Extended Unification Theorem) *Let* $\overline{E} = E_0, \ldots, E_n$, *either* $E_i \subseteq_{\mathscr{F}}$ $Term_{\mathscr{L}}$ *or* $E_i \subseteq_{\mathscr{F}} Atom_{\mathscr{L}}$ *or* $E_i \subseteq_{\mathscr{F}} QAtom_{\mathscr{L}}$ *or* $E_i \subseteq_{\mathscr{F}} OrdLit_{\mathscr{L}}$, $boundvars(\overline{E}) \subseteq$ $V \subseteq_{\mathscr{F}} Var_{\mathscr{L}}$. *If there exists a unifier of* $\mathscr{L}$ *for* $\overline{E}$, *then there exists* $\theta^* \in mgu_{\mathscr{L}}(\overline{E})$ *such that* $range(\theta^*|_{freevars(\overline{E})}) \cap V = \emptyset$.

Proof A straightforward consequence of Theorem 1. □

3.2 A Formal Treatment

Translation of a formula or a theory to *CNF* and clausal form, is based on the following lemma:

Lemma 1 *Let* $n_\phi, n_0 \in \mathbb{N}$, $\phi \in Form_{\mathscr{L}}$, $T \subseteq Form_{\mathscr{L}}$.

(I) *There exist either* $J_\phi = \emptyset$ *or* $J_\phi = \{(n_\phi, j) \mid j \leq n_{J_\phi}\}$, $J_\phi \subseteq \{(n_\phi, j) \mid j \in \mathbb{N}\}$, *a CNF* $\psi \in Form_{\mathscr{L} \cup \{\tilde{p}_j \mid j \in J_\phi\}}$, $S_\phi \subseteq_{\mathscr{F}} SimOrdCl_{\mathscr{L} \cup \{\tilde{p}_j \mid j \in J_\phi\}}$ *such that*

(a) $\|J_\phi\| \leq 2 \cdot |\phi|$;

(b) *either* $J_\phi = \emptyset$, $S_\phi = \{\Box\}$ *or* $J_\phi = S_\phi = \emptyset$ *or* $J_\phi \neq \emptyset$, $\Box \notin S_\phi \neq \emptyset$;

(c) *there exists an interpretation* $\mathfrak{A}$ *for* $\mathscr{L}$ *and* $\mathfrak{A} \models \phi$ *if and only if there exists an interpretation* $\mathfrak{A}'$ *for* $\mathscr{L} \cup \{\tilde{p}_j \mid j \in J_\phi\}$ *and* $\mathfrak{A}' \models \psi$, *satisfying* $\mathfrak{A} = \mathfrak{A}'|_{\mathscr{L}}$;

(d) *there exists an interpretation* $\mathfrak{A}$ *for* $\mathscr{L}$ *and* $\mathfrak{A} \models \phi$ *if and only if there exists an interpretation* $\mathfrak{A}'$ *for* $\mathscr{L} \cup \{\tilde{p}_j \mid j \in J_\phi\}$ *and* $\mathfrak{A}' \models S_\phi$, *satisfying* $\mathfrak{A} = \mathfrak{A}'|_{\mathscr{L}}$;

(e) $|\psi| \in O(|\phi|^2)$; *the number of all elementary operations of the translation of* ϕ *to* ψ, *is in* $O(|\phi|^2)$; *the time and space complexity of the translation of* ϕ *to* ψ, *is in* $O(|\phi|^2 \cdot (\log(1 + n_\phi) + \log |\phi|))$;

(f) $|S_\phi| \in O(|\phi|^2)$; *the number of all elementary operations of the translation of* ϕ *to* S_ϕ, *is in* $O(|\phi|^2)$; *the time and space complexity of the translation of* ϕ *to* S_ϕ, *is in* $O(|\phi|^2 \cdot (\log(1 + n_\phi) + \log |\phi|))$;

(g) if $\psi \notin Tcons_{\mathscr{L}}$, then $\psi = \bigwedge_{i \leq n_\psi} D_i$, D_i is a factor, $J_\phi \neq \emptyset$, for all $i \leq n_\psi$,
     $\emptyset \neq preds(D_i) \cap \tilde{\mathbb{P}} \subseteq \{\tilde{p}_{\mathrm{j}} \mid \mathrm{j} \in J_\phi\}$, for all $i < i' \leq n_\psi$, $lits(D_i) \neq lits(D_{i'})$;

(h) if $S_\phi \neq \emptyset, \{\square\}$, then $J_\phi \neq \emptyset$, for all $C \in S_\phi$, $\emptyset \neq preds(C) \cap \tilde{\mathbb{P}} \subseteq \{\tilde{p}_{\mathrm{j}} \mid \mathrm{j} \in J_\phi\}$;

(i) for all $a \in qatoms(\psi)$, there exists $\mathrm{j}^* \in J_\phi$ and $preds(a) = \{\tilde{p}_{\mathrm{j}^*}\}$;

(j) for all $\mathrm{j} \in J_\phi$, there exists a sequence $\bar{x}$ of variables of $\mathscr{L}$ and $\tilde{p}_{\mathrm{j}}(\bar{x}) \in atoms(\psi)$ satisfying, for all $a \in atoms(\psi)$ and $preds(a) = \{\tilde{p}_{\mathrm{j}}\}$, $a = \tilde{p}_{\mathrm{j}}(\bar{x})$; if there exists $a^* \in qatoms(\psi)$ and $preds(a^*) = \{\tilde{p}_{\mathrm{j}}\}$, then there exists $Qx\,\tilde{p}_{\mathrm{j}}(\bar{x}) \in qatoms(\psi)$ satisfying, for all $a \in qatoms(\psi)$ and $preds(a) = \{\tilde{p}_{\mathrm{j}}\}$, $a = Qx\,\tilde{p}_{\mathrm{j}}(\bar{x})$;

(k) for all $a \in qatoms(S_\phi)$, there exists $\mathrm{j}^* \in J_\phi$ and $preds(a) = \{\tilde{p}_{\mathrm{j}^*}\}$;

(l) for all $\mathrm{j} \in J_\phi$, there exists a sequence $\bar{x}$ of variables of $\mathscr{L}$ and $\tilde{p}_{\mathrm{j}}(\bar{x}) \in atoms(S_\phi)$ satisfying, for all $a \in atoms(S_\phi)$ and $preds(a) = \{\tilde{p}_{\mathrm{j}}\}$, $a = \tilde{p}_{\mathrm{j}}(\bar{x})$; if there exists $a^* \in qatoms(S_\phi)$ and $preds(a^*) = \{\tilde{p}_{\mathrm{j}}\}$, then there exists $Qx\,\tilde{p}_{\mathrm{j}}(\bar{x}) \in qatoms(S_\phi)$ satisfying, for all $a \in qatoms(S_\phi)$ and $preds(a) = \{\tilde{p}_{\mathrm{j}}\}$, $a = Qx\,\tilde{p}_{\mathrm{j}}(\bar{x})$;

(m) $tcons(\psi) = tcons(S_\phi) \subseteq tcons(\phi)$.

(II) There exist $J_T \subseteq \{(i,j) \mid i \geq n_0\}$ and $S_T \subseteq SimOrdCl_{\mathscr{L} \cup \{\tilde{p}_{\mathrm{j}} \mid \mathrm{j} \in J_T\}}$ such that

(a) either $J_T = \emptyset$, $S_T = \{\square\}$ or $J_T = S_T = \emptyset$ or $J_T \neq \emptyset$, $\square \notin S_T \neq \emptyset$;

(b) there exists an interpretation $\mathfrak{A}$ for $\mathscr{L}$ and $\mathfrak{A} \vDash T$ if and only if there exists an interpretation $\mathfrak{A}'$ for $\mathscr{L} \cup \{\tilde{p}_{\mathrm{j}} \mid \mathrm{j} \in J_T\}$ and $\mathfrak{A}' \vDash S_T$, satisfying $\mathfrak{A} = \mathfrak{A}'|_{\mathscr{L}}$;

(c) if $T \subseteq_{\mathscr{F}} Form_{\mathscr{L}}$, then $J_T \subseteq_{\mathscr{F}} \{(i,j) \mid i \geq n_0\}$, $\|J_T\| \leq 2 \cdot |T|$, $S_T \subseteq_{\mathscr{F}} SimOrdCl_{\mathscr{L} \cup \{\tilde{p}_{\mathrm{j}} \mid \mathrm{j} \in J_T\}}$, $|S_T| \in O(|T|^2)$; the number of all elementary operations of the translation of T to S_T, is in $O(|T|^2)$; the time and space complexity of the translation of T to S_T, is in $O(|T|^2 \cdot \log(1 + n_0 + |T|))$;

(d) if $S_T \neq \emptyset, \{\square\}$, then $J_T \neq \emptyset$, for all $C \in S_T$, $\emptyset \neq preds(C) \cap \tilde{\mathbb{P}} \subseteq \{\tilde{p}_{\mathrm{j}} \mid \mathrm{j} \in J_T\}$;

(e) for all $a \in qatoms(S_T)$, there exists $\mathrm{j}^* \in J_T$ and $preds(a) = \{\tilde{p}_{\mathrm{j}^*}\}$;

(f) for all $\mathrm{j} \in J_T$, there exists a sequence $\bar{x}$ of variables of $\mathscr{L}$ and $\tilde{p}_{\mathrm{j}}(\bar{x}) \in atoms(S_T)$ satisfying, for all $a \in atoms(S_T)$ and $preds(a) = \{\tilde{p}_{\mathrm{j}}\}$, $a = \tilde{p}_{\mathrm{j}}(\bar{x})$; if there exists $a^* \in qatoms(S_T)$ and $preds(a^*) = \{\tilde{p}_{\mathrm{j}}\}$, then there exists $Qx\,\tilde{p}_{\mathrm{j}}(\bar{x}) \in qatoms(S_T)$ satisfying, for all $a \in qatoms(S_T)$ and $preds(a) = \{\tilde{p}_{\mathrm{j}}\}$, $a = Qx\,\tilde{p}_{\mathrm{j}}(\bar{x})$;

(g) $tcons(S_T) \subseteq tcons(T)$.

Proof Cf. http://ii.fmph.uniba.sk/~guller/sci15.pdf, Sect. 3.3, Lemma 1, for a proof. In Table 1, for every form of $l \in Lit_{\mathscr{L}}$, an equisatisfiable $C \in SimOrdCl_{\mathscr{L}}$ is assigned. In Tables 2, 3, 4, 5 and 6, interpolation rules for all the connectives are proposed, which translation is based on. The lemma is proved. $\square$

Table 1 Translation of l to C

Case	l	C	l	C
1	a	$a \approx 1$	$\lvert a \rvert$	$\lvert a \rvert + 2 \leq 3 \cdot \lvert l \rvert$
2	$a \to 0$	$a \approx 0$	$\lvert a \rvert + 2$	$\lvert a \rvert + 2 \leq 3 \cdot \lvert l \rvert$
3	$\bar{c} \to b$	$\bar{c} < b \vee \bar{c} \approx b$	$\lvert b \rvert + 2$	$2 \cdot \lvert b \rvert + 4 \leq 3 \cdot \lvert l \rvert$
4	$a \to \bar{c}$	$a < \bar{c} \vee a \approx \bar{c}$	$\lvert a \rvert + 2$	$2 \cdot \lvert a \rvert + 4 \leq 3 \cdot \lvert l \rvert$
5	$a \to b$	$a < b \vee a \approx b$	$\lvert a \rvert + \lvert b \rvert + 1$	$2 \cdot \lvert a \rvert + 2 \cdot \lvert b \rvert + 2 \leq 3 \cdot \lvert l \rvert$
6	$(a \to 0) \to 0$	$0 < a$	$\lvert a \rvert + 4$	$\lvert a \rvert + 2 \leq 3 \cdot \lvert l \rvert$
7	$(a \to b) \to b$	$b < a \vee b \approx 1$	$\lvert a \rvert + 2 \cdot \lvert b \rvert + 2$	$\lvert a \rvert + 2 \cdot \lvert b \rvert + 3 \leq 3 \cdot \lvert l \rvert$
8	$a \to d$	$a < d \vee a \approx d$	$\lvert a \rvert + \lvert d \rvert + 1$	$2 \cdot \lvert a \rvert + 2 \cdot \lvert d \rvert + 2 \leq 3 \cdot \lvert l \rvert$
9	$d \to a$	$d < a \vee d \approx a$	$\lvert a \rvert + \lvert d \rvert + 1$	$2 \cdot \lvert a \rvert + 2 \cdot \lvert d \rvert + 2 \leq 3 \cdot \lvert l \rvert$
10	$a \to \Delta b$	$a \approx 0 \vee b \approx 1$	$\lvert a \rvert + \lvert b \rvert + 2$	$\lvert a \rvert + \lvert b \rvert + 4 \leq 3 \cdot \lvert l \rvert$
11	$\Delta b \to a$	$b < 1 \vee a \approx 1$	$\lvert a \rvert + \lvert b \rvert + 2$	$\lvert a \rvert + \lvert b \rvert + 4 \leq 3 \cdot \lvert l \rvert$

$a, b \in Atom_{\mathscr{L}} - Tcons_{\mathscr{L}}, \bar{c} \in \overline{C}_{\mathscr{L}}, d \in QAtom_{\mathscr{L}}$

The described translation produces order clausal theories in some restrictive form, which will be utilised in inference using our order hyperresolution calculus to get shorter deductions in average case. Let $P \subseteq \tilde{\mathbb{P}}$ and $S \subseteq OrdCl_{\mathscr{L} \cup P}$. S is admissible iff

(a) for all $a \in qatoms(S)$, $preds(a) \subseteq P$;

(b) for all $\tilde{p} \in P$, there exists a sequence $\bar{x}$ of variables of $\mathscr{L}$ and $\tilde{p}(\bar{x}) \in atoms(S)$ satisfying, for all $a \in atoms(S)$ and $preds(a) = \{\tilde{p}\}$, a is an instance of $\tilde{p}(\bar{x})$ of $\mathscr{L} \cup P$; if there exists $a^* \in qatoms(S)$ and $preds(a^*) = \{\tilde{p}\}$, then there exists $Qx\,\tilde{p}(\bar{x}) \in qatoms(S)$ satisfying, for all $a \in qatoms(S)$ and $preds(a) = \{\tilde{p}\}$, a is an instance of $Qx\,\tilde{p}(\bar{x})$ of $\mathscr{L} \cup P$.

(a) and (b) imply that for all $Qx\,a, Q'x'\,a' \in qatoms(S)$, if $preds(a) = preds(a')$, then $Q = Q'$, $x = x'$, $boundindset(Qx\,a) = boundindset(Q'x'\,a')$.

Theorem 3 *Let $n_0 \in \mathbb{N}$, $\phi \in Form_{\mathscr{L}}$, $T \subseteq Form_{\mathscr{L}}$. There exist $J_T^{\phi} \subseteq \{(i,j) \mid i \geq n_0\}$ and $S_T^{\phi} \subseteq SimOrdCl_{\mathscr{L} \cup \{\tilde{p}_j \mid j \in J_T^{\phi}\}}$ such that*

(i) *there exists an interpretation $\mathfrak{A}$ for $\mathscr{L}$ and $\mathfrak{A} \vDash T$, $\mathfrak{A} \therefore \vDash \phi$ if and only if there exists an interpretation $\mathfrak{A}'$ for $\mathscr{L} \cup \{\tilde{p}_j \mid j \in J_T^{\phi}\}$ and $\mathfrak{A}' \vDash S_T^{\phi}$, satisfying $\mathfrak{A} = \mathfrak{A}' \vert_{\mathscr{L}}$;*

(ii) *if $T \subseteq_{\mathscr{F}} Form_{\mathscr{L}}$, then $J_T^{\phi} \subseteq_{\mathscr{F}} \{(i,j) \mid i \geq n_0\}$, $\lVert J_T^{\phi} \rVert \in O(\lvert T \rvert + \lvert \phi \rvert)$, $S_T^{\phi} \subseteq_{\mathscr{F}} SimOrdCl_{\mathscr{L} \cup \{\tilde{p}_j \mid j \in J_T^{\phi}\}}$, $\lvert S_T^{\phi} \rvert \in O(\lvert T \rvert^2 + \lvert \phi \rvert^2)$; the number of all elementary operations of the translation of T and ϕ to S_T^{ϕ}, is in $O(\lvert T \rvert^2 + \lvert \phi \rvert^2)$; the time and space complexity of the translation of T and ϕ to S_T^{ϕ}, is in $O(\lvert T \rvert^2 \cdot \log(1 + n_0 + \lvert T \rvert) + \lvert \phi \rvert^2 \cdot (\log(1 + n_0) + \log \lvert \phi \rvert))$;*

(iii) *S_T^{ϕ} is admissible;*

(iv) *$tcons(S_T^{\phi}) \subseteq tcons(\phi) \cup tcons(T)$.*

Proof Cf. http://ii.fmph.uniba.sk/~guller/sci15.pdf, Sect. 3.3, Theorem 3, for a proof. ☐

Table 2 Binary interpolation rules for $\wedge$ and $\vee$

Case		Laws	
$\theta = \theta_1 \wedge \theta_2$			
Positive interpolation	$\dfrac{\tilde{p}_{\mathfrak{a}}(\bar{x}) \to \theta_1 \wedge \theta_2}{(\tilde{p}_{\mathfrak{a}}(\bar{x}) \to \tilde{p}_{\mathfrak{a}_1}(\bar{x})) \wedge (\tilde{p}_{\mathfrak{a}_1}(\bar{x}) \to \theta_1) \wedge (\tilde{p}_{\mathfrak{a}_2}(\bar{x}) \to \theta_2)}$	(6)	(13)
|Consequent| $= 9 + 4 \cdot |\bar{x}| + |\tilde{p}_{\mathfrak{a}_1}(\bar{x}) \to \theta_1| + |\tilde{p}_{\mathfrak{a}_2}(\bar{x}) \to \theta_2|$			
Positive interpolation	$\dfrac{\tilde{p}_{\mathfrak{a}}(\bar{x}) \to \theta_1 \wedge \theta_2}{\left\{\begin{array}{l}\tilde{p}_{\mathfrak{a}}(\bar{x}) < \tilde{p}_{\mathfrak{a}_1}(\bar{x}) \vee \tilde{p}_{\mathfrak{a}}(\bar{x}) < \tilde{p}_{\mathfrak{a}_2}(\bar{x}) \vee \tilde{p}_{\mathfrak{a}}(\bar{x}) \boxminus \tilde{p}_{\mathfrak{a}_2}(\bar{x}),\\ \tilde{p}_{\dot{\mathfrak{a}}}(\bar{x}) \to \theta_1 \cdot \tilde{p}_{\dot{\mathfrak{a}}_2}(\bar{x}) \to \theta_2\end{array}\right.}$		(14)
|Consequent| $= 12 + 8 \cdot |\bar{x}| + |\tilde{p}_{\mathfrak{a}_1}(\bar{x}) \to \theta_1| + |\tilde{p}_{\mathfrak{a}_2}(\bar{x}) \to \theta_2| \leq 15 \cdot (1 + |\bar{x}|) + |\tilde{p}_{\mathfrak{a}_1}(\bar{x}) \to \theta_1| + |\tilde{p}_{\mathfrak{a}_2}(\bar{x}) \to \theta_2|$			
Negative interpolation	$\dfrac{\theta_1 \wedge \theta_2 \to \tilde{p}_{\mathfrak{a}}(\bar{x})}{(\tilde{p}_{\mathfrak{a}_1}(\bar{x}) \to \tilde{p}_{\mathfrak{a}}(\bar{x}) \vee \tilde{p}_{\mathfrak{a}_2}(\bar{x}) \to \tilde{p}_{\mathfrak{a}}(\bar{x})) \wedge (\theta_1 \to \tilde{p}_{\mathfrak{a}_1}(\bar{x})) \wedge (\theta_2 \to \tilde{p}_{\mathfrak{a}_2}(\bar{x}))}$	(8)	(15)
|Consequent| $= 9 + 4 \cdot |\bar{x}| + |\theta_1 \to \tilde{p}_{\mathfrak{a}_1}(\bar{x})| + |\theta_2 \to \tilde{p}_{\mathfrak{a}_2}(\bar{x})| \leq 13 \cdot (1 + |\bar{x}|) + |\theta_1 \to \tilde{p}_{\mathfrak{a}_1}(\bar{x})| + |\theta_2 \to \tilde{p}_{\mathfrak{a}_2}(\bar{x})|$			
Negative interpolation	$\dfrac{\theta_1 \wedge \theta_2 \to \tilde{p}_{\dot{\mathfrak{a}}}(\bar{x})}{\left\{\begin{array}{l}\tilde{p}_{\mathfrak{a}_1}(\bar{x}) < \tilde{p}_{\mathfrak{a}}(\bar{x}) \vee \tilde{p}_{\mathfrak{a}_1}(\bar{x}) \boxminus \tilde{p}_{\mathfrak{a}}(\bar{x}) \vee \tilde{p}_{\mathfrak{a}_2}(\bar{x}) < \tilde{p}_{\mathfrak{a}}(\bar{x}) \vee \tilde{p}_{\mathfrak{a}_2}(\bar{x}) \boxminus \tilde{p}_{\mathfrak{a}}(\bar{x}),\\ \theta_1 \to \tilde{p}_{\mathfrak{a}_1}(\bar{x}),\, \theta_2 \to \tilde{p}_{\mathfrak{a}_2}(\bar{x})\end{array}\right.}$		(16)
|Consequent| $= 12 + 8 \cdot |\bar{x}| + |\theta_1 \to \tilde{p}_{\mathfrak{a}_1}(\bar{x})| + |\theta_2 \to \tilde{p}_{\mathfrak{a}_2}(\bar{x})| \leq 15 \cdot (1 + |\bar{x}|) + |\theta_1 \to \tilde{p}_{\mathfrak{a}_1}(\bar{x})| + |\theta_2 \to \tilde{p}_{\mathfrak{a}_2}(\bar{x})|$			
$\theta = \theta_1 \vee \theta_2$			
Positive interpolation	$\dfrac{\tilde{p}_{\mathfrak{a}}(\bar{x}) \to (\theta_1 \vee \theta_2)}{(\tilde{p}_{\mathfrak{a}}(\bar{x}) \to \tilde{p}_{\mathfrak{a}_1}(\bar{x}) \vee \tilde{p}_{\mathfrak{a}}(\bar{x}) \to \tilde{p}_{\mathfrak{a}_2}(\bar{x})) \wedge (\tilde{p}_{\mathfrak{a}_1}(\bar{x}) \to \theta_1) \wedge (\tilde{p}_{\mathfrak{a}_2}(\bar{x}) \to \theta_2)}$	(5)	(17)
|Consequent| $= 9 + 4 \cdot |\bar{x}| + |\tilde{p}_{\mathfrak{a}_1}(\bar{x}) \to \theta_1| + |\tilde{p}_{\mathfrak{a}_2}(\bar{x}) \to \theta_2| \leq 13 \cdot (1 + |\bar{x}|) + |\tilde{p}_{\mathfrak{a}_1}(\bar{x}) \to \theta_1| + |\tilde{p}_{\mathfrak{a}_2}(\bar{x}) \to \theta_2|$			

(continued)

Table 2 (continued)

Case		Laws
Positive interpolation	$\dfrac{\tilde{p}_{\bar{\mathbb{i}}}(\bar{x}) \to (\theta_1 \vee \theta_2)}{\tilde{p}_{\bar{\mathbb{i}}}(\bar{x}) < \tilde{p}_{\bar{\mathbb{i}}_1}(\bar{x}) \vee \tilde{p}_{\bar{\mathbb{i}}}(\bar{x}) \rightleftharpoons \tilde{p}_{\bar{\mathbb{i}}_1}(\bar{x}) \vee \tilde{p}_{\bar{\mathbb{i}}}(\bar{x}) < \tilde{p}_{\bar{\mathbb{i}}_2}(\bar{x}) \vee \tilde{p}_{\bar{\mathbb{i}}}(\bar{x}) \rightleftharpoons \tilde{p}_{\bar{\mathbb{i}}_2}(\bar{x}),\ \tilde{p}_{\bar{\mathbb{i}}_1}(\bar{x}) \to \theta_1, \tilde{p}_{\bar{\mathbb{i}}_2}(\bar{x}) \to \theta_2}$	(18)
	$|\text{Consequent}| = 12 + 8 \cdot |\bar{x}| + |\tilde{p}_{\bar{\mathbb{i}}_1}(\bar{x}) \to \theta_1| + |\tilde{p}_{\bar{\mathbb{i}}_2}(\bar{x}) \to \theta_2| \leq 15 \cdot (1 + |\bar{x}|) + |\tilde{p}_{\bar{\mathbb{i}}_1}(\bar{x}) \to \theta_1| + |\tilde{p}_{\bar{\mathbb{i}}_2}(\bar{x}) \to \theta_2|$	
Negative interpolation	$\dfrac{(\theta_1 \vee \theta_2) \to \tilde{p}_{\bar{\mathbb{i}}}(\bar{x})}{(\tilde{p}_{\bar{\mathbb{i}}_1}(\bar{x}) \to \tilde{p}_{\bar{\mathbb{i}}}(\bar{x})) \wedge (\tilde{p}_{\bar{\mathbb{i}}_2}(\bar{x}) \to \tilde{p}_{\bar{\mathbb{i}}}(\bar{x})) \wedge (\theta_1 \to \tilde{p}_{\bar{\mathbb{i}}_1}(\bar{x})) \wedge (\theta_2 \to \tilde{p}_{\bar{\mathbb{i}}_2}(\bar{x}))}$	(19) (7)
	$|\text{Consequent}| = 9 + 4 \cdot |\bar{x}| + |\theta_1 \to \tilde{p}_{\bar{\mathbb{i}}_1}(\bar{x})| + |\theta_2 \to \tilde{p}_{\bar{\mathbb{i}}_2}(\bar{x})| \leq 13 \cdot (1 + |\bar{x}|) + |\theta_1 \to \tilde{p}_{\bar{\mathbb{i}}_1}(\bar{x})| + |\theta_2 \to \tilde{p}_{\bar{\mathbb{i}}_2}(\bar{x})|$	
Negative interpolation	$\dfrac{(\theta_1 \vee \theta_2) \to \tilde{p}_{\bar{\mathbb{i}}}(\bar{x})}{\tilde{p}_{\bar{\mathbb{i}}_1}(\bar{x}) < \tilde{p}_{\bar{\mathbb{i}}}(\bar{x}) \vee \tilde{p}_{\bar{\mathbb{i}}_1}(\bar{x}) \rightleftharpoons \tilde{p}_{\bar{\mathbb{i}}}(\bar{x}),\ \tilde{p}_{\bar{\mathbb{i}}_2}(\bar{x}) < \tilde{p}_{\bar{\mathbb{i}}}(\bar{x}) \vee \tilde{p}_{\bar{\mathbb{i}}_2}(\bar{x}) \rightleftharpoons \tilde{p}_{\bar{\mathbb{i}}}(\bar{x}),\ \theta_1 \to \tilde{p}_{\bar{\mathbb{i}}_1}(\bar{x}),\ \theta_2 \to \tilde{p}_{\bar{\mathbb{i}}_2}(\bar{x})}$	(20)
	$|\text{Consequent}| = 12 + 8 \cdot |\bar{x}| + |\theta_1 \to \tilde{p}_{\bar{\mathbb{i}}_1}(\bar{x})| + |\theta_2 \to \tilde{p}_{\bar{\mathbb{i}}_2}(\bar{x})| \leq 15 \cdot (1 + |\bar{x}|) + |\theta_1 \to \tilde{p}_{\bar{\mathbb{i}}_1}(\bar{x})| + |\theta_2 \to \tilde{p}_{\bar{\mathbb{i}}_2}(\bar{x})|$	

Table 3 Binary interpolation rules for $\to$

Case		Laws	
$\theta = \theta_1 \to \theta_2,\ \theta_2 \neq 0$			
Positive interpolation	$\dfrac{\tilde{p}_{\tilde{i}}(\bar{x}) \to (\theta_1 \to \theta_2)}{(\tilde{p}_{\tilde{i}}(\bar{x}) \to \tilde{p}_{\tilde{i}_2}(\bar{x}) \vee \tilde{p}_{\tilde{i}_1}(\bar{x}) \to \tilde{p}_{\tilde{i}_2}(\bar{x})) \wedge (\theta_1 \to \tilde{p}_{\tilde{i}_1}(\bar{x}) \wedge (\tilde{p}_{\tilde{i}_2}(\bar{x}) \to \theta_2))}$	(9), (8)	(21)
	$\lvert\text{Consequent}\rvert = 9 + 4 \cdot \lvert\bar{x}\rvert + \lvert\theta_1 \to \tilde{p}_{\tilde{i}_1}(\bar{x})\rvert + \lvert\tilde{p}_{\tilde{i}_2}(\bar{x}) \to \theta_2\rvert \leq 13 \cdot (1 + \lvert\bar{x}\rvert) + \lvert\theta_1 \to \tilde{p}_{\tilde{i}_1}(\bar{x})\rvert + \lvert\tilde{p}_{\tilde{i}_2}(\bar{x}) \to \theta_2\rvert$		
Positive interpolation	$\dfrac{\tilde{p}_{\tilde{i}}(\bar{x}) \to (\theta_1 \to \theta_2)}{\left\{\begin{array}{l}\tilde{p}_{\tilde{i}}(\bar{x}) \prec \tilde{p}_{\tilde{i}_2}(\bar{x}) \vee \tilde{p}_{\tilde{i}}(\bar{x}) \boxplus \tilde{p}_{\tilde{i}_2}(\bar{x}) \vee \tilde{p}_{\tilde{i}_1}(\bar{x}) \prec \tilde{p}_{\tilde{i}_2}(\bar{x}) \vee \tilde{p}_{\tilde{i}_1}(\bar{x}) \boxplus \tilde{p}_{\tilde{i}_2}(\bar{x}),\\ \theta_1 \to \tilde{p}_{\tilde{i}_1}(\bar{x}), \tilde{p}_{\tilde{i}_2}(\bar{x}) \to \theta_2\end{array}\right.}$		(22)
	$\lvert\text{Consequent}\rvert = 12 + 8 \cdot \lvert\bar{x}\rvert + \lvert\theta_1 \to \tilde{p}_{\tilde{i}_1}(\bar{x})\rvert + \lvert\tilde{p}_{\tilde{i}_2}(\bar{x}) \to \theta_2\rvert \leq 15 \cdot (1 + \lvert\bar{x}\rvert) + \lvert\theta_1 \to \tilde{p}_{\tilde{i}_1}(\bar{x})\rvert + \lvert\tilde{p}_{\tilde{i}_2}(\bar{x}) \to \theta_2\rvert$		
Negative interpolation	$\dfrac{(\theta_1 \to \theta_2) \to \tilde{p}_{\tilde{i}}(\bar{x})}{((\tilde{p}_{\tilde{i}_1}(\bar{x}) \to \tilde{p}_{\tilde{i}_2}(\bar{x}) \to \tilde{p}_{\tilde{i}_2}(\bar{x}) \vee \tilde{p}_{\tilde{i}}(\bar{x})) \wedge (\tilde{p}_{\tilde{i}_2}(\bar{x}) \to \tilde{p}_{\tilde{i}}(\bar{x})) \wedge}{\quad (\tilde{p}_{\tilde{i}_1}(\bar{x}) \to \theta_1) \wedge (\theta_2 \to \tilde{p}_{\tilde{i}_2}(\bar{x}))}$	(11), (3), (1)	(23)
	$\lvert\text{Consequent}\rvert = 13 + 6 \cdot \lvert\bar{x}\rvert + \lvert\tilde{p}_{\tilde{i}_1}(\bar{x}) \to \theta_1\rvert + \lvert\theta_2 \to \tilde{p}_{\tilde{i}_2}(\bar{x})\rvert \leq 13 \cdot (1 + \lvert\bar{x}\rvert) + \lvert\tilde{p}_{\tilde{i}_1}(\bar{x}) \to \theta_1\rvert + \lvert\theta_2 \to \tilde{p}_{\tilde{i}_2}(\bar{x})\rvert$		
Negative interpolation	$\dfrac{(\theta_1 \to \theta_2) \to \tilde{p}_{\tilde{i}}(\bar{x})}{\left\{\begin{array}{l}\tilde{p}_{\tilde{i}_2}(\bar{x}) \prec \tilde{p}_{\tilde{i}_1}(\bar{x}) \vee \tilde{p}_{\tilde{i}_2}(\bar{x}) \boxplus 1 \vee \tilde{p}_{\tilde{i}}(\bar{x}) \boxplus 1,\\ \tilde{p}_{\tilde{i}_2}(\bar{x}) \prec \tilde{p}_{\tilde{i}_1}(\bar{x}) \vee \tilde{p}_{\tilde{i}_2}(\bar{x}) \boxplus \tilde{p}_{\tilde{i}_1}(\bar{x}), \tilde{p}_{\tilde{i}_1}(\bar{x}) \to \theta_1, \theta_2 \to \tilde{p}_{\tilde{i}_2}(\bar{x})\end{array}\right.}$		(24)
	$\lvert\text{Consequent}\rvert = 15 + 8 \cdot \lvert\bar{x}\rvert + \lvert\tilde{p}_{\tilde{i}_1}(\bar{x}) \to \theta_1\rvert + \lvert\theta_2 \to \tilde{p}_{\tilde{i}_2}(\bar{x})\rvert \leq 15 \cdot (1 + \lvert\bar{x}\rvert) + \lvert\tilde{p}_{\tilde{i}_1}(\bar{x}) \to \theta_1\rvert + \lvert\theta_2 \to \tilde{p}_{\tilde{i}_2}(\bar{x})\rvert$		

Table 4 Unary interpolation rules for $\rightarrow$

Case	Laws	
$\theta = \theta_1 \rightarrow 0$		
Positive interpolation $\dfrac{\tilde{p}_\mathbf{i}(\bar{x}) \rightarrow (\theta_1 \rightarrow 0)}{(\tilde{p}_\mathbf{i}(\bar{x}) \rightarrow 0 \vee \tilde{p}_{\mathbf{i}_1}(\bar{x}) \rightarrow 0) \wedge (\theta_1 \rightarrow \tilde{p}_{\mathbf{i}_1}(\bar{x}))}$	(9), (8)	(25)
|Consequent| $= 8 + 2 \cdot |\bar{x}| + |\theta_1 \rightarrow \tilde{p}_{\mathbf{i}_1}(\bar{x})| \leq 13 \cdot (1 + |\bar{x}|) + |\theta_1 \rightarrow \tilde{p}_{\mathbf{i}_1}(\bar{x})|$		
Positive interpolation $\dfrac{\tilde{p}_\mathbf{i}(\bar{x}) \rightarrow (\theta_1 \rightarrow 0)}{\{\tilde{p}_\mathbf{i}(\bar{x}) \doteq 0 \vee \tilde{p}_{\mathbf{i}_1}(\bar{x}) \doteq 0, \theta_1 \rightarrow \tilde{p}_{\mathbf{i}_1}(\bar{x})\}}$		(26)
|Consequent| $= 6 + 2 \cdot |\bar{x}| + |\theta_1 \rightarrow \tilde{p}_{\mathbf{i}_1}(\bar{x})| \leq 15 \cdot (1 + |\bar{x}|) + |\theta_1 \rightarrow \tilde{p}_{\mathbf{i}_1}(\bar{x})|$		
Negative interpolation $\dfrac{(\theta_1 \rightarrow 0) \rightarrow \tilde{p}_\mathbf{i}(\bar{x})}{((\tilde{p}_{\mathbf{i}_1}(\bar{x}) \rightarrow 0) \rightarrow 0 \vee \tilde{p}_\mathbf{i}(\bar{x})) \wedge (\tilde{p}_{\mathbf{i}_1}(\bar{x}) \rightarrow \theta_1)}$	(11)	(27)
|Consequent| $= 8 + 2 \cdot |\bar{x}| + |\tilde{p}_{\mathbf{i}_1}(\bar{x}) \rightarrow \theta_1| \leq 13 \cdot (1 + |\bar{x}|) + |\tilde{p}_{\mathbf{i}_1}(\bar{x}) \rightarrow \theta_1|$		
Negative interpolation $\dfrac{(\theta_1 \rightarrow 0) \rightarrow \tilde{p}_\mathbf{i}(\bar{x})}{\{0 < \tilde{p}_{\mathbf{i}_1}(\bar{x}) \vee \tilde{p}_\mathbf{i}(\bar{x}) \doteq 1, \tilde{p}_{\mathbf{i}_1}(\bar{x}) \rightarrow \theta_1\}}$		(28)
|Consequent| $= 6 + 2 \cdot |\bar{x}| + |\tilde{p}_{\mathbf{i}_1}(\bar{x}) \rightarrow \theta_1| \leq 15 \cdot (1 + |\bar{x}|) + |\tilde{p}_{\mathbf{i}_1}(\bar{x}) \rightarrow \theta_1|$		

Table 5 Unary interpolation rules for $\forall$ and $\exists$

Case	
$\theta = \forall x \, \theta_1$	
Positive interpolation $\dfrac{\tilde{p}_\mathbf{i}(\bar{x}) \rightarrow \forall x \, \theta_1}{(\tilde{p}_\mathbf{i}(\bar{x}) \rightarrow \forall x \, \tilde{p}_{\mathbf{i}_1}(\bar{x})) \wedge (\tilde{p}_{\mathbf{i}_1}(\bar{x}) \rightarrow \theta_1)}$	(29)
|Consequent| $= 6 + 2 \cdot |\bar{x}| + |\tilde{p}_{\mathbf{i}_1}(\bar{x}) \rightarrow \theta_1| \leq 13 \cdot (1 + |\bar{x}|) + |\tilde{p}_{\mathbf{i}_1}(\bar{x}) \rightarrow \theta_1|$	
Positive interpolation $\dfrac{\tilde{p}_\mathbf{i}(\bar{x}) \rightarrow \forall x \, \theta_1}{\{\tilde{p}_\mathbf{i}(\bar{x}) < \forall x \, \tilde{p}_{\mathbf{i}_1}(\bar{x}) \vee \tilde{p}_\mathbf{i}(\bar{x}) \doteq \forall x \, \tilde{p}_{\mathbf{i}_1}(\bar{x}), \tilde{p}_{\mathbf{i}_1}(\bar{x}) \rightarrow \theta_1\}}$	(30)
|Consequent| $= 10 + 4 \cdot |\bar{x}| + |\tilde{p}_{\mathbf{i}_1}(\bar{x}) \rightarrow \theta_1| \leq 15 \cdot (1 + |\bar{x}|) + |\tilde{p}_{\mathbf{i}_1}(\bar{x}) \rightarrow \theta_1|$	
Negative interpolation $\dfrac{\forall x \, \theta_1 \rightarrow \tilde{p}_\mathbf{i}(\bar{x})}{(\forall x \, \tilde{p}_{\mathbf{i}_1}(\bar{x}) \rightarrow \tilde{p}_\mathbf{i}(\bar{x})) \wedge (\theta_1 \rightarrow \tilde{p}_{\mathbf{i}_1}(\bar{x}))}$	(31)
|Consequent| $= 6 + 2 \cdot |\bar{x}| + |\theta_1 \rightarrow \tilde{p}_{\mathbf{i}_1}(\bar{x})| \leq 13 \cdot (1 + |\bar{x}|) + |\theta_1 \rightarrow \tilde{p}_{\mathbf{i}_1}(\bar{x})|$	
Negative interpolation $\dfrac{\forall x \, \theta_1 \rightarrow \tilde{p}_\mathbf{i}(\bar{x})}{\{\forall x \, \tilde{p}_{\mathbf{i}_1}(\bar{x}) < \tilde{p}_\mathbf{i}(\bar{x}) \vee \forall x \, \tilde{p}_{\mathbf{i}_1}(\bar{x}) \doteq \tilde{p}_\mathbf{i}(\bar{x}), \theta_1 \rightarrow \tilde{p}_{\mathbf{i}_1}(\bar{x})\}}$	(32)
|Consequent| $= 10 + 4 \cdot |\bar{x}| + |\theta_1 \rightarrow \tilde{p}_{\mathbf{i}_1}(\bar{x})| \leq 15 \cdot (1 + |\bar{x}|) + |\theta_1 \rightarrow \tilde{p}_{\mathbf{i}_1}(\bar{x})|$	
$\theta = \exists x \, \theta_1$	
Positive interpolation $\dfrac{\tilde{p}_\mathbf{i}(\bar{x}) \rightarrow \exists x \, \theta_1}{(\tilde{p}_\mathbf{i}(\bar{x}) \rightarrow \exists x \, \tilde{p}_{\mathbf{i}_1}(\bar{x})) \wedge (\tilde{p}_{\mathbf{i}_1}(\bar{x}) \rightarrow \theta_1)}$	(33)
|Consequent| $= 6 + 2 \cdot |\bar{x}| + |\tilde{p}_{\mathbf{i}_1}(\bar{x}) \rightarrow \theta_1| \leq 13 \cdot (1 + |\bar{x}|) + |\tilde{p}_{\mathbf{i}_1}(\bar{x}) \rightarrow \theta_1|$	

(continued)

226 D. Guller

Table 5 (continued)

Case	
Positive interpolation $\dfrac{\tilde{p}_{\tilde{\mathbf{i}}}(\bar{x}) \to \exists x\, \theta_1}{\{\tilde{p}_{\tilde{\mathbf{i}}}(\bar{x}) \prec \exists x\, \tilde{p}_{\tilde{\mathbf{i}}_1}(\bar{x}) \vee \tilde{p}_{\tilde{\mathbf{i}}}(\bar{x}) \approx \exists x\, \tilde{p}_{\tilde{\mathbf{i}}_1}(\bar{x}),\, \tilde{p}_{\tilde{\mathbf{i}}_1}(\bar{x}) \to \theta_1\}}$	(34)
$\lvert\text{Consequent}\rvert = 10 + 4 \cdot \lvert\bar{x}\rvert + \lvert\tilde{p}_{\tilde{\mathbf{i}}_1}(\bar{x}) \to \theta_1\rvert \le 15 \cdot (1 + \lvert\bar{x}\rvert) + \lvert\tilde{p}_{\tilde{\mathbf{i}}_1}(\bar{x}) \to \theta_1\rvert$	
Negative interpolation $\dfrac{\exists x\, \theta_1 \to \tilde{p}_{\tilde{\mathbf{i}}}(\bar{x})}{(\exists x\, \tilde{p}_{\tilde{\mathbf{i}}_1}(\bar{x}) \to \tilde{p}_{\tilde{\mathbf{i}}}(\bar{x})) \wedge (\theta_1 \to \tilde{p}_{\tilde{\mathbf{i}}_1}(\bar{x}))}$	(35)
$\lvert\text{Consequent}\rvert = 6 + 2 \cdot \lvert\bar{x}\rvert + \lvert\theta_1 \to \tilde{p}_{\tilde{\mathbf{i}}_1}(\bar{x})\rvert \le 13 \cdot (1 + \lvert\bar{x}\rvert) + \lvert\theta_1 \to \tilde{p}_{\tilde{\mathbf{i}}_1}(\bar{x})\rvert$	
Negative interpolation $\dfrac{\exists x\, \theta_1 \to \tilde{p}_{\tilde{\mathbf{i}}}(\bar{x})}{\{\exists x\, \tilde{p}_{\tilde{\mathbf{i}}_1}(\bar{x}) \prec \tilde{p}_{\tilde{\mathbf{i}}}(\bar{x}) \vee \exists x\, \tilde{p}_{\tilde{\mathbf{i}}_1}(\bar{x}) \approx \tilde{p}_{\tilde{\mathbf{i}}}(\bar{x}),\, \theta_1 \to \tilde{p}_{\tilde{\mathbf{i}}_1}(\bar{x})\}}$	(36)
$\lvert\text{Consequent}\rvert = 10 + 4 \cdot \lvert\bar{x}\rvert + \lvert\theta_1 \to \tilde{p}_{\tilde{\mathbf{i}}_1}(\bar{x})\rvert \le 15 \cdot (1 + \lvert\bar{x}\rvert) + \lvert\theta_1 \to \tilde{p}_{\tilde{\mathbf{i}}_1}(\bar{x})\rvert$	

Table 6 Unary interpolation rules for Δ

Case	
$\theta = \Delta\theta_1$	
Positive interpolation $\dfrac{\tilde{p}_{\tilde{\mathbf{i}}}(\bar{x}) \to \Delta\theta_1}{(\tilde{p}_{\tilde{\mathbf{i}}}(\bar{x}) \to \Delta\tilde{p}_{\tilde{\mathbf{i}}_1}(\bar{x})) \wedge (\tilde{p}_{\tilde{\mathbf{i}}_1}(\bar{x}) \to \theta_1)}$	(37)
$\lvert\text{Consequent}\rvert = 5 + 2 \cdot \lvert\bar{x}\rvert + \lvert\tilde{p}_{\tilde{\mathbf{i}}_1}(\bar{x}) \to \theta_1\rvert \le 13 \cdot (1 + \lvert\bar{x}\rvert) + \lvert\tilde{p}_{\tilde{\mathbf{i}}_1}(\bar{x}) \to \theta_1\rvert$	
Positive interpolation $\dfrac{\tilde{p}_{\tilde{\mathbf{i}}}(\bar{x}) \to \Delta\theta_1}{\{\tilde{p}_{\tilde{\mathbf{i}}}(\bar{x}) \approx 0 \vee \tilde{p}_{\tilde{\mathbf{i}}_1}(\bar{x}) \approx 1,\, \tilde{p}_{\tilde{\mathbf{i}}_1}(\bar{x}) \to \theta_1\}}$	(38)
$\lvert\text{Consequent}\rvert = 6 + 2 \cdot \lvert\bar{x}\rvert + \lvert\tilde{p}_{\tilde{\mathbf{i}}_1}(\bar{x}) \to \theta_1\rvert \le 15 \cdot (1 + \lvert\bar{x}\rvert) + \lvert\tilde{p}_{\tilde{\mathbf{i}}_1}(\bar{x}) \to \theta_1\rvert$	
Negative interpolation $\dfrac{\Delta\theta_1 \to \tilde{p}_{\tilde{\mathbf{i}}}(\bar{x})}{(\Delta\tilde{p}_{\tilde{\mathbf{i}}_1}(\bar{x}) \to \tilde{p}_{\tilde{\mathbf{i}}}(\bar{x})) \wedge (\theta_1 \to \tilde{p}_{\tilde{\mathbf{i}}_1}(\bar{x}))}$	(39)
$\lvert\text{Consequent}\rvert = 5 + 2 \cdot \lvert\bar{x}\rvert + \lvert\theta_1 \to \tilde{p}_{\tilde{\mathbf{i}}_1}(\bar{x})\rvert \le 13 \cdot (1 + \lvert\bar{x}\rvert) + \lvert\theta_1 \to \tilde{p}_{\tilde{\mathbf{i}}_1}(\bar{x})\rvert$	
Negative interpolation $\dfrac{\Delta\theta_1 \to \tilde{p}_{\tilde{\mathbf{i}}}(\bar{x})}{\{\tilde{p}_{\tilde{\mathbf{i}}_1}(\bar{x}) \prec 1 \vee \tilde{p}_{\tilde{\mathbf{i}}}(\bar{x}) \approx 1,\, \theta_1 \to \tilde{p}_{\tilde{\mathbf{i}}_1}(\bar{x})\}}$	(40)
$\lvert\text{Consequent}\rvert = 6 + 2 \cdot \lvert\bar{x}\rvert + \lvert\theta_1 \to \tilde{p}_{\tilde{\mathbf{i}}_1}(\bar{x})\rvert \le 15 \cdot (1 + \lvert\bar{x}\rvert) + \lvert\theta_1 \to \tilde{p}_{\tilde{\mathbf{i}}_1}(\bar{x})\rvert$	

Corollary 1 *Let $n_0 \in \mathbb{N}$, $\phi \in Form_{\mathscr{L}}$, $T \subseteq Form_{\mathscr{L}}$. There exist $J_T^{\phi} \subseteq \{(i,j) \mid i \ge n_0\}$ and $S_T^{\phi} \subseteq SimOrdCl_{\mathscr{L} \cup \{\tilde{p}_j \mid j \in J_T^{\phi}\}}$ such that*

(i) *$T \vDash \phi$ if and only if S_T^{ϕ} is unsatisfiable;*

(ii) *if $T \subseteq_{\mathscr{F}} Form_{\mathscr{L}}$, then $J_T^{\phi} \subseteq_{\mathscr{F}} \{(i,j) \mid i \ge n_0\}$, $\lVert J_T^{\phi}\rVert \in O(\lvert T\rvert + \lvert\phi\rvert)$, $S_T^{\phi} \subseteq_{\mathscr{F}} SimOrdCl_{\mathscr{L} \cup \{\tilde{p}_j \mid j \in J_T^{\phi}\}}$, $\lvert S_T^{\phi}\rvert \in O(\lvert T\rvert^2 + \lvert\phi\rvert^2)$; the number of all elementary operations of the translation of T and ϕ to S_T^{ϕ}, is in $O(\lvert T\rvert^2 + \lvert\phi\rvert^2)$; the time and space complexity of the translation of T and ϕ to $S_{T'}^{\phi}$, is in $O(\lvert T\rvert^2 \cdot \log(1 + n_0 + \lvert T\rvert) + \lvert\phi\rvert^2 \cdot (\log(1 + n_0) + \log\lvert\phi\rvert))$;*

(iii) *S_T^{ϕ} is admissible;*

(iv) *$tcons(S_T^{\phi}) \subseteq tcons(\phi) \cup tcons(T)$.*

Proof Cf. http://ii.fmph.uniba.sk/~guller/sci15.pdf, Sect. 3.3, Corollary 1, for a proof. □

4 Hyperresolution over Order Clauses

In this section, we propose an order hyperresolution calculus with truth constants operating over order clausal theories.

4.1 Order Hyperresolution Rules

At first, we introduce some basic notions and notation concerning chains of order literals. A chain $\varXi$ of $\mathscr{L}$ is a sequence $\varXi = \varepsilon_0 \diamond_0 v_0, \dots, \varepsilon_n \diamond_n v_n$, $\varepsilon_i \diamond_i v_i \in OrdLit_{\mathscr{L}}$, such that for all $i < n$, $v_i = \varepsilon_{i+1}$. ε_0 is the beginning element of $\varXi$ and v_n the ending element of $\varXi$. $\varepsilon_0 \varXi v_n$ denotes $\varXi$ together with its respective beginning and ending element. Let $\varXi = \varepsilon_0 \diamond_0 v_0, \dots, \varepsilon_n \diamond_n v_n$ be a chain of $\mathscr{L}$. $\varXi$ is an equality chain of $\mathscr{L}$ iff, for all $i \leq n$, $\diamond_i = {=}$. $\varXi$ is an increasing chain of $\mathscr{L}$ iff there exists $i^* \leq n$ such that $\diamond_{i^*} = {<}$. $\varXi$ is a contradiction of $\mathscr{L}$ iff $\varXi$ is an increasing chain of $\mathscr{L}$ of the form $\varepsilon_0 \varXi 0$ or $1 \varXi v_n$ or $\varepsilon_0 \varXi \varepsilon_0$. Let $S \subseteq OrdCl_{\mathscr{L}}$ be unit and $\varXi = \varepsilon_0 \diamond_0 v_0, \dots, \varepsilon_n \diamond_n v_n$ be a chain | an equality chain | an increasing chain | | a contradiction of $\mathscr{L}$. $\varXi$ is a chain | an equality chain | an increasing chain | a contradiction of S iff, for all $i \leq n$, $\varepsilon_i \diamond_i v_i \in S$. Let $\tilde{\mathbb{W}} = \{\tilde{w}_i \mid i \in \mathbb{I}\}$ such that $\tilde{\mathbb{W}} \cap (Func_{\mathscr{L}} \cup \{\tilde{f}_0\}) = \emptyset$; $\tilde{\mathbb{W}}$ is an infinite countable set of new function symbols. Let $\mathscr{L}$ contain a constant (nullary function) symbol. Let $P \subseteq \tilde{\mathbb{P}}$ and $S \subseteq OrdCl_{\mathscr{L} \cup P}$. We denote $GOrdCl_{\mathscr{L}} = \{C \mid C \in OrdCl_{\mathscr{L}}$ is closed$\} \subseteq OrdCl_{\mathscr{L}}$, $GInst_{\mathscr{L}}(S) = \{C \mid C \in GOrdCl_{\mathscr{L}}$ is an instance of S of $\mathscr{L}\} \subseteq GOrdCl_{\mathscr{L}}$, $ordtcons(S) = \{0 < 1\} \cup \{0 < \bar{c} \mid \bar{c} \in tcons(S) \cap \overline{C}_{\mathscr{L}}\} \cup \{\bar{c} < 1 \mid \bar{c} \in tcons(S) \cap \overline{C}_{\mathscr{L}}\} \cup \{\bar{c}_1 < \bar{c}_2 \mid \bar{c}_1, \bar{c}_2 \in tcons(S) \cap \overline{C}_{\mathscr{L}}, c_1 < c_2\} \subseteq GOrdCl_{\mathscr{L}}$.

A basic order hyperresolution calculus is defined as follows. The first rule is a central order hyperresolution one with obvious intuition.

$$(Basic\ order\ hyperresolution\ rule) \qquad (41)$$

$$\frac{l_0 \vee C_0, \dots, l_n \vee C_n \in S_{\kappa-1}}{\displaystyle\bigvee_{i=0}^{n} C_i \in S_\kappa};$$

$l_0, \dots, l_n$ is a contradiction of $\mathscr{L}_{\kappa-1}$.

We say that $\bigvee_{i=0}^{n} C_i$ is a basic order hyperresolvent of $l_0 \vee C_0, \dots, l_n \vee C_n$. The second rule is an auxiliary one which ensures a total order over derived atoms.

$$(Basic\ order\ trichotomy\ rule) \qquad (42)$$

$$\frac{a, b \in atoms(S_{\kappa-1}), \{a, b\} \nsubseteq Tcons_{\mathscr{L}}}{a \prec b \vee a \bumpeq b \vee b \prec a \in S_{\kappa}}.$$

$a \prec b \vee a \bumpeq b \vee b \prec a$ is a basic order trichotomy resolvent of a and b. The next two rules order a quantified atom and its ground instances.

$$(Basic\ order\ \forall\ \text{-quantification rule}) \qquad (43)$$

$$\frac{\forall x\, a \in qatoms^{\forall}(S_{\kappa-1})}{\forall x\, a \prec a\gamma \vee \forall x\, a \bumpeq a\gamma \in S_{\kappa}};$$

$$t \in GTerm_{\mathscr{L}_{\kappa-1}}, \gamma = x/t \in Subst_{\mathscr{L}_{\kappa-1}}, dom(\gamma) = \{x\} = vars(a).$$

$\forall x\, a \prec a\gamma \vee \forall x\, a \bumpeq a\gamma$ is a basic order $\forall$-quantification resolvent of $\forall x\, a$.

$$(Basic\ order\ \exists\ \text{-quantification rule}) \qquad (44)$$

$$\frac{\exists x\, a \in qatoms^{\exists}(S_{\kappa-1})}{a\gamma \prec \exists x\, a \vee a\gamma \bumpeq \exists x\, a \in S_{\kappa}};$$

$$t \in GTerm_{\mathscr{L}_{\kappa-1}}, \gamma = x/t \in Subst_{\mathscr{L}_{\kappa-1}}, dom(\gamma) = \{x\} = vars(a).$$

$a\gamma \prec \exists x\, a \vee a\gamma \bumpeq \exists x\, a$ is a basic order $\exists$-quantification resolvent of $\exists x\, a$. The last two rules introduce a witness with respect to infimum | supremum, as a ground term with a new function symbol, between a derived quantified atom and an atom | a quantified atom.

$$(Basic\ order\ \forall\text{-witnessing rule}) \qquad (45)$$

$$\frac{\forall x\, a \in qatoms^{\forall}(S_{\kappa-1}), b \in atoms(S_{\kappa-1}) \cup qatoms(S_{\kappa-1})}{a\gamma \prec b \vee b \bumpeq \forall x\, a \vee b \prec \forall x\, a \in S_{\kappa}};$$

$$\tilde{w} \in \tilde{\mathbb{W}} - Func_{\mathscr{L}_{\kappa-1}}, ar(\tilde{w}) = |freetermseq(\forall x\, a), freetermseq(b)|,$$
$$\gamma = x/\tilde{w}(freetermseq(\forall x\, a), freetermseq(b)) \in Subst_{\mathscr{L}_{\kappa}}, dom(\gamma) = \{x\} = vars(a).$$

$a\gamma \prec b \vee b \bumpeq \forall x\, a \vee b \prec \forall x\, a$ is a basic order $\forall$-witnessing resolvent of $\forall x\, a$ and b.

$$(Basic\ order\ \exists\text{-witnessing rule}) \qquad (46)$$

$$\frac{\exists x\, a \in qatoms^{\exists}(S_{\kappa-1}), b \in atoms(S_{\kappa-1}) \cup qatoms(S_{\kappa-1})}{b \prec a\gamma \vee \exists x\, a \bumpeq b \vee \exists x\, a \prec b \in S_{\kappa}};$$

$$\tilde{w} \in \tilde{\mathbb{W}} - Func_{\mathscr{L}_{\kappa-1}}, ar(\tilde{w}) = |freetermseq(\exists x\, a), freetermseq(b)|,$$
$$\gamma = x/\tilde{w}(freetermseq(\exists x\, a), freetermseq(b)) \in Subst_{\mathscr{L}_{\kappa}}, dom(\gamma) = \{x\} = vars(a).$$

$b \prec a\gamma \vee \exists x\, a \bumpeq b \vee \exists x\, a \prec b$ is a basic order $\exists$-witnessing resolvent of $\exists x\, a$ and b.

The basic order hyperresolution calculus can be generalised to an order hyperresolution one. Intuition behind rules is similar to that in the basic case.

$$(\text{Order hyperresolution rule}) \qquad (47)$$

$$\frac{\bigvee_{j=0}^{k_0} \varepsilon_j^0 \diamond_j^0 v_j^0 \vee \bigvee_{j=1}^{m_0} l_j^0, \ldots, \bigvee_{j=0}^{k_n} \varepsilon_j^n \diamond_j^n v_j^n \vee \bigvee_{j=1}^{m_n} l_j^n \in S_{\kappa-1}^{Vr}}{\left(\bigvee_{i=0}^{n} \bigvee_{j=1}^{m_i} l_j^i\right)\theta \in S_\kappa};$$

for all $i < i' \leq n$,

$\quad freevars(\bigvee_{j=0}^{k_i} \varepsilon_j^i \diamond_j^i v_j^i \vee \bigvee_{j=1}^{m_i} l_j^i) \cap freevars(\bigvee_{j=0}^{k_{i'}} \varepsilon_j^{i'} \diamond_j^{i'} v_j^{i'} \vee \bigvee_{j=1}^{m_{i'}} l_j^{i'}) = \emptyset$,

$\theta \in mgu_{\mathscr{L}_{\kappa-1}}\left(\bigvee_{j=0}^{k_0} \varepsilon_j^0 \diamond_j^0 v_j^0, l_1^0, \ldots, l_{m_0}^0, \ldots, \bigvee_{j=0}^{k_n} \varepsilon_j^n \diamond_j^n v_j^n, l_1^n, \ldots, l_{m_n}^n, \right.$

$\qquad \left. \{v_0^0, \varepsilon_0^1\}, \ldots, \{v_0^{n-1}, \varepsilon_0^n\}, \{a, b\}\right)$,

$dom(\theta) = freevars\left(\{\varepsilon_j^i \diamond_j^i v_j^i \mid j \leq k_i, i \leq n\}, \{l_j^i \mid 1 \leq j \leq m_i, i \leq n\}\right)$,

$a = \varepsilon_0^0, b = 1$ *or* $a = v_0^n, b = 0$ *or* $a = \varepsilon_0^0, b = v_0^n$,

there exists $i^* \leq n$ *suchthat* $\diamond_0^{i^*} = \prec$.

$\left(\bigvee_{i=0}^{n} \bigvee_{j=1}^{m_i} l_j^i\right)\theta$ is an order hyperresolvent of $\bigvee_{j=0}^{k_0} \varepsilon_j^0 \diamond_j^0 v_j^0 \vee \bigvee_{j=1}^{m_0} l_j^0, \ldots, \bigvee_{j=0}^{k_n} \varepsilon_j^n \diamond_j^n v_j^n \vee \bigvee_{j=1}^{m_n} l_j^n$.

$$(\text{Order trichotomy rule}) \qquad (48)$$

$$\frac{a, b \in atoms(S_{\kappa-1}^{Vr}), \{a, b\} \nsubseteq Tcons_{\mathscr{L}}}{a \prec b \vee a \eqcirc b \vee b \prec a \in S_\kappa};$$

$vars(a) \cap vars(b) = \emptyset$.

$a \prec b \vee a \eqcirc b \vee b \prec a$ is an order trichotomy resolvent of a and b.

$$(\text{Order } \forall\text{-quantification rule}) \qquad (49)$$

$$\frac{\forall x \, a \in qatoms^\forall(S_{\kappa-1})}{\forall x \, a \prec a \vee \forall x \, a \eqcirc a \in S_\kappa}.$$

$\forall x \, a \prec a \vee \forall x \, a \eqcirc a$ is an order $\forall$-quantification resolvent of $\forall x \, a$.

$$(\text{Order } \exists\text{-quantification rule}) \qquad (50)$$

$$\frac{\exists x \, a \in qatoms^\exists(S_{\kappa-1})}{a \prec \exists x \, a \vee a \eqcirc \exists x \, a \in S_\kappa}.$$

$a \prec \exists x \, a \vee a \eqcirc \exists x \, a$ is an order $\exists$-quantification resolvent of $\exists x \, a$.

$$(\text{Order } \forall\text{-witnessing rule}) \qquad (51)$$

$$\frac{\forall x \, a \in qatoms^\forall(S_{\kappa-1}^{Vr}), b \in atoms(S_{\kappa-1}^{Vr}) \cup qatoms(S_{\kappa-1}^{Vr})}{a\gamma \prec b \vee b \eqcirc \forall x \, a \vee b \prec \forall x \, a \in S_\kappa};$$

$freevars(\forall x\, a) \cap freevars(b) = \emptyset,$

$\tilde{w} \in \tilde{\mathbb{W}} - Func_{\mathscr{L}_{\kappa-1}}, ar(\tilde{w}) = \lfloor freetermseq(\forall x\, a), freetermseq(b)\rfloor,$

$\gamma = x/\tilde{w}(freetermseq(\forall x\, a), freetermseq(b)) \cup id|_{vars(a)-\{x\}} \in Subst_{\mathscr{L}_{\kappa}},$

$dom(\gamma) = \{x\} \cup (vars(a) - \{x\}) = vars(a).$

$a\gamma \prec b \vee b \bumpeq \forall x\, a \vee b \prec \forall x\, a$ is an order $\forall$-witnessing resolvent of $\forall x\, a$ and b.

$$(\text{Order } \exists\text{-witnessing rule}) \qquad (52)$$

$$\frac{\exists x\, a \in qatoms^{\exists}(S^{Vr}_{\kappa-1}), b \in atoms(S^{Vr}_{\kappa-1}) \cup qatoms(S^{Vr}_{\kappa-1})}{b \prec a\gamma \vee \exists x\, a \bumpeq b \vee \exists x\, a \prec b \in S_{\kappa}};$$

$freevars(\exists x\, a) \cap freevars(b) = \emptyset,$

$\tilde{w} \in \tilde{\mathbb{W}} - Func_{\mathscr{L}_{\kappa-1}}, ar(\tilde{w}) = \lfloor freetermseq(\exists x\, a), freetermseq(b)\rfloor,$

$\gamma = x/\tilde{w}(freetermseq(\exists x\, a), freetermseq(b)) \cup id|_{vars(a)-\{x\}} \in Subst_{\mathscr{L}_{\kappa}},$

$dom(\gamma) = \{x\} \cup (vars(a) - \{x\}) = vars(a).$

$b \prec a\gamma \vee \exists x\, a \bumpeq b \vee \exists x\, a \prec b$ is an order $\exists$-witnessing resolvent of $\exists x\, a$ and b.

Let $\mathscr{L}_0 = \mathscr{L} \cup P$, a reduct of $\mathscr{L} \cup \tilde{\mathbb{W}} \cup P$, and $S_0 = \emptyset \subseteq GOrdCl_{\mathscr{L}_0} \mid OrdCl_{\mathscr{L}_0}$. Let $\mathscr{D} = C_1, \dots, C_n, C_{\kappa} \in GOrdCl_{\mathscr{L}\cup\tilde{\mathbb{W}}\cup P} \mid OrdCl_{\mathscr{L}\cup\tilde{\mathbb{W}}\cup P}, n \geq 1$. $\mathscr{D}$ is a deduction of C_n from S by basic order hyperresolution iff, for all $1 \leq \kappa \leq n$, $C_{\kappa} \in ordtcons(S) \cup GInst_{\mathscr{L}_{\kappa-1}}(S)$, or there exist $1 \leq j_k^* \leq \kappa - 1, k = 1, \dots, m$, such that C_{κ} is a basic order resolvent of $C_{j_1^*}, \dots, C_{j_m^*} \in S_{\kappa-1}$ using Rule (41)–(46) with respect to $\mathscr{L}_{\kappa-1}$ and $S_{\kappa-1}$; $\mathscr{D}$ is a deduction of C_n from S by order hyperresolution iff, for all $1 \leq \kappa \leq n$, $C_{\kappa} \in ordtcons(S) \cup S$, or there exist $1 \leq j_k^* \leq \kappa - 1, k = 1, \dots, m$, such that C_{κ} is an order resolvent of $C'_{j_1^*}, \dots, C'_{j_m^*} \in S^{Vr}_{\kappa-1}$ using Rule (47)–(52) with respect to $\mathscr{L}_{\kappa-1}$ and $S_{\kappa-1}$ where $C'_{j_k^*}$ is a variant of $C_{j_k^*} \in S_{\kappa-1}$ of $\mathscr{L}_{\kappa-1}$; $\mathscr{L}_{\kappa}$ and S_{κ} are defined by recursion on $1 \leq \kappa \leq n$ as follows:

$$\mathscr{L}_{\kappa} = \begin{cases} \mathscr{L}_{\kappa-1} \cup \{\tilde{w}\} & \text{in case of Rule (45),(46) | (51),(52),} \\ \mathscr{L}_{\kappa-1} & \text{else,} \end{cases} \quad \text{a reduct of } \mathscr{L} \cup \tilde{\mathbb{W}} \cup P;$$

$$S_{\kappa} = S_{\kappa-1} \cup \{C_{\kappa}\} \subseteq GOrdCl_{\mathscr{L}_{\kappa}} \mid OrdCl_{\mathscr{L}_{\kappa}},$$

$$S^{Vr}_{\kappa} = Vrnt_{\mathscr{L}_{\kappa}}(S_{\kappa}) \subseteq OrdCl_{\mathscr{L}_{\kappa}}.$$

$\mathscr{D}$ is a refutation of S iff $C_n = \square$. We denote

$$clo^{\mathscr{BH}}(S) = \{C \mid \text{there exists a deduction of } C \text{ from } S$$
$$\text{by basic order hyperresolution}\} \subseteq GOrdCl_{\mathscr{L}\cup\tilde{\mathbb{W}}\cup P},$$

$$clo^{\mathscr{H}}(S) = \{C \mid \text{there exists a deduction of } C \text{ from } S$$
$$\text{by order hyperresolution}\} \subseteq OrdCl_{\mathscr{L}\cup\tilde{\mathbb{W}}\cup P}.$$

4.2 Refutational Soundness and Completeness

We are in position to prove the refutational soundness and completeness of the order hyperresolution calculus. At first, we list some auxiliary lemmata.

Lemma 2 (Lifting Lemma) *Let $\mathscr{L}$ contain a constant symbol. Let $P \subseteq \tilde{\mathbb{P}}$ and $S \subseteq OrdCl_{\mathscr{L} \cup P}$. Let $C \in clo^{\mathscr{B}\mathscr{H}}(S)$. There exists $C^* \in clo^{\mathscr{H}}(S)$ such that C is an instance of C^* of $\mathscr{L} \cup \tilde{\mathbb{W}} \cup P$.*

Proof Technical, analogous to the standard one. □

Lemma 3 (Reduction Lemma) *Let $\mathscr{L}$ contain a constant symbol. Let $P \subseteq \tilde{\mathbb{P}}$ and $S \subseteq OrdCl_{\mathscr{L} \cup P}$. Let $\{ \bigvee_{j=0}^{k_i} \varepsilon_j^i \diamond_j^i v_j^i \vee C_i \mid i \leq n \} \subseteq clo^{\mathscr{B}\mathscr{H}}(S)$ such that for all $\mathscr{S} \in \mathscr{S}el(\{ \{ j \mid j \leq k_i \}_i \mid i \leq n \})$, there exists a contradiction of $\{ \varepsilon_{\mathscr{S}(i)}^i \diamond_{\mathscr{S}(i)}^i v_{\mathscr{S}(i)}^i \mid i \leq n \} \subseteq GOrdCl_{\mathscr{L} \cup \tilde{\mathbb{W}} \cup P}$. There exists $\emptyset \neq I^* \subseteq \{ i \mid i \leq n \}$ such that $\bigvee_{i \in I^*} C_i \in clo^{\mathscr{B}\mathscr{H}}(S)$.*

Proof Technical, analogous to the one of Proposition 2, [22]. □

Lemma 4 (Unit Lemma) *Let $\mathscr{L}$ contain a constant symbol. Let $P \subseteq \tilde{\mathbb{P}}$ and $S \subseteq OrdCl_{\mathscr{L} \cup P}$. Let $\square \notin clo^{\mathscr{B}\mathscr{H}}(S) = \{ \bigvee_{j=0}^{k_\iota} \varepsilon_j^\iota \diamond_j^\iota v_j^\iota \mid \iota < \gamma \}$, $\gamma \leq \omega$. There exists $\mathscr{S}^* \in \mathscr{S}el(\{ \{ j \mid j \leq k_\iota \}_\iota \mid \iota < \gamma \})$ such that there does not exist a contradiction of $\{ \varepsilon_{\mathscr{S}^*(\iota)}^\iota \diamond_{\mathscr{S}^*(\iota)}^\iota v_{\mathscr{S}^*(\iota)}^\iota \mid \iota < \gamma \} \subseteq GOrdCl_{\mathscr{L} \cup \tilde{\mathbb{W}} \cup P}$.*

Proof Technical, a straightforward consequence of König's Lemma and Lemma 3. □

Let $\{0, 1\} \subseteq X \subseteq Tcons_{\mathscr{L}}$. X is admissible with respect to suprema and infima iff, for all $\emptyset \neq Y_1, Y_2 \subseteq \overline{X}$ and $\bigvee Y_1 = \bigwedge Y_2$, $\bigvee Y_1 \in Y_1$, $\bigwedge Y_2 \in Y_2$.

Theorem 4 (Refutational Soundness and Completeness) *Let $\mathscr{L}$ contain a constant symbol. Let $P \subseteq \tilde{\mathbb{P}}$, $S \subseteq OrdCl_{\mathscr{L} \cup P}$, tcons(S) be admissible with respect to suprema and infima. $\square \in clo^{\mathscr{H}}(S)$ if and only if S is unsatisfiable.*

Proof Cf. http://ii.fmph.uniba.sk/~guller/sci15.pdf, Sect. 4.2, Theorem 4, for a proof. □

Consider $S = \{0 \prec a\} \cup \{a \prec \frac{1}{n} \mid n \geq 2\} \subseteq OrdCl_{\mathscr{L}}$, $a \in Pred_{\mathscr{L}} - Tcons_{\mathscr{L}}$, $ar_{\mathscr{L}}(a) = 0$. $tcons(S)$ is not admissible with respect to suprema and infima; for $\{0\}$ and $\{ \frac{1}{n} \mid n \geq 2 \}$, $\bigvee \{0\} = \bigwedge \{ \frac{1}{n} \mid n \geq 2 \} = 0$, $0 \notin \{ \frac{1}{n} \mid n \geq 2 \}$. S is unsatisfiable; both the cases $\|a\|^{\mathfrak{A}} = 0$ and $\|a\|^{\mathfrak{A}} > 0$ lead to $\mathfrak{A} \nvDash S$ for every interpretation $\mathfrak{A}$ for $\mathscr{L}$. However, $\square \notin clo^{\mathscr{H}}(S)$. So, the condition on $tcons(S)$ being admissible with respect to suprema and infima, is necessary.

Corollary 2 *Let $\mathscr{L}$ contain a constant symbol. Let $n_0 \in \mathbb{N}$, $\phi \in Form_{\mathscr{L}}$, $T \subseteq Form_{\mathscr{L}}$, tcons(T) be admissible with respect to suprema and infima. There exist $J_T^\phi \subseteq \{ (i,j) \mid i \geq n_0 \}$ and $S_T^\phi \subseteq SimOrdCl_{\mathscr{L} \cup \{ \tilde{p}_j \mid j \in J_T^\phi \}}$ such that $tcons(S_T^\phi)$ is admissible with respect to suprema and infima; $T \vDash \phi$ if and only if $\square \in clo^{\mathscr{H}}(S_T^\phi)$.*

Proof By Corollary 1 for n_0, ϕ, T, there exist

$$J_T^\phi \subseteq \{ (i,j) \mid i \geq n_0 \}, S_T^\phi \subseteq SimOrdCl_{\mathscr{L} \cup \{ \tilde{p}_j \mid j \in J_T^\phi \}}$$

Table 7 An example: $\phi = \forall x\,(q_1(x) \to \overline{0.3}) \to (\exists x\,q_1(x) \to \overline{0.5})$

$\phi = \forall x\,(q_1(x) \to \overline{0.3}) \to (\exists x\,q_1(x) \to \overline{0.5})$

$$\left\{\tilde{p}_0(x) \prec 1, \underbrace{(\forall x\,(q_1(x) \to \overline{0.3})}_{\tilde{p}_1(x)} \to \underbrace{(\exists x\,q_1(x) \to \overline{0.5}))}_{\tilde{p}_2(x)} \to \tilde{p}_0(x)\right\} \tag{24}$$

$$\left\{\tilde{p}_0(x) \prec 1, \tilde{p}_2(x) \prec \tilde{p}_1(x) \vee \tilde{p}_2(x) = 1 \vee \tilde{p}_0(x) = 1, \tilde{p}_2(x) \prec \tilde{p}_0(x) \vee \tilde{p}_2(x) = \tilde{p}_0(x), \tilde{p}_1(x) \to \forall x\,\underbrace{(q_1(x) \to \overline{0.3})}_{\tilde{p}_3(x)}, (\underbrace{\exists x\,q_1(x)}_{\tilde{p}_4(x)} \to \underbrace{\overline{0.5}}_{\tilde{p}_5(x)}) \to \tilde{p}_2(x)\right\} \tag{30}, (24)$$

$$\left\{\tilde{p}_0(x) \prec 1, \tilde{p}_2(x) \prec \tilde{p}_1(x) \vee \tilde{p}_2(x) = 1 \vee \tilde{p}_0(x) = 1, \tilde{p}_2(x) \prec \tilde{p}_0(x) \vee \tilde{p}_2(x) = \tilde{p}_0(x), \tilde{p}_1(x) \prec \forall x\,\tilde{p}_3(x) \vee \tilde{p}_1(x) = \forall x\,\tilde{p}_3(x), \tilde{p}_3(x) \to \underbrace{(q_1(x) \to \overline{0.3})}_{\tilde{p}_6(x)\;\; \tilde{p}_7(x)},\right.$$
$$\left. \tilde{p}_5(x) \prec \tilde{p}_4(x) \vee \tilde{p}_5(x) = 1 \vee \tilde{p}_2(x) = 1, \tilde{p}_5(x) \prec \tilde{p}_2(x) \vee \tilde{p}_5(x) = \tilde{p}_2(x), \tilde{p}_4(x) \to \underbrace{\exists x\,q_1(x)}_{\tilde{p}_8(x)}, \overline{0.5} \prec \tilde{p}_5(x) \vee \overline{0.5} = \tilde{p}_5(x)\right\} \tag{22}, (34)$$

$S^\phi = \left\{ \boxed{\tilde{p}_0(x) \prec 1} \right.$ [1]

$\tilde{p}_2(x) \prec \tilde{p}_1(x) \vee \tilde{p}_2(x) = 1 \vee \boxed{\tilde{p}_0(x) = 1}$ [2]

$\boxed{\tilde{p}_2(x) \prec \tilde{p}_0(x)} \vee \tilde{p}_2(x) = \tilde{p}_0(x)$ [3]

$\boxed{\tilde{p}_1(x) \prec \forall x\,\tilde{p}_3(x)} \vee \tilde{p}_1(x) = \forall x\,\tilde{p}_3(x)$ [4]

$\boxed{\tilde{p}_3(x) \prec \tilde{p}_7(x)} \vee \tilde{p}_3(x) = \tilde{p}_7(x) \vee \boxed{\tilde{p}_6(x) \prec \tilde{p}_7(x)} \vee$
$\tilde{p}_6(x) = \tilde{p}_7(x)$ [5]

$\boxed{q_1(x) \prec \tilde{p}_6(x) \vee q_1(x) = \tilde{p}_6(x)}$ [6]

$\boxed{\tilde{p}_7(x) \prec \overline{0.3} \vee \tilde{p}_7(x) = \overline{0.3}}$ [7]

$\tilde{p}_5(x) \prec \tilde{p}_4(x) \vee \tilde{p}_5(x) = 1 \vee \boxed{\tilde{p}_2(x) = 1}$ [8]

$\boxed{\tilde{p}_5(x) \prec \tilde{p}_2(x)} \vee \tilde{p}_5(x) = \tilde{p}_2(x)$ [9]

$\boxed{\tilde{p}_4(x) \prec \exists x\,\tilde{p}_8(x) \vee \tilde{p}_4(x) = \exists x\,\tilde{p}_8(x)}$ [10]

$\boxed{\tilde{p}_8(x) \prec q_1(x) \vee \tilde{p}_8(x) = q_1(x)}$ [11]

$\left. \boxed{\overline{0.5} \prec \tilde{p}_5(x) \vee \overline{0.5} = \tilde{p}_5(x)} \right\}$ [12]

Rule (47): [1][2]:

$\tilde{p}_2(x) \prec \tilde{p}_1(x) \vee \boxed{\tilde{p}_2(x) = 1}$ [13]

Rule (47): [3][13]:

$\boxed{\tilde{p}_2(x) = \tilde{p}_0(x)} \vee \tilde{p}_2(x) \prec \tilde{p}_1(x)$ [14]

Rule (47): [1][13][14]:

$\boxed{\tilde{p}_2(x) \prec \tilde{p}_1(x)}$ [15]

Rule (47): [8][15]:

$\tilde{p}_5(x) \prec \tilde{p}_4(x) \vee \boxed{\tilde{p}_5(x) = 1}$ [16]

Rule (47): [9][16]:

$\boxed{\tilde{p}_5(x) = \tilde{p}_2(x)} \vee \tilde{p}_5(x) \prec \tilde{p}_4(x)$ [17]

Rule (47): [15][16][17]:

$\boxed{\tilde{p}_5(x) \prec \tilde{p}_4(x)}$ [18]

Rule (49): $\forall x\,\tilde{p}_3(x)$:

$\boxed{\forall x\,\tilde{p}_3(x) \prec \tilde{p}_3(x) \vee \forall x\,\tilde{p}_3(x) = \tilde{p}_3(x)}$ [19]

$\overline{0.3} \prec \overline{0.5} \in ordtcons(S^\phi)$

$\boxed{\overline{0.3} \prec \overline{0.5}}$ [20]

repeatedly **Rule (47)**: [4][5][7][9][12][15][19][20]:

$\boxed{\tilde{p}_6(x) \prec \tilde{p}_7(x) \vee \tilde{p}_6(x) = \tilde{p}_7(x)}$ [21]

Rule (52): $\exists x\,\tilde{p}_8(x), \overline{0.5}$:

$\overline{0.5} \prec \tilde{p}_8(\tilde{w}_{(0,0)}) \vee \boxed{\exists x\,\tilde{p}_8(x) \prec \overline{0.5} \vee \exists x\,\tilde{p}_8(x) = \overline{0.5}}$ [22]

repeatedly **Rule (47)**: [10][12][18][22]:

$\boxed{\overline{0.5} \prec \tilde{p}_8(\tilde{w}_{(0,0)})}$ [23]

repeatedly **Rule (47)**: [6][7][11][21]; $\tilde{w}_{(0,0)}$: [20][23]:

$\square$ [24]

and Corollary 1(i,iv) hold for ϕ, T, S_T^ϕ; we have $tcons(T)$ is admissible with respect to suprema and infima, $tcons(S_T^\phi) \subseteq tcons(\phi) \cup tcons(T)$; $tcons(\phi) \subseteq_{\mathscr{F}} Tcons_{\mathscr{L}}$, $tcons(S_T^\phi)$ is admissible with respect to suprema and infima; we have $T \vDash \phi$ if and only if S_T^ψ is unsatisfiable; by Theorem 4 for $\{\tilde{p}_{\mathrm{j}} \mid \mathrm{j} \in J_T^\phi\}$, S_T^ϕ, S_T^ϕ is unsatisfiable if and only if $\square \in clo^{\mathscr{H}}(S_T^\phi)$; $T \vDash \phi$ if and only if $\square \in clo^{\mathscr{H}}(S_T^\phi)$. The corollary is proved. $\square$

In Table 7, we show that $\phi = \forall x\, (q_1(x) \to \overline{0.3}) \to (\exists x\, q_1(x) \to \overline{0.5}) \in Form_{\mathscr{L}}$ is logically valid using the proposed translation to order clausal form and the order hyperresolution calculus.

5 Conclusions

In the paper, we have proposed a modification of the hyperresolution calculus from [2, 19] which is suitable for automated deduction with explicit partial truth. The first-order Gödel logic is expanded by a countable set of intermediate truth constants $\bar{c}$, $c \in (0, 1)$. We have modified translation of a formula to an equivalent satisfiable finite order clausal theory, consisting of order clauses. An order clause is a finite set of order literals of the form $\varepsilon_1 \diamond \varepsilon_2$ where ε_i is an atom or a quantified atom, and $\diamond$ is the connective $\boldsymbol{\pi}$ or $\prec$. $\boldsymbol{\pi}$ and $\prec$ are interpreted by the equality and standard strict linear order on $[0, 1][0, 1]$, respectively. We have investigated the so-called canonical standard completeness, where the semantics of Gödel logic is given by the standard G-algebra and truth constants are interpreted by 'themselves'. The modified hyperresolution calculus is refutation sound and complete for a countable order clausal theory if the set of truth constants occurring in the theory, is admissible with respect to suprema and infima. This condition covers the case of finite order clausal theories. As an interesting consequence, we get an affirmative solution to the open problem of recursive enumerability of unsatisfiable formulae in Gödel logic with truth constants and the projection operator $\boldsymbol{\Delta}$.

Corollary 3 *The set of unsatisfiable formulae of $\mathscr{L}$ is recursively enumerable.*

Proof Let $\phi \in Form_{\mathscr{L}}$. Then ϕ contains a finite number of truth constants and $tcons(\{\phi\})$ is admissible with respect to suprema and infima. The statement ϕ is unsatisfiable, is equivalent to $\{\phi\} \vDash 0$. Hence, the problem that ϕ is unsatisfiable can be reduced to the deduction problem $\{\phi\} \vDash 0$ after a constant number of steps. Let $n_0 \in \mathbb{N}$. By Corollary 2 for n_0, 0, $\{\phi\}$, there exist $J^0_{\{\phi\}} \subseteq \{(i,j) \mid i \geq n_0\}$, $S^0_{\{\phi\}} \subseteq SimOrdCl_{\mathscr{L} \cup \{\bar{p}_j \mid j \in J^0_{\{\phi\}}\}}$ and $tcons(S^0_{\{\phi\}})$ is admissible with respect to suprema and infima, $\{\phi\} \vDash 0$ if and only if $\square \in clo^{\mathscr{H}}(S^0_{\{\phi\}})$; if $\{\phi\} \vDash 0$, then $\square \in clo^{\mathscr{H}}(S^0_{\{\phi\}})$ and we can decide it after a finite number of steps. This straightforwardly implies that the set of unsatisfiable formulae of $\mathscr{L}$ is recursively enumerable. The corollary is proved. $\qquad\square$

References

1. Baaz, M., Ciabattoni, A., Fermüller, C.G.: Theorem proving for prenex Gödel logic with Delta: checking validity and unsatisfiability. Log. Methods Comput. Sci. **8** (2012)
2. Guller, D.: An order hyperresolution calculus for Gödel logic—General first-order case. In: Rosa, A.C., Correia, A.D., Madani, K., Filipe, J., Kacprzyk, J. (eds.) IJCCI 2012—Proceedings of the 4th International Joint Conference on Computational Intelligence, Barcelona, Spain, 5–7 October 2012, pp. 329–342. SciTePress (2012)

3. Davis, M., Putnam, H.: A computing procedure for quantification theory. J. ACM **7**, 201–215 (1960)
4. Davis, M., Logemann, G., Loveland, D.: A machine program for theorem-proving. Commun. ACM **5**, 394–397 (1962)
5. Robinson, J.A.: A machine-oriented logic based on the resolution principle. J. ACM **12**, 23–41 (1965)
6. Robinson, J.A.: Automatic deduction with hyper-resolution. Int. J. Comput. Math. **1**, 227–234 (1965)
7. Gallier, J.H.: Logic for Computer Science: Foundations of Automatic Theorem Proving. Harper and Row Publishers Inc, New York (1985)
8. Biere, A., Heule, M.J., van Maaren, H., Walsh, T.: Handbook of Satisfiability. Frontiers in Artificial Intelligence and Applications, vol. 185. IOS Press, Amsterdam (2009)
9. Hájek, P.: Metamathematics of Fuzzy Logic. Trends in Logic. Springer (2001)
10. Pavelka, J.: On fuzzy logic I, II, III. Semantical completeness of some many-valued propositional calculi. Math. Logic Q. **25**, 45–52, 119–134, 447–464 (1979)
11. Novák, V., Perfilieva, I., Močkoř, J.: Mathematical Principles of Fuzzy Logic. The Springer International Series in Engineering and Computer Science. Springer, US (1999)
12. Esteva, F., Godo, L., Montagna, F.: The ŁΠ and ŁΠ $\frac{1}{2}$ logics: two complete fuzzy systems joining Łukasiewicz and product logics. Arch. Math. Log. **40**, 39–67 (2001)
13. Savický, P., Cignoli, R., Esteva, F., Godo, L., Noguera, C.: On product logic with truth-constants. J. Log. Comput. **16**, 205–225 (2006)
14. Esteva, F., Godo, L., Noguera, C.: On completeness results for the expansions with truth-constants of some predicate fuzzy logics. In: Stepnicka, M., Novák, V., Bodenhofer, U. (eds.) New Dimensions in Fuzzy Logic and Related Technologies. Proceedings of the 5th EUSFLAT Conference, Ostrava, Czech Republic, September 11–14, 2007, Volume 2: Regular Sessions, Universitas Ostraviensis, pp. 21–26 (2007)
15. Esteva, F., Gispert, J., Godo, L., Noguera, C.: Adding truth-constants to logics of continuous t-norms: axiomatization and completeness results. Fuzzy Sets Syst. **158**, 597–618 (2007)
16. Esteva, F., Godo, L., Noguera, C.: First-order t-norm based fuzzy logics with truth-constants: distinguished semantics and completeness properties. Ann. Pure Appl. Logic **161**, 185–202 (2009)
17. Esteva, F., Godo, L., Noguera, C.: Expanding the propositional logic of a t-norm with truth-constants: completeness results for rational semantics. Soft Comput. **14**, 273–284 (2010)
18. Esteva, F., Godo, L., Noguera, C.: On expansions of WNM t-norm based logics with truth-constants. Fuzzy Sets Syst. **161**, 347–368 (2010)
19. Guller, D.: A generalisation of the hyperresolution principle to first order Gödel logic. In: Computational Intelligence—International Joint Conference, IJCCI 2012 Barcelona, Spain, October 5–7, 2012 Revised Selected Papers. Studies in Computational Intelligence, vol. 577, pp. 159–182. Springer (2015)
20. Guller, D.: An order hyperresolution calculus for Gödel logic with truth constants. In: Rosa, A.C., Dourado, A., Correia, K.M., Filipe, J., Kacprzyk, J. (eds.) FCTA 2014—Proceedings of the 6th International Joint Conference on Computational Intelligence, Rome, Italy, 22–24 October 2014, pp. 37–52. SciTePress (2014)
21. Apt, K.R.: Introduction to logic programming. Technical Report CS-R8826, Centre for Mathematics and Computer Science, Amsterdam, The Netherlands (1988)
22. Guller, D.: On the refutational completeness of signed binary resolution and hyperresolution. Fuzzy Sets Syst. **160**, 1162–1176 (2009). Featured Issue: Formal Methods for Fuzzy Mathematics. Approximation and Reasoning, Part II

A Fuzzy Approach for Performance Appraisal: The Evaluation of a Purchasing Specialist

Hatice Esen, Tuğçen Hatipoğlu and Ali İhsan Boyacı

Abstract One of the most important issues faced in organizations is the objective performance measure of employees who have a crucial role in the success of the production/service processes. Performance appraisal is vital since it helps to clarify whether the company is going in line with the predetermined objectives. Performance appraisal is the evaluation of the general performance level of personnel according to the previously determined targets and/or performance factors. The main purpose of performance appraisal is not only to increase the performance of employees, but also to unite the individual targets with the company's targets. Performance appraisal requires evaluation and decision making in uncertain environments involving multiple factors. In this paper, a performance appraisal model is constructed to deal with the uncertainty and also to objectively measure the employees' performances. The criteria used in the model are defined in terms of the fuzzy numbers and linguistic variables. The three main criteria of the performance appraisal model are decision making and leadership, communication and relations, and technical skills. The model is applied in the purchasing department of a company operating in the automotive industry. The developed model has the flexibility to be applied in different departments with a modification in the criteria under the title of "Technical skills".

Keywords Human resources management · Performance appraisal · Fuzzy logic

H. Esen (✉) · T. Hatipoğlu · A.İ. Boyacı
Department of Industrial Engineering, Kocaeli University, Umuttepe, Kocaeli, Turkey
e-mail: hatice.eris@kocaeli.edu.tr

T. Hatipoğlu
e-mail: tugcen.hatipoglu@kocaeli.edu.tr

A.İ. Boyacı
e-mail: ali.ihsan@kocaeli.edu.tr

© Springer International Publishing Switzerland 2016
J.J. Merelo et al. (eds.), *Computational Intelligence*,
Studies in Computational Intelligence 620, DOI 10.1007/978-3-319-26393-9_14

1 Introduction

Human resources management is a discipline involving the auditing, directing, organizing, planning and developing of strategies about the improvement, employment and supply of human resources which is essential for a competitive advantage in companies [1]. Since successful human resources management leads to better performance, it brings a competitive advantage by increasing the efficiency within organizations [2]. The first step of successful human resources management is the objective performance appraisal of employees which is the most dynamic one among all the company resources.

Setting salaries and making promotion decisions should not be the only purposes of an effective Performance appraisal programme. It should help in the construction of a performance improvement plan which includes leading from the department manager to improve and develop employee skills. Thus, it can be put in the same category with training which aims to improve the performance by considering the future and developing efficient programmes [3].

There are two important and difficult tasks in organizations; to determine the achievement of the personnel in their jobs and their capabilities. For each employee, the expectations, senses of duty, capabilities, knowledge/talent, and working disciplines are different. Because of these natural differences, their performances are also different. All workers may not satisfy their duties. So, managers naturally want to learn the abilities and the success of the personnel working for them. Performance appraisal criteria are needed to be able to measure the satisfaction level of the targets by the personnel [4].

Generally, performance appraisal is based on the individual's characteristics, behavioral criteria, and the results and aims about the job. However, two common important errors are made about the criteria of performance appraisal. The first one is wrongly assuming that the criteria are only related to the job. Some indicators should be defined to represent the objectives of the work. A universal criteria bundle doesn`t exist. The work analysis should be used in the identification of performance criteria of a job title. Secondly, a selected criterion of work performance should be measured correctly and precisely.

Existing methods like graphic rating scale, group order ranking or individual ranking don`t consider the uncertainty and imprecision of factors used in service evaluation [5]. Performance appraisal generally consists of assessment and decision making under uncertainty, based on multiple factors of a quantitative and qualitative nature, temporal and resource constraints, varying tactics and strategies, domain-specific knowledge and information asymmetries, etc. Fuzziness generally exists in most human perception and thinking [3].

The evaluation process should be as objective as possible to prevent mental anguish and satisfy the expectations of employees. In most cases, the impossibility of complete objectivity leads the evaluators to be completely subjective which causes a mistrust in their authority [6].

In this study, a fuzzy modeling approach is employed to deal with the problems mentioned above and to evaluate the employees' performance objectively. A performance appraisal model involving three main and sixteen sub-criteria was developed. The main criteria are Decision Making and Leadership, Communication and Relations, and Technical skills. The proposed model can be used for different departments with a modification in the job title which is under the main criteria of "Technical skills". The model can objectively define and give weights to the criteria for all job titles.

The main objectives of the study can be summarized as follows;

(1) Evaluating and ranking the employees based on Decision Making and Leadership, Communication and Relations, and Technical skills criteria
(2) Supporting the managers in determination of training needs, career planning, promotions, and fair payoff by providing information about the employees' performance and capabilities/skills.
(3) The model allows flexibility to be used for different job titles by changing the criteria under the main criterion of "Technical skills".
(4) The model gives weights to each criterion which is a lacking feature in the current performance appraisal models.
(5) Since performance appraisal is a decision making process, it involves uncertainty. To deal with this uncertainty and objectively measure the employees' performances, the criteria of the model are defined as fuzzy numbers and linguistic variables.

The purchasing department can be counted as one of the most difficult areas of the evaluation. The competitive environment forces the companies to evaluate their operations and decrease the costs by continuously improving them. The purchasing cost is one of the biggest sources of expenditure in companies. It usually amounts to 40–70 % of a firm's sales volume (depending upon the degree of vertical integration in the industry) which makes it a possible source to increase the competitiveness of companies [7]. The duty of the purchasing department is to improve the efficiency of the supply chain by communicating with the suppliers. The performance measure of the purchasing department is quite important because of its critical role [8].

Due to the reasons explained above and to understand the model better, the criteria of the purchasing specialist position are defined and weighted.

The paper is organized as follows; a literature survey is given in the second part entitled performance appraisal. In the third part of the study the methodology, Fuzzy Analytic Hierarchy Process (AHP), is explained. In the fourth section, the proposed performance appraisal model is detailed. The last section of the study is comprised of the results and comments about the application.

2 Performance Appraisal

Performance appraisal is the assessment of employees' achievement in their work according to the previously determined reachable level specific to the employee and/or the performance criteria. It is an important source of continuous improvement for companies which assess the performance of the employees in the right manner. As a social rule states in organizations, organizational goals should direct the individual behavior of the employees. In the same direction, performance appraisal should also be done with the aim of increasing the performance of employees in line with the organization's policies [9].

Cleveland et al. stated that there are four main purposes of performance appraisal; making a distinction among workers, distinguishing the strong and weak sides of employees, evaluating the human resources system of the organization, and setting a base for the management decisions [10].

The points that should be considered during the performance appraisal studies are;

- The purposes and aims of the appraisal should be clearly explained to employees
- The performance targets expected from employees should be described to them.
- The performance targets should be reachable, understandable and objective.
- The appraisal should be fair and balanced. Enough a sufficient level of objectivity and limited subjectivity should be guaranteed in the appraisal to be able to satisfy the employees' demands [11].

Several methods exist in the literature for human performance appraisal. The most popular ones are graphic rating scale, force distribution, behaviorally anchored rating scale, management by objectives, 360° evaluation method, etc. But these methods don't provide a good analysis of data and they lack the ability to handle the uncertainties and ambiguities in the data [12].

In the paper by Shaout and Shammari (1998), a good application of fuzzy set theory is represented for multi-attribute performance appraisal for faculty members [13]. The idea of using acceptability as a criterion in the evaluation of performance appraisal methods is tested in Hedge and Teachout's study (2000). They compare four different performance appraisal techniques in terms of the user acceptability and the differences in rater acceptance [14]. In Lefkowitz's (2000) study, a literature review is conducted about the supervisor interpersonal effect and performance appraisals [15]. An examination of the effects of social context on the appraisal process and a relevant literature review is given in the study by Levy and Williams (2004) [16]. A fuzzy group decision making system- fuzzy group decision support system (FGDSS) is developed in the study by Chang et al. (2007) to find a solution for the appraisal problem of military officers' performance [17]. By using the Decision Making Theory, a flexible 360-degree performance appraisal model is proposed in the study by Andrés et al. (2010). In their model, the management team is allowed to know how to combine the individual opinions [18]. Moon et al. (2010) present a new ranking procedure to rank order the performance of candidates applying for promotion in a military organization in Korea. The metric distance and fuzzy mean value are used in the ranking procedure which allows the combination of the scores of each evaluator [19].

Manoharan et al. (2011) show that an integrated tool like fuzzy multi-attribute decision making (FMADM), with fuzzy analytic hierarchy process (FAHP), and fuzzy quality function deployment (FQFD) can be used as a supporting tool for the performance appraisal system [3]. Spence and Keeping (2011) discuss a phenomenon labeled as conscious rating distortion which occurs when managers evaluate the employees with motives other than accuracy. They apply regulatory focus theory to study performance ratings [20]. Barone and DeCarlo (2012) investigate the factors of the effect of performance trends on the salesperson evaluations of practicing sales managers [21]. Min-peng et al. (2012) create performance indicators based on the factors like morality, ability, diligence, and performance. Then, AHP is used to find the weight of every indicator. After that, fuzzy evaluation method is applied in the construction of a performance appraisal model to be able to measure the Research and Development performance of engineers [22]. Subjective evaluations from different appraisal sources are weighted and combined with a mathematical model developed in the study by Sepehrirad et al. (2012) in a 360-degree performance appraisal. They categorize the performance appraisal criteria by using Delphi method and according to the characteristics of the National Iranian Productivity Organization (NIPO). The importance of each appraisal criterion is calculated with fuzzy AHP technique [23].

Espinilla et al. (2013) develop an integrated model for 360-degree performance appraisal which can handle the heterogeneous information and give a final verbal evaluation for each employee. The model aggregated the interaction among criteria and the relevance of reviewers by using the averages of weights [24]. In Gürbüz and Albayrak's study (2014), a new hybrid method including Analytical Network Process and Choquet Integral is used to evaluate the employees' performances doing the same task. They employ the proposed method for marketing department employees and show the performance appraisal model. The criteria of their model are experience, marketing ability, salary/satisfaction, social power, educational level, marketing politics, management politics, salary politics and ability of managing the changes [6]. An online fuzzy based decision support system is developed by Samuel et al. (2014) to evaluate human resource performance. The delays and biases present in the orthodox performance appraisal systems in organizations are managed with a computational method in their study [25].

3 Fuzzy Analytic Hierarchy Process Method

In Analytical Hierarchy Process, a flexible and structured methodology, complex decision variables are structured into a hierarchical framework to solve and analyze them [26]. Human thinking and judgment are ambiguous and it is not meaningful to

Table 1 Triangular fuzzy scale of preference

Relative importance	Definition	Fuzzy scale	Fuzzy reciprocal scale
1	Equal importance	(1, 1, 1)	(1, 1, 1)
3	Moderate importance	(1, 3, 5)	(1/5,1/3,1)
5	Strong importance	(3, 5, 7)	(1/7,1/5,1/3)
7	Demonstrated importance	(5, 7, 9)	(1/9,1/7,1/5)
9	Extreme importance	(7, 9, 9)	(1/9,1/9,1/7)

represent them with point numbers. Interval judgments better represent them than precise value judgments. Thus, the priority between decision variables is decided according to the triangular fuzzy numbers. The final priority weights are found with synthetic extent analysis method and this is called fuzzy extended AHP [27]. Thus, using fuzzy theory in AHP is more reasonable and effective than classic AHP.

There are various AHP methods related to the fuzziness, while the most popular one is Chang's approach. Chang established the extent analysis method (EAM) for synthetic values of pair wise comparisons with the use of triangular fuzzy numbers (TFNs) [28]. The triangular fuzzy conversion scale, given in Table 1, is used in the appraisal model of this paper.

Let $X = \{x_1, x_2, \ldots, x_n\}$ be an object set, and $U = \{u_1, u_2, \ldots, u_m\}$ be a goal set. According to the method of Chang's extent analysis, each object is taken and extent analysis for each goal, g_i, is performed, respectively. Therefore, m extent analysis values for each object can be obtained, with the following signs:

$$M_{g_i}^1, M_{g_i}^2, \ldots, M_{g_i}^m \; i = 1, 2, \ldots, n \tag{1}$$

where all the $M_{g_i}^j$ (j = 1, 2, ..., m) are triangular fuzzy numbers.

The steps of Chang's extent analysis can be given as in the following:

Step 1: The value of fuzzy synthetic extent with respect to the *i*th object is defined as

$$S_i = \sum_{j=1}^m M_{g_i}^j \otimes \left[\sum_{i=1}^n \sum_{j=1}^m M_{g_i}^j \right]^{-1} \tag{2}$$

To obtain $\sum_{j=1}^m M_{g_i}^j$, perform the fuzzy addition operation of m extent analysis values for a particular matrix such that

$$\sum_{j=1}^m M_{g_i}^j = \left(\sum_{j=1}^m l_j, \; \sum_{j=1}^m m_j, \; \sum_{j=1}^m u_j \right) \tag{3}$$

and to obtain $\left[\sum_{i=1}^n \sum_{j=1}^m M_{g_i}^j \right]^{-1}$, perform the fuzzy addition operation of $M_{g_i}^j$ (j = 1, 2, ..., m) values such that

$$\sum_{i=1}^{n} \sum_{j=1}^{m} M_{g_i}^{j} = \left(\sum_{i=1}^{n} l_i, \ \sum_{i=1}^{n} m_i, \ \sum_{i=1}^{n} u_i \right) \tag{4}$$

and then compute the inverse of the vector in Eq. (4) such that

$$\left[\sum_{i=1}^{n} \sum_{j=1}^{m} M_{g_i}^{j} \right]^{-1} = \left(\frac{1}{\sum_{i=1}^{n} u_i}, \ \frac{1}{\sum_{i=1}^{n} m_i}, \ \frac{1}{\sum_{i=1}^{n} l_i} \right) \tag{5}$$

Step 2: The degree of possibility of $M_2 = (l_2, m_2, u_2) \geq M_1 = (l_1, m_1, u_1)$ is defined as

$$V\ (M_2 \geq M_1) = sup_{y \geq x} \lfloor \min \mu_{M_1}(x), \mu_{M_2}(y) \rfloor \tag{6}$$

and can be equivalently expressed as follows:

$$V(M_2 \geq M_1) = \text{hgt}(M_1 \cap M_2)$$

$$= \mu_{M_2}(d) \begin{cases} 1, & \text{if } m_2 \geq m_1 \\ 0, & \text{if } l_1 \geq u_2 \\ \frac{l_1 - u_2}{(m_2 - u_2) - (m_1 - l_1)} & \text{otherwise} \end{cases} \tag{7}$$

where d is the ordinate of the highest intersection point D between μ_{M_1} and μ_{M_2} (see Fig. 1).

To compare M_1 and M_2, we need both the values of $V(M_2 \geq M_1)$ and $V(M_1 \geq M_2)$.

Step 3: The degree possibility for a convex fuzzy number to be greater than k convex fuzzy numbers M_i $(i = 1, 2,..., k)$ can be defined by.

$$V(M \geq M_1, M_2, \ldots, M_k) = V[(M \geq M_1) \text{ and } (M \geq M_2) \text{ and } \ldots \text{ and } (M \geq M_k)]$$
$$= \min V\ (M \geq M_i), \ i = 1, 2, \ldots, k$$

$$\tag{8}$$

Fig. 1 The intersection between M_1 and M_2

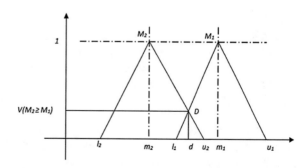

Assume that

$$d'(A_i) = \min V(S_i \geq S_k). \tag{9}$$

For $k = 1, 2, \ldots, n; k \neq i$. Then the weight vector is given by

$$W' = \left(d'(A_1), d'(A_2), \ldots, d'(A_n)\right)^T \tag{10}$$

where A_i $(i = 1, 2, \ldots, n)$ are n elements.

Step 4: Via normalization, the normalized weight vectors are

$$W = (d(A_1), d(A_2), \ldots, d(A_n))^T \tag{11}$$

where W is a non fuzzy number [29].

4 Proposed Performance Appraisal Model

Since business environments have become more complicated and are subject to fast change nowadays, managers should be aware of the importance of allocating enough time and attention to observe and measure their subordinates' performance [30]. Performance appraisal is concerned with the measurement and control of individual performances and integrates them with the organization's objectives [31].

Performance appraisal is a decision making process which involves uncertainty. To overcome the uncertainty and evaluate the workers performance objectively, a performance appraisal model is developed of which the criteria are defined as the fuzzy numbers and the linguistic variables. The scope of the study is to identify and objectively weight the criteria for the performance appraisal of each job position. To explain the model better, the criteria for the purchasing specialist position are identified and weighted.

The application of the model was done at a company in the automotive supplier industry in Kocaeli. After meeting with the executives of the company, the boundaries of the study were determined and a team was created including an academic expert, a production manager, a human resources manager, a purchase manager, and a logistics and supply manager.

First of all, the aim, importance and framework of the study was discussed, then the criteria found by brainstorming was noted in the team's meeting that was organized to discuss the criteria to be used in the performance measure. The criteria draft was evaluated again; whether they can be measured, or they can be represented by another criterion and whether there are any missing criteria were discussed. After this evaluation, the criteria were finalized. Every criterion was described in detail to prevent any misunderstanding about their meaning during their usage.

The proposed model makes the performance appraisal by using three main criteria. These main criteria are Decision Making and Leadership, Communication and Relations, and Technical skills.

The Decision Making and Leadership, and Communication and Relations criteria and their sub-criteria can be mutually used in the evaluation of all job titles. The sub-criteria under the main criterion of "Technical skills" changes according to the job title. To make the model more understandable, the weights used in the performance appraisal of the job title "Purchasing Specialist" are identified. The sub-criteria under "Technical skills" are flexible enough to be used for another job title. The hierarchy belonging to the proposed performance appraisal of "Purchasing Specialist" in the model can be seen in Fig. 2.

The decision structure has two levels;

First level (Level of determinants): determinants of the performance appraisal are determined as Decision Making and Leadership (DML), Communication and Relations (CR) and Technical Skills (TS).

Second level: this level consists of 16 sub-criteria. Six sub-criteria about Decision Making and Leadership are Problem solving and result orientation (PS), Agility (AG), Adaptability (AD), Team building and management (TB), Project management (PM) and Strategy Elaboration (SE). Four sub-criteria about Communication and Relations are dealing with organization (DO), Communication (C), International Mindset (IM) and Interpersonal Skills (IS). Six sub-criteria about Technical Skills are Negotiation Practice (NP), purchasing tools practice (PT), Financial awareness

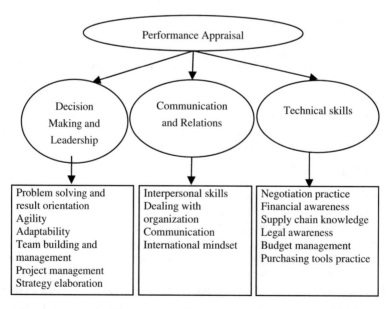

Fig. 2 Hierarchies in the AHP

(FA), Supply chain knowledge (SK), Legal awareness (LA) and Budget management (BM). The explanation of the sub-criteria is given below:

Decision Making and Leadership

(1) Problem solving and result orientation (PS): Ability to detect, design and implement solutions adapted to situations and people (evaluate, diagnose) applying QRQC (Quick Response Quality Control). Identification and weighting of important parameters, identification of causes, priorities and development of solutions. Ability to achieve results regardless of circumstances, but not at any cost. Focus on pragmatic and practical tasks and ability to act in the field. Willingness and ability to meet commitments.

(2) Agility (AG): Ability to combine speed and rationality in decision making followed by the implementation of an action plan. Ability to react to requests and situations within the required time constraints. Capability to manage a heavy work load in a stressed environment and ability to deal with urgencies.

(3) Adaptability (AD): Ability to adapt easily to different working environments. Functioning in a matrix environment or with occasional apparently contradictory issues. Analysis of problems from several points of view, including the points of view of others.

(4) Team building and management (TB): Ability to manage and coach people, to create a team spirit, to establish (common and individual) objectives and to assess performance and competences. Efficient delegation of appropriate tasks. Management of structural conflicts. Ability to select and recruit team members. Integration of the team inside the structure. Focus on people development. Gathering of different personalities and mobilization of them towards a common objective.

(5) Project management (PM): Ability to plan resources in order to manage a project successfully according to standards (quality, cost, delivery) for customer satisfaction. Establishment of clear, realistic timeframes for goal accomplishment.

(6) Strategy Elaboration (SE): Ability to anticipate future evolutions (helicopter view), to define vision, objectives, strategic action plans and milestones in order to meet objectives and to implement their strategic plans.

Communication and Relations

(1) Dealing with organization (DO): Ability to deal with organizational complexity, internal and external stakeholders.

(2) Communication (C): Ability "get the point across" and to get the "buy-in" of the target audience. Selection of the most pertinent information, reliable sources, appropriate population, the best media. Definition and organization of the content (according to different cultures and levels) at the appropriate time. Enhancement of information sharing and feedback. Openness to listen to other options and to take them into consideration. Ability to analyze complicated situations.

(3) International Mindset (IM): Ability to work with people from different cultures. Integration of other cultural values and systems and development of cultural of open mindedness. Ability to think globally, act locally.

(4) Interpersonal Skills (IS): Ability to develop interactive listening skills and to give constructive reinforcement. Ability to deal with interpersonal conflict. Ability to influence in a positive way the work of colleagues through enthusiastic communication.

Technical Skills

(1) Negotiation practice (NP): Ability to prepare the negotiation strategy and tactics based on assessment of company's levers versus suppliers. Knowledge of his/her limits and supplier's decision drivers. Ability to conduct the negotiation in a professional manner ensuring a win-win deal. Ability to obtain closure.

(2) Purchasing tools practice (PT): Knowledge of purchasing tools: methodology tools (commodity matrix, system audit, process audit, initial assessment, etc.), analysis tools (market surveys, etc.), transaction tools (weekly report, EDI, MRP system, etc.)

(3) Financial awareness (FA): Understanding of financial reports. Understanding of management accounting and different methods to establish a price and to evaluate the financial health of a supplier.

(4) Supply chain knowledge (SK): Ability to use supply chain knowledge to contribute to the optimization of the supply chain together with the logistics department and supplier.

(5) Legal awareness (LA): Ability to formalize the contractual relationship in all domains (confidentiality, development, supplies, equipment etc.). Ability to manage supplier disputes and major crises with the help of the legal counsel. Ability to use the law to elaborate his/her commodity strategy.

(6) Budget management (BM): Ability to evaluate all necessary resources and their costs in order to achieve a target. Ability to manage a budget. Evaluation and anticipation of the financial impact of new decisions. Management of situation changes and minimization of excess costs.

The triangular fuzzy conversion scale, given in Table 1, is used in the appraisal model of this study. Fuzzy pair-wise comparisons of three main decision criteria on "Performance Appraisal" are presented in Table 2. Then the sub-criteria are pair wise compared in Tables 3, 4 and 5 respectively and weights are calculated.

As seen in Table 2; $S_{DML} = (0.060, 0.103, 0.273)$, $S_{CR} = (0.099, 0.291, 0.819)$ and $S_{TS} = (0.224, 0.605, 1.522)$ are calculated. Then $W' = (0.089, 0.655, 1.000)$ is

Table 2 Fuzzy linguistic preference relation decision matrix of three main criteria

	DML			C			TS		
DML	1	1	1	1/5	1/3	1	1/7	1/5	1/3
C	1	3	5	1	1	1	1/5	1/3	1/1
TS	3	5	7	1	3	5	1	1	1

Table 3 Fuzzy linguistic preference relation decision matrix of decision making and leadership

	PS			AG			AD			TB			PM			SE		
PS	1	1	1	1/7	1/5	1/3	1/5	1/3	1	3	5	7	1	3	5	5	7	9
AG	3	5	7	1	1	1	1	3	5	3	5	7	3	5	7	7	9	9
AD	1	3	5	1/5	1/3	1	1	1	1	3	5	7	1	3	5	5	7	9
TB	1/7	1/5	1/3	1/7	1/5	1/3	1/7	1/5	1/3	1	1	1	1/5	1/3	1	3	5	7
PM	1/5	1/3	1	1/7	1/5	1/3	1/5	1/3	1	1	3	5	1	1	1	1	3	5
SE	1/9	1/7	1/5	1/9	1/7	1/5	1/9	1/7	1/5	1/7	1/5	1/3	1/5	1/3	1	1	1	1

Table 4 Fuzzy linguistic preference relation decision matrix of communication and relations

	DO			C			IM			IS		
DO	1	1	1	1/5	1/3	1	3	5	7	1	3	5
C	1	3	5	1	1	1	5	7	9	3	5	7
IM	1/7	1/5	1/3	1/9	1/7	1/5	1	1	1	1/5	1/3	1
IS	1/5	1/3	1	1/7	1/5	1/3	1	3	5	1	1	1

Table 5 Fuzzy linguistic preference relation decision matrix of technical skills

	NP			PT			FA			SK			LA			BM		
NP	1	1	1	1	3	5	3	5	7	5	7	9	7	9	9	3	5	7
PT	1/5	1/3	1	1	1	1	3	5	7	3	5	7	5	7	9	1	3	5
FA	1/7	1/5	1/3	1/7	1/5	1/3	1	1	1	1/5	1/3	1	5	7	9	1/7	1/5	1/3
SK	1/9	1/7	1/5	1/7	1/5	1/3	1	3	5	1	1	1	1	3	5	1/5	1/3	1
LA	1/9	1/9	1/7	1/9	1/7	1/5	1/9	1/7	1/5	1/5	1/3	1	1	1	1	1/7	1/5	1/3
BM	1/7	1/5	1/3	1/5	1/3	1	3	5	7	1	3	5	3	5	7	1	1	1

obtained and priority weights vector of each main criteria is $W = (0.051, 0.376, 0.573)^T$. Similarly the priority weights vector of each main sub-criteria is seen in Table 3, $W = (0.2347, 0.3411, 0.2700, 0.0491, 0.1052, 0)^T$, in Table 4 $W = (0.351, 0.493, 0, 0.156)^T$ and in Table 5 $W = (0.368, 0.291, 0.065, 0.072, 0, 0.204)^T$.

Table 6 shows the overall or global importance levels of for the main criteria and sub-criteria. According to these results, the performance appraisal of a purchasing expert is evaluated as following:

It is further observed that the priority of the main criteria "Technical Skills" with %57 is highest followed by "Communication and Relations" with %38 while "Decision Making and Leadership" is just %5.

In the case of the sub-criteria the priority is highest for "Negotiation practice", "Purchasing tools practice" and "Budget management" respectively under "Technical Skills"; "Communication" and "Dealing with organization" among "Communication and relations"; "Agility", "Adaptability" and "Problem Solving and Result Orientation" among "Decision Making and Leadership".

Table 6 Global importance levels of sub-criteria

Global importance of three main criteria	Global importance of sub-criteria	Weights
Decision making and leadership (0.051)	PS (0.235)	0.012
	AG (0.341)	0.017
	AD (0.270)	0.014
	TB (0.049)	0.002
	PM (0.105)	0.005
	SE (0)	0.000
Communication and relations (0.376)	DO (0.351)	0.132
	C (0.493)	0.185
	IM (0)	0.000
	IS (0.156)	0.059
Technical skills (0.573)	NP (0.368)	0.211
	PT (0.291)	0.167
	FA (0.065)	0.037
	SK (0.072)	0.041
	LA (0)	0.000
	BM (0.204)	0.117

5　Conclusions

For all organizations, it is vital to use a measurement system that gives the performance levels of individuals. It is important not only to maximize the usage of employees' capabilities but also to integrate the employees' performance with the aims of the organization [32].

Besides being the most important part of an effective human resources management strategy, performance appraisal is one of the most important elements to reach the objectives of organizational management. Performance appraisal should be used as a tool to direct, stimulate, and increase the motivation and the trust of workers through the organization. The most important part of the evaluation is to make it as objective as possible.

The contributions of this study into the literature can be summarized as follows; 1. The proposed model can be used for other job titles in companies by allowing flexibility in the sub-criteria under the main criterion of "Technical skills".

The current performance appraisal models don't weight the criterion assuming that they all have the same importance. The proposed model has such a structure and objectivity to satisfy this lack of current models. 3. Performance appraisal is a decision making process which involves uncertainty. To overcome the uncertainty and evaluate the workers performance objectively, a performance appraisal model is developed of which the criteria are defined as the fuzzy numbers and the linguistic variables.

The evaluation process of the performance can be thought of as a complex multi-criteria decision making problem considering multiple factors and sub factors affecting the evaluation. Fuzzy AHP method enables decision-makers to realize a hierarchical structure and an effective vague assessment of the main and sub factors' weights. Hence, we used a fuzzy approach for the evaluation of personnel performance. By utilizing fuzzy AHP method, the weights of sub factors are determined subsequently.

By applying the model in a company operating as an automotive supplier, it has been shown that the model can be used in practice without any difficulty. During the application, the weighing of the specific criteria used in the performance appraisal of a purchasing specialist was done. The main criteria are ordered as Technical Skills, Communication and Relations, and Decision Making and Leadership as a result of the evaluation. The most important sub-criteria are identified as Negotiation practice, Communication and Purchasing tools practice.

To explain the model better with the help of an example, an application was conducted in the purchasing department of a company. In future studies, performance appraisals of other job titles in the organization will also be developed by modifying the "Technical skills" criteria. Performance appraisal studies require sustainability. With the software implementation that is planned to be developed, usage efficiency of the model is expected to be increased. As a result, company managers will be able to assess the employees' performance in an objective, faster and easier way.

References

1. Doğan, S., Demiral, Ö.: İnsan Kaynaklari Yönetiminde Çalişanlarin Kendilerine Doğru Yolculuk Yöntemi: Yetenek Yönetimi. Ç.Ü. Sosyal Bilimler Enstitüsü Dergisi 17(3):145–166 (2008)
2. Stavrou, E.T., Charalambous, C., Spiliotis, S., et al.: Human resource management and performance: a neural network analysis. Eur. J. Oper. Res. 181, 453–467 (2007)
3. Manoharan, T.R., Muralidharan, C., Deshmukh, S.G., et al.: An integrated fuzzy multi-attribute decision-making model for employees' performance appraisal. Int. J. Hum. Resour. Manage. 22, 722–745 (2011)
4. Kılıç, S.: Performans değerlendirme sisteminin kariyer planlamasıyla ilişkisinin analizi ve bir uygulama. Dissertation, Uludağ University (2011)
5. Manoharan, T.R., Muralidharan, C., Deshmugh, S.G., et al.: Employees Performance Appraisal Using Data Envelopment Analysis: A Case Study. Res. Pract. Hum. Res. Manage. 17, 92–111 (2009)
6. Gürbüz, T., Albayrak, Y.E.: An engineering approach to human resources performance evaluation: Hybrid MCDM application with interactions. Appl. Soft Comput. 21, 365–375 (2014)
7. Saranga, H., Moser, R.: Performance evaluation of purchasing and supply management using value chain DEA approach. Eur. J. Oper. Res. 207, 197–205 (2010)
8. Easton, L., Murphy, D.J., Pearson, J.N., et al.: Purchasing performance evaluation: with data envelopment analysis. Eur. J. Purchasing Supply Manage 8, 123–134 (2002)

9. Bourguignon, A.: Performance management and management control: evaluated managers' point of view. Eur. Account. Rev. **13**(4), 659–687 (2004). doi:10.1080/09638180 42000216875

10. Cleveland, J.N., Mohammed, S., Skattebo, A.L., Sin, H.P. et al.: Multiple purposes of performance appraisal: a replication and extension. In: Annual Conference for the Society for Industrial and Organizational Psychology. Orlando, FL, 2003

11. Gürbüz, T.: Multiple criteria human performance evaluation using choquet integral. Int. J. Comput. Intell. Syst. **3**(3), 290–300 (2010). doi:10.1080/18756891.2010.9727700

12. Li, Y., Jiang, D., Li, F., et al.: The application of generating fuzzy ID3 algorithm in performance evaluation. Procedia Eng. **29**, 229–234 (2012)

13. Shaout, A., Al-Shammari, M.: Fuzzy logic modeling for performance appraisal systems a framework for empirical evaluation. Expert Syst. Appl. **14**, 323–328 (1998)

14. Hedge, J.W., Teachout, M.S.: Exploring the concept of acceptability as a criterion for evaluating performance measures. Group Org. Manage. **25**, 22–44 (2000)

15. Lefkowitz, J.: The role of interpersonal affective regard in supervisory performance ratings: a literature review and proposed causal model. J. Occup. Organ. Psychol. **73**, 67–85 (2000)

16. Levy, P.E., Williams, J.R.: The social context of performance appraisal: a review and framework for the future. J. Manage. **30**(6), 881–905 (2004)

17. Chang, J.R., Cheng, C.H., Chen, L.S., et al.: A fuzzy-based military officer performance appraisal system. Appl. Soft Comput. **7**, 936–945 (2007)

18. De Andrés, R., García-Lapresta, J.L., González-Pachón, J., et al.: Performance appraisal based on distance function methods. Eur. J. Oper. Res. **207**, 1599–1607 (2010)

19. Moon, C., Lee, J., Lim, S., et al.: A performance appraisal and promotion ranking system based on fuzzy logic: An implementation case in military organizations. Appl. Soft Comput. **10**, 512–519 (2010)

20. Spence, J.R., Keeping, L.: Conscious rating distortion in performance appraisal: a review, commentary, and proposed framework for research. Hum. Res. Manage. Rev. **21**, 85–95 (2011)

21. Barone, M.J., DeCarlo, T.E.: Performance trends and salesperson evaluations: the moderating roles of evaluation task, managerial risk propensity, and firm strategic orientation. J. Pers. Selling Sales Manage. **32**(2), 207–223 (2012)

22. Min-peng, X., Xiao-hu, Z., Xin, D. et al.: Modeling of engineering R&D staff performance appraisal model based on fuzzy comprehensive evaluation. In: International Conference on Complexity Science and Information Engineering. Systems Engineering Procedia, vol. 4, pp. 236–242 (2012)

23. Sepehrirad, R., Azar, A., Sadeghi, A., et al.: Developing a hybrid mathematical model for 360-degree performance appraisal: a case study. Procedia—Soc. Behav. Sci. **62**, 844–848 (2012)

24. Espinilla, M., De Andrés, R., Martínez, F.J., Martínez, L., et al.: A 360-degree performance appraisal model dealing with heterogeneous information and dependent criteria. Inf. Sci. **222**, 459–471 (2013)

25. Samuel, O.W., Omisore, M.O., Atajeromavwo, E.J., et al.: Online fuzzy based decision support system for human resource performance appraisal. Measurement **55**, 452–461 (2014)

26. Boroushaki, S., Malczewski, J.: Implementing an extension of the analytical hierarchy process using ordered weighted averaging operators with fuzzy quantifiers in ArcGIS. Comput. Geosci. **34**, 399–410 (2008)

27. Chan, F.T.S., Kumar, N.: Global supplier development considering risk factors using fuzzy extended AHP-based approach. Omega **35**(4), 417–431 (2007)

28. Heo, E., Kim, J., Cho, S., et al.: Selecting hydrogen production methods using fuzzy analytic hierarchy process with opportunities, costs, and risks. Int. J. Hydrogen Energy **37**, 17655–17662 (2012)

29. Kahraman, C., Cebeci, U., Ruan, D., et al.: Multi-attribute comparison of catering service companies using fuzzy AHP: the case of Turkey. Int. J. Prod. Econ. **87**, 171–184 (2004)

30. Çelik, D.A.: Enabling more objective performance appraisals: A training program model of pinpointing. Soc. Behav. Sci. **150**, 794–802 (2014)
31. Sanyal, M.K., Biswas, S.B.: Employee motivation from performance appraisal implications: test of a theory in the software industry in West Bengal (India). Procedia Econ. Finan. **11**, 182–196 (2014)
32. Jefferson, A.L.: Performance appraisal applied to leadership. Educ. Stud. **36**(1), 111–114 (2010)

M-valued Measure of Roughness for Approximation of *L*-fuzzy Sets and Its Topological Interpretation

Sang-Eon Han and Alexander Šostak

Abstract We develop a scheme allowing to measure the "quality" of rough approximation of fuzzy sets. This scheme is based on what we call "an approximation quadruple" (L, M, φ, ψ) where L and M are cl-monoids (in particular, $L = M = [0, 1]$) and $\psi : L \to M$ and $\varphi : M \to L$ are satisfying certain conditions mappings (in particular, they can be the identity mappings). In the result of realization of this scheme we get measures of upper and lower rough approximation for L-fuzzy subsets of a set equipped with a reflexive transitive M-fuzzy relation R. In case the relation R is also symmetric, these measures coincide and we call their value by the measure of roughness of rough approximation. Basic properties of such measures are studied. A realization of measures of rough approximation in terms of L-fuzzy topologies is presented.

Keywords L-fuzzy set · M-relation · L-fuzzy rough set · Cl-monoid · M-valued measure of inclusion · (L, M)-fuzzy topology · (L, M)-fuzzy co-topology

1 Introduction

The concept of a rough subset of a set equipped with an equivalence relation was introduced by Pawlak [24]. Rough sets found important applications in real-world problems, and also arouse interest among "pure" mathematicians as an interesting mathematical notion having deep relations with other fundamental mathemat-

---

The support of the ESF project 2013/0024/1DP/1.1.1.2.0/13/APIA/VIAA/045 is kindly announced.

S.-E. Han
Department of Mathematics Education, Institute of Pure and Applied Mathematics,
Chonbuk National University, Jeonju, Jeonbuk 561-756, Republic of Korea

A. Šostak (✉)
Institute of Mathematics and Computer Science, University of Latvia,
Raina Boulevard 29, Riga 1459, Latvia
e-mail: sostaks@latnet.lv

J.J. Merelo et al. (eds.), *Computational Intelligence*,
Studies in Computational Intelligence 620, DOI 10.1007/978-3-319-26393-9_15

ical concepts, in particular, with topology. Soon after Pawlak's work, the concept of roughness was extended to the context of fuzzy sets; Dubois' and Prade's paper [7] was the first work in this direction. At present there is a vast literature where fuzzy rough sets are investigated and applied. In particular, fuzzy rough sets are studied and used in [6, 12, 13, 19, 23, 25–27, 36, 38–40] just to mention a few of numerous works dealing with (fuzzy) rough sets. In our paper [30] we undertook the first, as far as we know, attempt to measure the degree of roughness of a fuzzy set, or to state it in another way, to measure, "how much rough" is a given (fuzzy) subset of a set equipped with a (fuzzy) relation. Here we develop further the approach of measuring the roughness initiated in [30]. Namely we propose a scheme allowing to measure the "quality" of rough approximation of fuzzy sets. This scheme is based on what we call "an approximation quadruple" (L, M, φ, ψ) where L and M are cl-monoids (in particular, $L = M = [0, 1]$) and $\varphi : L \to M$ and $\psi : M \to L$ are satisfying certain conditions mappings (in particular, they can be the identity mappings.) In the result of realization of this scheme we get measures of upper and lower rough approximation for L-fuzzy subsets of a set equipped with a reflexive transitive M-fuzzy relation R. In case when the relation R is also symmetric, these measures coincide and we call their value by the measure of roughness of rough approximation. Basic properties of such measures are studied. In addition we present an interpretation of measures of rough approximation in terms of L-fuzzy topologies.

The structure of the paper is as follows. In the next section we recall two notions which are fundamental for our work, namely a cl-*monoid* and an *M-relation*. In the third section we introduce the measure of inclusion of one fuzzy set into another, and describe the behavior of this measure.

In Sect. 4 we define operators of upper and lower rough approximation for an L-fuzzy subset of a set endowed with an M-relation. These operators, as special cases, contain various rough approximation-type operators used by different authors, see e.g. [8, 16, 17, 25, 32, 35].

In Sect. 5 we define the measures of upper $\mathcal{K}(A)$ and lower $\mathcal{T}(A)$ rough approximation for an L-fuzzy subset A of a set endowed with an M-relation. Essentially, $\mathcal{K}(A)$ is the measure of inclusion of the upper approximation of an L-fuzzy set A into A, while $\mathcal{T}(A)$ is the measure of inclusion of A into its lower approximation. By showing $\mathcal{K}(A) = \mathcal{T}(A)$ whenever R is symmetric, we come to the measure of roughness $\mathcal{RA}(A)$ of an L-fuzzy set A.

In Sect. 6 we interpret the operator of measuring roughness of rough approximation as an (L, M)-fuzzy ditopology (that is a pair of an (L, M)-fuzzy topology $\mathcal{T}$ and an (L, M)-fuzzy $\mathcal{K}$ co-topology) on a set X and discuss some issues of this interpretation.

In the last, Conclusion, section we discuss some directions for the prospective work.

2 Preliminaries

2.1 Cl-Monoids

Let $(L, \leq, \wedge, \vee)$ denote a complete lattice, that is a lattice in which arbitrary suprema (joins) and infima (meets) exist, see e.g. [9]. In particular, the top 1_L and the bottom 0_L elements in L exist and $0_L \neq 1_L$. In our work, the concept of a cl-monoid, introduced by G. Birkhoff will play a fundamental role:

Definition 1 [1] A tuple $(L, \leq, \wedge, \vee, *)$ is called a cl-monoid if $(L, \leq, \wedge, \vee)$ is a complete lattice and the binary operation $*: L \times L \to L$ satisfies conditions:

(0$*$) $*$ is monotone: $\alpha \leq \beta \implies \alpha * \gamma \leq \beta * \gamma$ for all $\alpha, \beta, \gamma \in L$;
(1$*$) $*$ is commutative: $\alpha * \beta = \beta * \alpha$ for all $\alpha, \beta \in L$;
(2$*$) $*$ is associative: $(\alpha * \beta) * \gamma = \alpha * (\beta * \gamma)$ for all $\alpha, \beta, \gamma \in L$;
(3$*$) $*$ distributes over arbitrary joins:
  $\alpha * \left(\bigvee_{i \in I} \beta_i \right) = \bigvee_{i \in I} (\alpha * \beta_i)$ for all $\alpha \in L$, for all $\{\beta_i \mid i \in I\} \subseteq L$,
(4$*$) $\alpha * 1_L = \alpha$, $\quad \alpha * 0_L = 0_L$ for all $\alpha \in L$.

Remark 1 Note, that a cl-monoid can be defined also as an integral commutative quantale in the sense of Rosenthal [28].

Remark 2 In case $L = [0, 1]$ the operation $*: [0, 1] \times [0, 1] \to [0, 1]$ satisfying properties (0$*$), (1$*$), (2$*$) and (4$*$) (defined in a slightly different form) for the first time appeared in Menger's papers, see e.g. [22] under the name *a triangular norm*, or a *t-norm*. Later *t*-norms were thoroughly studied by different authors, see e.g. [18, 29]. A *t*-norm satisfying property (3$*$) is called lower semicontinuous.

In a cl-monoid a further binary operation $\mapsto$, residium, is defined:

$$\alpha \mapsto \beta = \bigvee \{\lambda \in L \mid \lambda * \alpha \leq \beta\}.$$

Residuation is connected with the operation $*$ by the Galois connection:

$$\alpha * \beta \leq \gamma \iff \alpha \leq (\beta \mapsto \gamma),$$

see e.g. [14]. In the following proposition we collect well-known properties of the residium:

Proposition 1 (see [14, 15])

(1$\mapsto$) $\left(\bigvee_i \alpha_i \right) \mapsto \beta = \bigwedge_i \left(\alpha_i \mapsto \beta \right)$ *for all* $\{\alpha_i \mid i \in I\} \subseteq L$, *for all* $\beta \in L$;
(2$\mapsto$) $\alpha \mapsto \left(\bigwedge_i \beta_i \right) = \bigwedge_i (\alpha \mapsto \beta_i)$ *for all* $\alpha \in L$, *for all* $\{\beta_i \mid i \in I\} \subseteq L$;
(3$\mapsto$) $1_L \mapsto \alpha = \alpha$ *for all* $\alpha \in L$;
(4$\mapsto$) $\alpha \mapsto \beta = 1_L$ *whenever* $\alpha \leq \beta$;
(5$\mapsto$) $\alpha * (\alpha \mapsto \beta) \leq \beta$ *for all* $\alpha, \beta \in L$;

$(6\mapsto)$ $(\alpha \mapsto \beta) * (\beta \mapsto \gamma) \le \alpha \mapsto \gamma$ for all $\alpha, \beta, \gamma \in L$;
$(7\mapsto)$ $\alpha \mapsto \beta \le (\alpha * \gamma \mapsto \beta * \gamma)$ for all $\alpha, \beta, \gamma \in L$.

In the sequel we will need also the following two lemmas:

Lemma 1 Let $(L, \le, \wedge, \vee, *)$ be a cl-monoid. Then for every $\{\alpha_i \mid i \in I\} \subseteq L$ and every $\{\beta_i \mid i \in I\} \subseteq L$ it holds $(\bigwedge_i \alpha_i) \mapsto (\bigwedge_i \beta_i) \ge \bigwedge_i (\alpha_i \mapsto \beta_i)$.

Proof Applying Proposition 1 we have:

$$(\bigwedge_i \alpha_i) \mapsto (\bigwedge_i \beta_i) = \bigwedge_i \alpha_i \mapsto \bigwedge_j \beta_j = \bigwedge_j (\bigwedge_i \alpha_i \mapsto \beta_j) \ge \bigwedge_j (\alpha_j \mapsto \beta_j). \qquad \square$$

Lemma 2 Let $(L, \le, \wedge, \vee, *)$ be a cl-monoid. Then for every $\{\alpha_i \mid i \in I\} \subseteq L$, and every $\{\beta_i \mid i \in I\} \subseteq L$ it holds: $(\bigvee_i \alpha_i) \mapsto (\bigvee_i \beta_i) \ge \bigwedge_i (\alpha_i \mapsto \beta_i)$.

Proof Applying Proposition 1 we have $(\alpha_i \mapsto \beta_i) * \alpha_i \le \beta_i$ for each $i \in I$. Let $c = \bigwedge_i (\alpha_i \mapsto \beta_i)$. Then $c * \alpha_i \le \beta_i$ for each $i \in I$. Taking suprema on the both sides of the above inequality over $i \in I$ we get $c * \bigvee_i \alpha_i \le \bigvee_i \beta_i$ and hence, by the Galois connection, $\bigwedge_i (\alpha_i \mapsto \beta_i) \le \bigvee_i \alpha_i \mapsto \bigvee_i \beta_i$. $\qquad \square$

2.2 The Framework of Our Research

In our work we use two apriori independent cl-monoids $L = (L, \le_L, \wedge_L, \vee_L, *_L)$ and $M = (M, \le_M, \wedge_M, \vee_M, *_M)$.[1] The first one of the cl-monoids, L is the background for L-fuzzy sets, that is for the objects which we aim to approximate, while the second cl-monoid, M is used as the set of values taken by the measure to estimate the precision of the approximation. Apriory these cl-monoids are unrelated. However, in order to get substantial results about the measure of approximation we need some connections between cl-monoids L and M. Therefore we assume that there are fixed mappings $\varphi : L \to M$ and $\psi : M \to L$ preserving bottom and top elements of the lattices and such that

$$\varphi(\bigwedge_i \alpha_i) = \bigwedge_i (\varphi(\alpha_i)) \text{ and } \psi(\bigvee_i \lambda_i) = \bigvee_i (\psi(\lambda_i))$$

for all $\{\alpha_i : i \in I\} \subseteq L$, $\{\lambda_i : i \in I\} \subseteq M$. Besides we require that

$$\varphi(\alpha *_L \beta) = \varphi(\alpha) *_M \varphi(\beta) \; \forall \alpha, \beta \in L \text{ and } \psi(\lambda *_M \mu) = \psi(\lambda) *_L \psi(\mu) \; \forall \lambda, \mu \in M.$$

In the result we come to a quadruple (L, M, φ, ψ) which will be referred to as *an approximation quadruple*. Below we give some examples of approximation quadruples (L, M, φ, ψ).

1. L and M are arbitrary homomorphic cl-monoids and $\varphi : L \to M$ and $\psi : M \to L$ are the corresponding homomorphisms. In particular, ψ may be defined as the inverse of φ.

---

[1] The subscripts $_L$ and $_M$ will be usually omitted as soon as it is clear from the context in which monoid we are working.

2. $L = ([0, 1], *)$ is the unit interval endowed with a lower semicontinuous *t*-norm $*$ and $M = \{0, 1\} \subset [0, 1]$. We define $\psi : M \to L$ as inclusion and $\varphi : L \to M$ by setting $\varphi(1_L) = 1_M$ and $\varphi(\alpha) = 0_M$ whenever $\alpha < 1$.

3. $M = ([0, 1], *)$ is the unit interval endowed with a lower semicontinuous *t*-norm $*$ and $L = \{0, 1\} \subset [0, 1]$. We define $\varphi : L \to M$ as the inclusion map and $\psi : L \to M$ by setting $\psi(0_L) = 1_M$ and $\psi(\alpha) = 1_M$ whenever $\alpha > 0$.

4. $L = ([0, 1], *)$ is the unit interval endowed with a lower semicontinuous *t*-norm $*$ and $a \in L$, $a \neq 0_L, 1_L$ is an idempotent element; $M = (\{0, a, 1\}, *) \subseteq L$. We define mapping $\psi : M \to L$ as an embedding and mapping $\varphi : L \to M$ by setting $\varphi(1_L) = 1_M$; $\varphi(\alpha) = a$ if $a \leq \alpha < 1_L$ and $\varphi(\alpha) = 0_M$ if $\alpha < a$.

5. $M = ([0, 1], *)$ is the unit interval endowed with a lower semicontinuous *t*-norm $*$ and $a \in M$, $a \neq 0_M, 1_M$ is an idempotent element and $L = (\{0, a, 1\}, *) \subseteq L$. We define the mapping $\varphi : L \to M$ an the embedding and the mapping $\psi : M \to L$ by setting $\psi(\lambda) = 1_L$ for $\lambda > a$; $\psi(\lambda) = a$ if $0_M < \lambda \leq a$ and $\psi(0_M) = 0_L$.

2.3 M-relations

Definition 2 Let a cl-monoid M and a set X be given. An M-valued relation, or just an M-relation on a set X is a mapping $R : X \times X \to M$.

(r) M-relation R is called reflexive if $R(x, x) = 1_M$ for each $x \in X$;
(s) M-relation R is called symmetric if $R(x, y) = R(y, x)$ for all $x, y \in X$;
(t) M-relation R is called transitive if $R(x, y) * R(y, z) \leq R(x, z)$ $\forall x, y, z \in X$.

A reflexive symmetric transitive M-relation is called an M-equivalence, or a similarity M-relation.

Let **REL**(M) be the category whose objects are pairs (X, R), where X is a set and $R : X \times X \to M$ is a transitive reflexive M-relation on it. Morphisms in **REL**(M) are mappings $f : (X, R_X) \to (Y, R_Y)$ such that

$$R_X(x, x') \leq R_Y(f(x), f(x')) \text{ for all } x, x' \in X.$$

Remark 3 In particular cases, notions similar to our concept of an M-valued relation appear in many papers written by different authors and often under different names (see e.g. [7, 37]) etc. For the first time it appeared in Zadeh's paper [41] under the name "a fuzzy relation". Zadeh's fuzzy relation can be described as our M-relation where $M = ([0, 1], \leq, \wedge, \vee, \wedge)$, that is when $M = [0, 1]$ is viewed as a lattice with usual order and φ, ψ are the identity mappings.

2.4 L-fuzzy Sets

For completeness we recall that given a complete lattice, in particular a cl-monoid L, and a set X an L-fuzzy subset, or just an L-subset of X is a mapping $A : X \to L$ (see e.g. [10, 11]). The family of all L-fuzzy subsets of X is denoted L^X. Union and intersections of a family $\{A_i : i \in I\}$ are defined as $\bigvee_{i \in I} A_i$ and $\bigwedge_{i \in I} A_i$. Further, we recall that given a mapping $f : X \to Y$ and L-subsets $A \in L^X$ and $B \in L^Y$ the image $f(A) \in L^Y$ of A is defined by $f(A)(y) = \sup_{x \in f^{-1}(y)} A(x)$ and the preimage $f^{-1}(B) \in L^X$ of B is defined by $f^{-1}(B)(x) = B(f(x))$.

If $(L, \leq, \wedge, \vee, *)$ is a cl-monoid, and X is a set, then the lattice and the monoidal structures of L can be pointwise lifted to the L-powerset L^X of X. Namely, given $A, B \in L^X$ we set $A \leq B$ iff $A(x) \leq B(x)$ for all $x \in X$, and define operations on L^X by setting

$$(A \wedge B)(x) = A(x) \wedge B(x), \ (A \vee B)(x) = A(x) \vee B(x),$$

$$(A * B)(x) = A(x) * B(x) \ \forall x \in X.$$

One can easily notice that in this way $(L^X, \leq, \wedge, \vee, *)$ becomes a cl-monoid.

3 M-valued Measure of Inclusion of L-fuzzy Sets

Let L, M be cl-monoids and let $\varphi : M \to L$ satisfy the conditions stated in Sect. 2.2, that is $\varphi(\bigwedge_i \alpha_i) = \bigwedge_i (\varphi(\alpha_i))$, $\varphi(0_M) = 0_L$, $\varphi(1_M) = 1_L$ and $\varphi(\alpha *_M \beta) = \varphi(\alpha) *_L \varphi(\beta)$ for all $\alpha, \beta \in M$

Definition 3 By setting

$$A \hookrightarrow B = \inf_{x \in X} \varphi(A(x) \mapsto B(x))$$

for all $A, B \in L^X$ we obtain a mapping $\hookrightarrow : L^X \times L^X \to M$. Equivalently, $\hookrightarrow$ can be defined by

$$A \hookrightarrow B = \varphi(\inf(A \mapsto B)),$$

where the infimum of the L-fuzzy set $A \mapsto B$ is taken in the lattice L^X. We call $A \hookrightarrow B$ by the M-valued measure of inclusion of the L-fuzzy set A into the L-fuzzy set B.

As the next proposition shows, the measure of inclusion $\hookrightarrow : L^X \times L^X \to M$ has properties in a certain sense resembling the properties of the residuation:

Proposition 2 *Mapping* $\hookrightarrow : L^X \times L^X \to M$ *defined above satisfies the following properties:*

$(1\hookrightarrow)$ $\left(\bigvee_i A_i\right) \hookrightarrow B = \bigwedge_i \left(A_i \hookrightarrow B\right)$ *for all* $\{A_i \mid i \in I\} \subseteq L^X$, *for all* $B \in L^X$;

$(2\hookrightarrow)$ $A \hookrightarrow \left(\bigwedge_i B_i\right) = \bigwedge_i (A \hookrightarrow B_i)$ *for all* $A \in L^X$, *for all* $\{B_i \mid i \in I\} \subseteq L^X$;

$(3\hookrightarrow)$ $A \hookrightarrow B = 1_L$ *whenever* $A \le B$;

$(4\hookrightarrow)$ $1_X \hookrightarrow A = \varphi(\inf_x A(x))$ *for all* $A \in L^X$;

$(5\hookrightarrow)$ $(A \hookrightarrow B) \le (A * C \hookrightarrow B * C)$ *for all* $A, B, C \in L^X$;

$(6\hookrightarrow)$ $(A \hookrightarrow B) * (B \hookrightarrow C) \subseteq (A \hookrightarrow C)$ *for all* $A, B, C \in L^X$;

$(7\hookrightarrow)$ $\left(\bigwedge_i A_i\right) \hookrightarrow \left(\bigwedge_i B_i\right) \ge \bigwedge_i (A_i \hookrightarrow B_i)$ *for all* $\{A_i : i \in I\}, \{B_i : i \in I\} \subseteq L^X$.

$(8\hookrightarrow)$ $\left(\bigvee_i A_i\right) \hookrightarrow \left(\bigvee_i B_i\right) \ge \bigwedge_i (A_i \hookrightarrow B_i)$ *for all* $\{A_i : i \in I\}, \{B_i : i \in I\} \subseteq L^X$.

Proof

(1) $\left(\bigvee_i A_i\right) \hookrightarrow B = \inf_x \left(\varphi\left(\bigvee_i A_i(x) \mapsto B(x)\right)\right) = \inf_x(\varphi(\bigwedge_i(A_i(x) \mapsto B(x)))) = \bigwedge_i$
$\left(\varphi(\inf_x\left((A_i(x) \mapsto B(x)))\right)\right) = \bigwedge_i(A_i \hookrightarrow B)$.

(2) $A \hookrightarrow \bigwedge_i B_i = \inf(\varphi(A(x) \mapsto \bigwedge_i B_i(x)) = \inf_x \left(\bigwedge_i \left(\varphi(A(x) \mapsto B_i(x))\right)\right) = \bigwedge_i(\inf_x$
$\varphi(A(x) \mapsto B_i(x)))) = \bigwedge_i(A \hookrightarrow B_i)$.

(3) Let $A \le B$. Then $A \hookrightarrow B = \inf_x \varphi(A(x) \mapsto B(x)) = \inf_x (\varphi(1_M)) = 1_L$.

(4) $1_X \hookrightarrow A = \inf_x(\varphi(1_X \mapsto A(x)) = \varphi(\inf_x(1_X \mapsto A(x)) = \varphi(\inf_x A(x)) = \inf_x(\varphi(A(x)))$.

(5) $A \hookrightarrow B = \inf_x (\varphi(A(x) \mapsto B(x))) \le \inf_x (\varphi(A(x) * C(x) \mapsto B(x) * C(x))) = A * C \hookrightarrow B * C$.

(6) $A \hookrightarrow B) * (B \hookrightarrow C) = \inf_x ((\varphi(A(x) \mapsto B(x))) * \varphi((B(x) \mapsto C(x)))) = \inf_x(\varphi(A(x) \mapsto B(x)) * ((B(x) \mapsto C(x))) \le \inf_x (\varphi(A(x) \mapsto C(x))) = A \hookrightarrow C$.

(7) Referring to Lemma 1 for every $x \in X$ we have

$$\left(\bigwedge_i A_i(x) \mapsto \bigwedge_i B_i(x)\right) \ge (\bigwedge_i (A_i(x) \mapsto B_i(x))).$$

Taking infimum by x and an image by φ and recalling that φ preserves meets, we exchange $\bigwedge$ and φ in the right side of the inequality and get the requested

$$\left(\bigwedge_i A_i\right) \hookrightarrow \left(\bigwedge_i B_i\right) \ge \bigwedge_i (A_i \hookrightarrow B_i).$$

(8) Referring to Lemma 2 for every $x \in X$ we have

$$\left(\bigvee_i A_i(x) \mapsto \bigvee_i B_i(x)\right) \ge (\bigwedge_i (A_i(x) \mapsto B_i(x))).$$

Taking infimum by x and image by φ and recalling that φ preserves meets we exchange $\bigwedge$ and φ on the left side of the inequality, and get the requested

$$\left(\bigvee_i A_i\right) \hookrightarrow \left(\bigvee_i B_i\right) \ge \bigwedge_i (A_i \hookrightarrow B_i).$$

$\square$

4 Rough Approximation of an *L*-fuzzy Set Induced by a Reflexive Transitive *M*-relation

Let L, M be cl-monoids and let $\psi : M \to L$ satisfy the conditions stated in Sect. 2.2, that is $\psi(\bigvee_i \lambda_i) = \bigvee(\psi \lambda_i)$, $\psi(0_M) = 0_L$, $\psi(1_M) = 1_L$ and $\psi(\lambda *_L \mu) = \psi(\lambda) *_M \psi(\mu)$ for all $\lambda, \mu \in M$.

Further, let $R : X \times X \to M$ be a reflexive transitive *M*-relation on a set X and A be an *L*-fuzzy subset of the set X, that is $A \in L^X$. By the *M*-rough approximation of the *L*-fuzzy set A we call the pair $(u_R(A), l_R(A))$ where $u_R : L^X \to L^X$ and $l_R : L^X \to L^X$ are respectively operators of upper and lower *M*-rough approximations of A defined below.

4.1 Upper Rough Approximation of an L-fuzzy Set Induced by a Reflexive Transitive M-relation

Given a reflexive transitive *M*-relation $R : X \times X \to M$, we define the upper rough approximation operator $u_R : L^X \to L^X$ by

$$u_R(A)(x) = \sup_{x'} \left(\psi(R(x, x')) * A(x') \right) \ \forall A \in L^X, \ \forall x \in X.$$

Theorem 1 *The upper rough approximation operator satisfies the following properties:*

(1u) $u_R(0_X) = 0_X$;
(2u) $A \leq u_R(A) \ \forall A \in L^X$;
(3u) $u_R(\bigvee_i A_i) = \bigvee_i u_R(A_i) \ \forall \{A_i \mid i \in I\} \subseteq L^X$;
(4u) $u_R(u_R(A)) = u_R(A) \ \forall A \in L^X$.

Proof Statement (1u) is obvious. To prove (2u) notice that taking into account reflexivity of the *M*-relation R we have:

$u_R(A)(x) = \sup_{x'}(\psi(R(x, x')) * A(x')) \geq \psi(R(x, x)) * A(x) = A(x)$.

We prove property (3u) as follows:

$$u_R(\bigvee_i A_i)(x) = \sup_{x'}(\psi(R(x, x')) * (\bigvee_i A_i(x'))) = \sup_{x'} \left(\bigvee_i \psi(R(x, x')) * A_i(x') \right) = \bigvee_i \left(\sup_{x'} \psi((R(x, x')) * A_i(x')) \right) = \bigvee_i (u_R(A_i)(x)) = \left(\bigvee_i (u_R(A_i)) \right)(x).$$

To prove property (4u) we take into account transitivity of the *M*-relation and get the following chain of inequalities:

$$u_R(u_R(A))(x) = \sup_{x'} \left(u_R(A)(x') * \psi(R(x, x')) \right)$$
$$= \sup_{x''} \sup_{x'}(A(x'') * ((\psi(R(x, x')) * (\psi R(x', x''))))$$
$$\leq \sup_{x''} \sup_{x'}(A(x'') * \psi(R(x, x') * R(x', x'')))$$
$$\leq \sup_{x''} A(x'') * \psi(R(x, x'')) = u_R(A)(x).$$

Since the converse inequality follows from (2u), we get property (4u).

4.2 Lower Rough Approximation of an L-fuzzy Set Induced by a Reflexive Transitive M-relation

Given a reflexive transitive M-relation $R : X \times X \rightarrow M$, we define a lower rough approximation operator $l_R : L^X \rightarrow L^X$ by

$$l_R(A)(x) = \inf_{x'} \left(\psi(R(x, x')) \mapsto A(x') \right) \ \forall A \in L^X \ \forall x \in X.$$

Theorem 2 *The lower rough approximation operator satisfies the following properties:*

(1l) $l_R(1_X) = 1_X$;
(2l) $A \geq l_R(A) \ \forall A \in L^X$;
(3l) $l_R(\bigwedge_i A_i) = \bigwedge_i l_R(A_i) \ \forall \{A_i \mid i \in I\} \subseteq L^X$;
(4l) $l_R(l_R(A)) = l_R(A) \ \forall A \in L^X$.

Proof Statement (1l) is obvious. We obtain property (2l) as follows:

$$l_R(A)(x) = \inf_{x'}(\psi(R(x, x') \mapsto A(x'))$$
$$\leq \psi(R(x, x)) \mapsto A(x) = 1_M \mapsto A(x) = A(x).$$

We prove property (3l) as follows:

$$l_R(\bigwedge_i A_i)(x) = \inf_{x'} \left(\psi(R(x, x')) \mapsto \bigwedge_i A_i(x') \right)$$
$$= \inf_{x'} \bigwedge_i \left(\psi(R(x, x')) \mapsto A_i(x') \right)$$
$$= \bigwedge_i \inf_{x'} \left(\psi R(x, x') \mapsto A_i(x') \right) = \bigwedge_i l_R(A_i).$$

To prove property (4l) we take into account transitivity of the M-relation and, applying Proposition 1, get the following chain of inequalities:

$$l_R(l_R(A))(x) = \inf_{x'}(\psi R(x, x') \mapsto l_R(A)(x'))$$
$$= \inf_{x'}(\psi R(x, x') \mapsto \inf_{x'}(\psi R(x', x'') \mapsto A(x'')))$$
$$= \inf_{x'}(\inf_{x''}(\psi R(x, x') * \psi R(x', x'') \mapsto A(x'')))$$
$$\geq \inf_{x''}(\psi R(x, x'') \mapsto A(x'')) = l_R(A)(x).$$

Since the converse inequality follows from (2l), we get property (4l). $\qquad\square$

Remark 4 Operators similar to our upper and lower approximation operators in special cases, in particular when $M = L$ and $\varphi = \psi$ is the identity mapping can be found in the works of several authors, see e.g. [19, 25, 32, 35].

5 M-valued Measure of Roughness of an L-fuzzy Set

5.1 Definitions and Basic Properties of M-valued Measure of Roughness of an L-fuzzy Set

Let (L, M, φ, ψ) be an approximation quadruple and let R be a reflexive transitive M-relation on a set X.

Given an L-fuzzy set $A \in L^X$ we define the measure $\mathcal{K}(A)$ of its upper rough approximation by $\mathcal{K}(A) = u_R(A) \hookrightarrow A$ and the measure $\mathcal{T}(A)$ of its lower rough approximation by $\mathcal{T}(A) = A \hookrightarrow l_R(A)$.

Theorem 3 *If R is also symmetric, that is an equivalence M-relation, then $\mathcal{K}(A) = \mathcal{T}(A)$ for every L-fuzzy set A.*

Proof Note that for the measure of the upper rough approximation we have

$$\mathcal{K}(A) = u_R(A) \hookrightarrow A = \varphi \left(\inf_x (u_R(A)(x) \mapsto A(x)) \right)$$
$$= \varphi \left(\inf_x \left(\sup_{x'} (A(x) * \psi(R(x,x')) \mapsto A(x)) \right) \right)$$
$$= \varphi \left(\inf_x \inf_{x'} \left(A(x) * \psi(R(x,x')) \mapsto A(x) \right) \right)$$
$$= \varphi \left(\inf_{x,x'} \left(A(x) * \psi(R(x,x')) \mapsto A(x) \right) \right).$$

On the other hand, for the lower rough approximations we have

$$\mathcal{T}(A) = A \hookrightarrow l_R(A) = \varphi \left(\inf_x (A(x) \mapsto l_R(A)(x)) \right)$$
$$= \varphi \left(\inf_x (A(x) \mapsto \inf_{x'} \left(\psi R(x,x') \mapsto A(x) \right)) \right)$$
$$= \varphi \left(\inf_x \inf_{x'} (A(x) \mapsto (\psi(R(x,x')) \mapsto A(x))) \right)$$
$$= \varphi \left(\inf_x \inf_{x'} (A(x) * \psi R(x,x') \mapsto A(x)) \right)$$
$$= \varphi \left(\inf_{x,x'} (A(x) * \psi R(x,x') \mapsto A(x)) \right).$$

Since $R(x,x') = R(x',x)$ in case R is symmetric, to complete the proof it is sufficient to notice that $(\alpha * \beta) \mapsto \gamma = \alpha \mapsto (\beta \mapsto \gamma)$ for any $\alpha, \beta, \gamma \in L$. Indeed,

$$(\alpha * \beta) \mapsto \gamma = \bigvee \{\lambda \mid \lambda * (\alpha * \beta) \leq \gamma\}, \text{ and}$$
$$\alpha \mapsto (\beta \mapsto \gamma) = \bigvee \{\lambda \mid \lambda * \alpha \leq \beta \mapsto \gamma\} = \bigvee \{\lambda \mid (\lambda * \alpha) * \beta \mapsto \gamma\};$$

the last equality is justified by Galois connection between $\mapsto$ and $*$. $\qquad \square$

The previous theorem allows us to introduce the following definition:

Definition 4 Let (L, M, φ, ψ) be an approximation quadruple and let R be an equivalence M-relation on a set X and $A \in L^X$. The measure of rough approximation of A is defined by

$$\mathcal{RA}(A) = u_R(A) \hookrightarrow A = A \hookrightarrow l_R(A).$$

In the next theorem we collect the main properties of the operators $\mathcal{K} : L^X \to L$ and $\mathcal{T} : L^X \to L$, and hence also of the operator $\mathcal{RA} : L^X \to L$ in case the relation R is symmetric.

Theorem 4 *M-valued measures of roughness of upper and lower rough approximation $\mathcal{K}, \mathcal{T} : L^X \to L$ have the following properties:*

1. *$\mathcal{K}(0_X) = 1_L$ where $0_X : X \to L$ is the constant function: $0_X(x) = 0_L$;*
2. *$\mathcal{T}(1_X) = 1_L$ where $1_X : X \to L$ is the constant function: $1_X(x) = 1_L$;*
3. *$\mathcal{K}(u(A)) = 1_L$ for every $A \in L^X$;*
4. *$\mathcal{T}(l(A)) = 1_L$ for every $A \in L^X$;*
5. *$\mathcal{K}(\bigvee_i A_i) \geq \bigwedge_i \mathcal{K}(A_i)$ for every family of L-fuzzy sets $\{A_i \mid i \in I\} \subseteq L^X$;*

6. $\mathcal{T}(\bigwedge_i A_i) \geq \bigwedge_i \mathcal{T}(A_i)$ *for every family of L-fuzzy sets* $\{A_i \mid i \in I\} \subseteq L^X$.

Proof (1) Referring to Theorem 1 and applying Proposition 2 (3 $\hookrightarrow$) we have $\mathcal{K}(A) = u_R(0_X) \hookrightarrow 0_X = 0_X \hookrightarrow 0_X = 1_M$. (2) Referring to Theorem 2 and applying Proposition 2 (3 $\hookrightarrow$), we have $\mathcal{T}(A) = 1_X \hookrightarrow l_R(1_X) = 1_X \hookrightarrow 1_X = 1_M$. (3) Referring to Theorem 1 and applying Proposition 2 (3 $\hookrightarrow$), we have

$$\mathcal{K}(u(A))) = u_R(u_R(A) \hookrightarrow u_R(A) = u_R(A) \hookrightarrow u_R(A) = 1_M.$$

(4) Referring to Theorem 2 and applying Proposition 2 (3 $\hookrightarrow$), we have

$$\mathcal{T}(l(A))) = l_R(A) \hookrightarrow l_R(l_R(A)) = l_R(A) \hookrightarrow l_L(A) = 1_M$$

(5) Referring to Theorem 1 and applying Proposition 2 (8 $\hookrightarrow$), we have

$$\mathcal{K}\left(\bigvee_i A_i\right) = u_R\left(\bigvee_i A_i\right) \hookrightarrow \bigvee_i A_i$$
$$= \bigvee_i u_R(A_i) \hookrightarrow \bigvee_i A_i \geq \bigwedge_i (u(A_i) \hookrightarrow A_i) = \bigwedge_i \mathcal{K}(A_i).$$

(6) Referring to Theorem 2 and applying Proposition 2 (7 $\hookrightarrow$), we have

$$\mathcal{T}(\bigwedge_i A_i) = (\bigwedge_i A_i) \hookrightarrow l(\bigwedge_i A_i) = \bigwedge_i A_i \hookrightarrow \bigwedge_i l(A_i) \geq \bigwedge_i A_i \hookrightarrow l(\bigwedge_i A_i).$$

$\square$

Theorem 5 *Let* $R_X : X \times X \to L$ *and* $R_Y : Y \times Y \to L$ *be reflexive transitive L-relations on sets X and Y respectively. Further, let f* : $X \to Y$ *be a mapping such that* $R_X(x,x') \leq R_Y(f(x),f(x'))$ *for every* $x,x' \in X$. *Then* $\mathcal{K}_X(f^{-1}(B)) \geq \mathcal{K}_Y(B)$ *and* $\mathcal{T}_X(f^{-1}(B)) \geq \mathcal{T}_Y(B)$ *for every* $B \in L^Y$.

The proof follows from the next sequences of (in)equalities:

$$\mathcal{K}_X(f^{-1}(B)) = u_{R_X}(f^{-1}(B)) \hookrightarrow f^{-1}(B))$$
$$= \varphi\left(\inf_x(u_{R_X}(f^{-1}(B))(x) \mapsto f^{-1}(B)(x))\right)$$
$$= \varphi\left(\inf_x\left(\sup_{x'} B(f(x)) * \psi R(x,x')\right) \mapsto B(f(x)))\right)$$
$$\geq \varphi\left(\inf_x\left(\sup_{x'} B(f(x') * \psi R(f(x),f(x'))))\right) \mapsto B(f(x))\right)$$
$$\geq \varphi\left(\inf_y\left(\sup'_y(B(y') * \psi R(y,y')) \mapsto B(y))\right)\right) = \mathcal{K}_Y(B)$$

and

$$\mathcal{T}_X(f^{-1}(B)) = (f^{-1}(B) \hookrightarrow l_{R_X}(f^{-1}(B)))$$
$$= \varphi\left(\inf_x(f^{-1}(B)(x) \mapsto l_{R_X}(f^{-1}(B)(x)))\right)$$
$$= \varphi\left(\inf_x(B(f(x)) \mapsto \inf_{x'}(\psi R(x,x') \mapsto f^{-1}(x')))\right)$$
$$= \varphi\left(\inf_x \inf_{x'}(B(f(x)) \mapsto (\psi R(x,x')) \mapsto B(f(x'))))\right)$$
$$\geq \varphi\left(\inf_x \inf_{x'}(B(f(x)) \mapsto (\psi R(f(x),f(x')) \mapsto B(f(x'))))\right)$$
$$\geq \varphi\left(\inf_y \inf_{y'}(B(y) \mapsto (\psi R(y,y') \mapsto B(y'))))\right) = B \hookrightarrow l_{R_Y}(B) = \mathcal{T}_Y(B).$$

$\square$

5.2 Examples of M-valued Measure of Roughness of Rough Approximation of L-fuzzy Sets

Here we illustrate M-valued measure of approximation in case when $L = M = ([0, 1], *)$ where $*$ is one of the basic t-norms: minimum $(*= \wedge)$, Łukasiewicz $(*_L)$ and product $(*= \cdot)$ and $\varphi = \psi = id$ are the identity mappings. A reader can easily construct the modifications of these examples for other approximation quadruples (L, M, φ, ψ), in particular, for the ones described in Sect. 2.2, in Examples (1)–(3) and in case of the t-norm $\wedge$, also in Examples (4)–(5).

Example 1 Let $*_L$ be the Łukasiewicz t-norm on the unit interval $L = [0, 1]$, that is

$$\alpha * \beta = \min(\alpha + \beta - 1, 1)$$

and $\mapsto_L : L \times L \to L$ be the corresponding residium, that is

$$\alpha \mapsto_L \beta = \max\{1 - \alpha + \beta, 0\}.$$

Then, given an equivalence L-relation R on a set X and $A \in L^X$ we have:

$$\mathcal{K}(A) = \inf_x \inf_{x'} (2 - A(x) + A(x') - R(x, x'));$$

$$\mathcal{T}(A) = \inf_x \inf_{x'} (2 - A(x) + A(x') - R(x', x)).$$

In particular, if $R : X \times X \to [0, 1]$ is the discrete relation, that is

$$R(x, x') = \begin{cases} 1 & \text{if } x = x' \\ 0 & \text{otherwise,} \end{cases}$$

we have

$$\mathcal{R}\mathcal{A}(A) = 1 \text{ for every } A \in L^X.$$

On the other hand for the indiscrete relation (that is $R(x, x') = 1$ for all $x, x' \in X$)

$$\mathcal{R}\mathcal{A}(A) = 1 - \inf_{x, x'} | A(x) - A(x') | \text{ for all } A \in L^X.$$

Example 2 Let $*= \wedge$ be the minimum t-norm on the unit interval $L = [0, 1]$, and $\mapsto : L \times L$ be the corresponding residium, that is

$$\alpha \mapsto \beta = \begin{cases} 1 & \text{if } \alpha \leq \beta \\ \beta & \text{otherwise} \end{cases}.$$

Then given a reflexive transitive *L*-relation R on a set X and $A \in L^X$ we have:

$$\mathcal{K}(A) = \inf_{x} \inf_{x'} (A(x') \wedge R(x, x') \mapsto A(x));$$

$$\mathcal{T}(A) = \inf_{x} \inf_{x'} (A(x') \wedge R(x', x) \mapsto A(x)).$$

In particular, $\mathcal{R}\mathcal{A}(A) = 1$ for every $A \in L^X$ in case the relation R is symmetric.

Example 3 Let $* = \cdot$ be the product t-norm on the unit interval $[0,1]$, and $\mapsto: L \times L$ be the corresponding residium, that is

$$\alpha \mapsto \beta = \begin{cases} 1 & \text{if } \alpha \leq \beta \\ \frac{\beta}{\alpha} & \text{if } \alpha > \beta \end{cases}.$$

Then, for a reflexive transitive *L*-relation we have:

$$\mathcal{K}(A) = \inf_{x} \inf_{x'} (A(x') \cdot R(x, x') \mapsto A(x));$$

$$\mathcal{T}(A) = \inf_{x} \inf_{x'} (A(x') \cdot R(x', x) \mapsto A(x)).$$

In particular

$$\mathcal{R}\mathcal{A}(A) = \inf_{x} \inf_{x'} (A(x') \cdot R(x, x') \mapsto A(x))$$

in case R is symmetric.

6 Measure of Roughness of a Fuzzy Set: Ditopological Interpretation

In this section we briefly discuss an alternative view on the concepts studied in the work. A reader not interested in the topological aspects of approximation, may omit this section. A detailed topological analysis of the M-valued measuring of rough approximation of L-fuzzy sets will be developed in the subsequent paper.

Notice that conditions (2), (4), and (6) of Theorem 4 actually mean that the mapping $\mathcal{T} : L^X \to M$ is an (L, M)-fuzzy topology on the set X, while conditions (1), (3), and (5) of this theorem mean that the mapping $\mathcal{K} : L^X \to M$ is an (L, M)-co-topology on this set (see e.g. [20, 21, 31, 33, 34]). Since the mappings $\mathcal{T}$ and $\mathcal{K}$ are not mutually related via complementation on the lattices L and M (which even need not exist on the lattice) we may interpret the pair $(\mathcal{T}, \mathcal{K})$ as an (L, M)-fuzzy ditopology on the set X [3].

Let $\alpha \in M$ be fixed and let

$$\mathcal{K}_\alpha = \{A \mid \mathcal{K}(A) \geq \alpha\} \text{ and } \mathcal{T}_\alpha = \{A \mid \mathcal{T}(A) \geq \alpha\}.$$

Then, applying again Theorem 4, we easily conclude that $\mathcal{T}_\alpha$ satisfies the axioms of a Chang-Goguen L-fuzzy topology, see [4, 11], and $\mathcal{K}_\alpha$ satisfies the axioms of a Chang-Goguen L-fuzzy co-topology. Hence for each $\alpha \in L$ the pair $(\mathcal{T}_\alpha, \mathcal{K}_\alpha)$ can be realized as a Chang-Goguen L-fuzzy ditopology on X [2].

From Theorem 5 we conclude that if $f : (X, R_X) \to (Y, R_Y)$ is a morphism in the category **REL**(M) of sets endowed with reflexive transitive M-relations, then

$$f : (X, \mathcal{T}_X, \mathcal{K}_X) \to (Y, \mathcal{T}_Y, \mathcal{K}_Y)$$

is continuous mapping of the corresponding (L, M)-fuzzy ditopological spaces. Thus we come to the following

Theorem 6 *By assigning to every object (X, R_X) from the category **REL**(M) (see Sect. 2.2) an (L, M)-fuzzy ditopological space $(X, \mathcal{T}_X, \mathcal{K}_X)$, and interpreting a morphism $f : (X, R_X) \to (Y, R_Y)$ of **REL**(M) as a mapping $f : (X, \mathcal{T}_X, \mathcal{K}_X) \to (Y, \mathcal{T}_Y, \mathcal{K}_Y)$ we obtain a functor*

$$\varPhi : \mathbf{REL}(M) \to \mathbf{DiTop}(L, M),$$

*where **DiTop**(L, M) is the category of (L, M)-fuzzy ditopological spaces and their continuous mappings [3].*

Corollary 1 *Let $\alpha \in M$ be fixed. By assigning to every object (X, R_X) from the category **REL**(M) a Chang-Goguen L-ditopological space $(X, \mathcal{T}_X{}^\alpha, \mathcal{K}_X{}^\alpha)$, and realizing a morphism $f : (X, R_X) \to (Y, R_Y)$ from **REL**(M) as a mapping $f : (X, \mathcal{T}_X{}^\alpha, \mathcal{K}_X{}^\alpha) \to (Y, \mathcal{T}_Y{}^\alpha, \mathcal{K}_Y{}^\alpha)$ we obtain a functor*

$$\varPhi_\alpha : \mathbf{REL}(M) \to \mathbf{DiTop}(L),$$

*where **DiTop**(L, M) is the category of Chang-Goguen L-ditopological spaces and their continuous mappings.*

7 Conclusion

In this paper we proposed an approach allowing to measure the roughness of lower and upper rough approximation for L-fuzzy subsets of a set endowed with a reflexive transitive M-relation. The basics of the theory of roughness measure were developed here. Besides, a natural interpretation of the operator of measure of rough approximation as a fuzzy ditopology was sketched here. However, several crucial issues concerning this theory remain untouched in this work. As one of the first goals for the further work we see the development of a consistent categorical viewpoint on the measure of rough approximation. In particular, it is important to study the behavior of the measure of approximation under operations of products, direct sums, quotients, etc., and to research the behavior of the measure of roughness under images and preimages of special mappings between sets endowed with reflexive transitive *M*-relations.

Another interesting, in our opinion, direction of the research is to develop the topological model of this theory sketched in Sect. 6. The restricted volume of this work does not allow us to linger on this subject. However, in our opinion the topological interpretation of the theory could be helpful for further studies.

Besides we hope that the concept of an *M*-valued measure of rough approximation will be helpful also in some problems of practical nature, since it allows in a certain sense to measure the quality of the rough approximation.

References

1. Birkhoff, G.: Lattice Theory. AMS Providence, RI (1995)
2. Brown, L.M., Ertürk, R., Dost, S.: Ditopological texture spaces and fuzzy topology, I. Basic concepts. Fuzzy Sets Syst. **110**, 227–236 (2000)
3. Brown, L.M., Šostak, A.: Categories of fuzzy topologies in the context of graded ditopologies. Iran. J. Fuzzy Syst. **11**(6), 1–20 (2014)
4. Chang, C.L.: Fuzzy topological spaces. J. Math. Anal. Appl. **24**, 182–190 (1968)
5. Chen, P., Zhang, D.: Alexandroff *L*-cotopological spaces. Fuzzy Sets Syst. **161**, 2505–2514 (2010)
6. Ciucci, D.: Approximation algebra and framework. Fund. Inform. **94**(2), 147–161 (2009)
7. Dubois, D., Prade, H.: Rough fuzzy sets and fuzzy rough sets. Int. J. Gen. Syst. **17**(2–3), 191–209 (1990)
8. Eļkins, A., Šostak, A.: On some categories of approximate systems generated by *L*-relations. In: 3rd Rough Sets Theory Workshop, pp. 14–19. Italy (2011)
9. Gierz, G., Hoffman, K.H., Keimel, K., Lawson, J.D., Mislove, M.W., Scott, D.S.: Continuous Lattices and Domains. Cambridge University Press, Cambridge (2003)
10. Goguen, J.A.: *L*-fuzzy sets. J. Math. Anal. Appl. **18**, 145–174 (1967)
11. Goguen, J.A.: The fuzzy Tychonoff theorem. J. Math. Anal. Appl. **43**, 734–742 (1973)
12. Han, S.-E., Kim, I.-S., Šostak, A.: On approximate-type systems generated by L-relations. Inf. Sci. **281**, 8–20 (2014)
13. Hao, J., Li, Q.: The relation between *L*-topology. Fuzzy Sets Syst. **178** (2011)
14. Höhle, U.: *M*-valued sets and sheaves over integral commutative CL-monoids. In: Rodabaugh, S.E., Höhle abd, U., Klement, E.P. (eds.) Applications of Category Theory to Fuzzy Subsets, pp. 33–72. Kluwer Academic Publishers, Docrecht (1992)
15. Höhle, U.: Commutative residuated *l*-monoids. In: Höhle abd, U., Klement, E.P. (eds.) Non-classical Logics and their Appl. to Fuzzy Subsets, pp. 53–106. Kluwer Academic Publishers, Docrecht (1995)
16. Järvinen, J.: On the structure of rough approximations. Fund. Inform. **53**, 135–153 (2002)
17. Järvinen, J., Kortelainen, J.: A unified study between modal-like operators, topologies and fuzzy sets. Fuzzy Sets Syst. **158**, 1217–1225 (2007)
18. Klement, E.P., Mesiar, R., Pap, E.: Triangular Norms. Kluwer Academic Publishers, Dordrecht (2000)
19. Kortelainen, J.: On relationship between modified sets, topological spaces and rough sets. Fuzzy Sets Syst. **61**, 91–95 (1994)

20. Kubiak, T.: On fuzzy topologies, PhD Thesis, Adam Mickiewicz University Poznan, Poland (1985)
21. Kubiak, T., Šostak, A.: A fuzzification of the category of M-valued L-topological spaces. Appl. Gen. Topol. **5**, 137–154 (2004)
22. Menger, K.: Geometry and positivism - a probabilistic microgeometry. In: Selected Papers in Logic, Foundations, Didactics, Economics. Reidel, Dodrecht (1979)
23. Mi, J.S., Hu, B.Q.: Topological and lattice structure of L-fuzzy rough sets determined by upper and lower sets. Inf. Sci. **218**, 194–204 (2013)
24. Pawlak, Z.: Rough sets. Int. J. Comp. Inf. Sci. **11**, 341–356 (1982)
25. Qin, K., Pei, Z.: On the topological properties of fuzzy rough sets. Fuzzy Sets Syst. **151**, 601–613 (2005)
26. Qin, K., Pei, Z.: Generalized rough sets based on reflexive and transitive relations. Inf. Sci. **178**, 4138–4141 (2008)
27. Radzikowska, A.M., Kerre, E.E.: A comparative study of fuzzy rough sets. Fuzzy Sets Syst. **126**, 137–155 (2002)
28. Rosenthal, K.I.: Quantales and their applications. In: Pitman Research Notes in Mathematics 234. Longman Scientific and Technical (1990)
29. Schweitzer, B., Sklar, A.: Probabilistic Metric Spaces. North Holland, New York (1983)
30. Šostak, A.: Measure of roughness of rough approximation of fuzzy sets and its topological interpretation. In: Proceedings of the International Conference on Fuzzy Computation Theory and Applications (FCTA-2014), pp. 61–67 (2014)
31. Šostak, A.: On a fuzzy topological structure. Suppl. Rend. Circ. Matem. Palermo, Ser II **11**, 125–186 (1985)
32. Šostak, A.: Towards the theory of approximate systems: variable range categories. In: Proceedings of ICTA2011, Islamabad, Pakistan, pp. 265–284. Cambridge University Publ. (2012)
33. Šostak, A.: Two decades of fuzzy topology: basic ideas, notions and results. Russ. Math. Surv. **44**, 125–186 (1989)
34. Šostak, A.: Basic structures of fuzzy topology. J. Math. Sci. **78**, 662–701 (1996)
35. Šostak, A.: Towards the theory of M-approximate systems: fundamentals and examples. Fuzzy Sets Syst. **161**, 2440–2461 (2010)
36. Tiwari, S.P., Srivastava, A.K.: Fuzzy rough sets, fuzzy pre oders and fuzzy topoloiges. Fuzzy Sets Syst. **210**, 63–68 (2013)
37. Valverde, L.: On the structure of F-indistinguishability operators. Fuzzy Sets Syst. **17**, 313–328 (1985)
38. Yao, Y.Y.: On generalizing Pawlak approximation operators. In: Proceedings First International Conference Rough Sets and Current Trends in Computing, pp. 298–307 (1998)
39. Yao, Y.Y.: A comparative study of fuzzy sets and rough sets. Inf. Sci. **109**, 227–242 (1998)
40. Yu, H., Zhan, W.R.: On the topological properties of generalized rough sets. Inf. Sci. **263**, 141–152 (2014)
41. Zadeh, L.: Similarity relations and fuzzy orderings. Inf. Sci. **3**, 177–200 (1971)

Fuzzy Control of a Sintering Plant Using the Charging Gates

Marco Vannocci, Valentina Colla, Piero Pulito, Michele Zagaria,
Vincenzo Dimastromatteo and Marco Saccone

Abstract The industrial priorities in the automation of the sinter plant comprise stable production rate at the highest productivity level specially within an integrated steelwork and classical control scheme may fail due to the complexity of the sinter process. The paper describes an approach exploiting a fuzzy rule- based expert system to control the charging gates of a sinter plant. Two different control strategies are presented and discussed in details within an innovative advisory system that supports the plant operators in the choice of the most promising action to do on each gate. A third strategy that combines the strong points of the two detailed ones is presented and studied in feasibility. Through a suitable exploitation of real-time data, the advisory system suggests the most promising action to do by reproducing the knowledge of the most expert operators, supporting the technicians in the control of the plant. Thus, this approach can also be used to train new plant operator before involving them in the actual plant operations. The performance of the detailed strategies and the goodness of the system have been evaluated for long time in the sinter plant of one of the biggest integrated steelworks in Europe, namely the ILVA Taranto Works in Italy.

Keywords Fuzzy control · Expert systems · Sinter plant

M. Vannocci · V. Colla (✉)
PERCRO Laboratory, Scuola Superiore Sant'Anna,
Via Alamanni 13D, San Giuliano Terme, Pisa, Italy
e-mail: colla@sssup.it

M. Vannocci
e-mail: m.vannocci@sssup.it

P. Pulito · M. Zagaria · V. Dimastromatteo · M. Saccone
ILVA S.p.A., Strada Statale Appia Km 648, Taranto, Italy
e-mail: piero.pulito@gruppoilva.com

M. Zagaria
e-mail: michele.zagaria@gruppoilva.com

V. Dimastromatteo
e-mail: vincenzo.dimastromatteo@gruppoilva.com

M. Saccone
e-mail: automazionepma.taranto@gruppoilva.com

© Springer International Publishing Switzerland 2016
J.J. Merelo et al. (eds.), *Computational Intelligence*,
Studies in Computational Intelligence 620, DOI 10.1007/978-3-319-26393-9_16

1 Introduction

Within a steelwork, the sintering process is a central operation in the production cycle: the treatment is basically a high temperature process that starts from raw materials mixture (such as fine iron ores) and produces a particular form of agglomerate material known as *sinter*, which is one of the material fed to the blast furnace in order to produce the pig iron, which is subsequently refined in the steel shop to produce the liquid steel.

The sintering process is articulated into a series of standard operations. A first preliminary task is the acceptation and the storage of the iron-bearing raw materials in the ore stockyard followed by the crashing and the screening of these raw materials. Then, the following phases which are more specific of the sintering process can be pointed out: (I) raw materials are mixed together with water and then granulated into a pseudo-particles in a rotary mixer drum and then stocked in a feed hopper; (II) after the hopper, the moistened mix passes through the charging gates and it is accumulated just before a leveler that strips out the exceeding material; thus the moistened mixture is charging as a layer onto continuously moving pallet-cars called "strand"; (III) after the ignition of the material close to the charging zone, the burning process is propagated by chemical reaction thanks to the air sucked through the strand by the so-called *wind boxes*, that are depressurised air ducts mounted below the strand; (IV) at the end of the strand the solidified agglomerate is broken within a crusher and cooled within a cooler strand; (V) finally the cooled material is conveyed to a second crusher in order to obtain a suitable size of the particles of the final sinter.

The overall process must be controlled in order to ensure that all the mix is burned just earlier than being discharged into the crusher. The points at which the flame front reaches the base of the strand are called "burn-through points" (BTPs). Thus, among the aims of the control process of the plant, two of them are of considerable interest and can be summarized as follows: to ensure that the BTPs are aligned in the transversal direction of the strand and to ensure that this alignment happens just earlier than the discharge. As a matter of fact, a uniform flame front is guaranteed to the former condition while the latter one optimizes the production capacity of the plant. In fact there is an evident waste of productivity of the plant if the BTPs occur too early compared to the end of the process; analogously, the quality of the sinter lowers if the burning process is not completed before the discharge and this fact negatively affects the production rate of the overall steel plant and the following production stages.

Predictive capabilities have been used to develop control schemes controlling the speed of the strand. In [1] the prediction of the waste gas temperature is used to manipulate the strand speed, while in [2] the same variable is controlled using a process model identified from the observed data. A different perspective is presented in [3], where the control scheme tries to keep the temperature distribution at the end of the plant on a pre-defined curve in order to yields a target BTP. Also in this case the manipulated variable is the strand speed.

The new approach presented in the paper is based on fuzzy rule-based expert systems and exploits the charging gates as controlled variables. Two different control strategies will be presented and discussed as well as an innovative advisory system that supports the plant operators in the control of the plant.

The paper is organized as follows. Section 2 provides a brief overview on the techniques to extract the fuzzy rules from the real data and to build the rule-basis. Section 3 describes the two control strategies while the advisory system is presented in Sect. 4. Section 5 presents the results of the on-the-plant tests and Sect. 6 proposed a feasibility study for a strategy which, on the basis on the previously achieved results, integrates the two proposed control approaches. Finally Sect. 7 provides some concluding remarks.

2 Rules Extraction

When dealing with fuzzy rule-based systems the most important component of the entire system is the rule basis. In fact, this is the component conveying most of the system's knowledge and it contains also the relationships between the variables coming from the field of application. The rule basis is of the utmost importance as it is the core of the fuzzy inference system: different rule bases can produce antithetical results although exploiting the same input data and the same mechanism, starting from the same fuzzy sets and membership functions and procedures for fuzzyfication of the crisp input data and defuzzification to produce numerical results.

Thus, the most important and exciting task during the design of the system, is the formalization of a correct and representative rule-basis for the considered application, starting from the available knowledge of the phenomenon or process under consideration and/or from the field expertise and practical knowledge of technical personnel who continuously handles the system to control.

In the last years, several methods and techniques have been proposed in the literature to cope with this issue. Heuristic approaches can be found in [4–6], while examples of clustering-based technique are presented in [7–9]. Other methods are based on Support Vector Machines (SVM) [10], Genetic Algorithms (GA) [11–13] or bacterial evolutionary algorithms [14]. Memetic algorithms belong to evolutionary algorithms too and they try to emulate cultural evolutions instead of the biological ones. An application of such techniques to rules extraction can be found in [15, 16].

Apart from evolutionary algorithms and clustering techniques, neuro-fuzzy methods [17] as well as data-mining techniques [18] can be applied.

A further approach, to be eventually integrated with the previously cited ones, is based on the direct interviews of the experts in order to acquire the knowledge and to understand the best practices that are commonly in use. Starting from the experts' knowledge and the best practices, a selection, management and aggregation of the information must be performed. This task is usually completed taking into account a trade-off between the dimension of the rule base (usually a small one is preferable in order to gain computational efficiency as well as easy interpretation for further adaptation) and the goodness of the information to embed into the rules. This approach has been successfully applied by some of the authors in past works [19, 20].

A positive side-effect of this approach consist in the fact that, when heuristic procedures or automatic techniques (for instance neural networks in the case of the particular category of the so-called Adaptive Neuro-Fuzzy Inference Systems—ANFIS [21]) are applied to tune some of the parameters of the FIS on the basis of the achieved experimental results or of the available data, the new FIS can be easily interpreted and the tuning operation helps the technicians to improve their own knowledge of the system under consideration, as it exploits the same kind of formalization they suggested. In the present case, the rules have been heuristically tuned in order to achieve final satisfactory performance of the process.

3 Control Strategies

Some mathematical models have been developed in order to cope with the dynamic of the sintering process in an analytical way. A first attempt has been made in [22] while a different perspective has been developed in [23, 24]. A different approach based on multiple-valued logic is the core of the present paper and concepts of fuzzy sets [25], fuzzy control [26–28] and expert systems [29, 30] are used to develop the strategies and to build the advisory system.

3.1 Overview of the Sintering Machine

The transversal direction of the strand can be divided into four segments denoted by A, B, C and D. Each of them covers the overall length of the machine and it is about one meter wide, so that they cover the overall width of the strand. Regarding the longitudinal direction, three different macro-zones can be pointed out as depicted in Fig. 1 and described in the following:

- charging zone ($\mathcal{A}$) at the very beginning of the strand;
- permeabilities zone ($\mathcal{B}$) after the ignition hood at the 3rd wind box;
- burn-through points zone (C) at the end of the bed covering a wide area of about 48 m^2.

Within the macro-zone $\mathcal{A}$ the feed hopper, 6 charging gates and 6 infrared sensors can be found, while the permeability sensors, which take 4 different permeability measurements along the transversal direction of the strand and that are indicated in the following as K_A, K_B, K_C, K_D, are located within the macro-zone $\mathcal{B}$. Finally, in the last zone (C) a regular grid of thermocouples has been installed, such as depicted in Fig. 2 that measures 24 temperature values: among them, the maximum one of

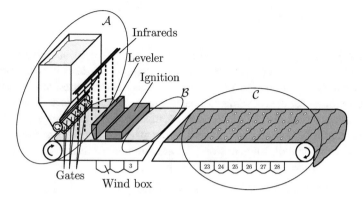

Fig. 1 Sketch of the macro-zones $\mathcal{A}, \mathcal{B}$ and $\mathcal{C}$ of the plant

Fig. 2 Sketch of the grid of thermocouples

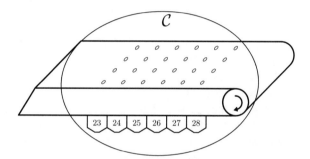

each segment is the associated BTP, thus finally there are 4 BTP values indicated as $BTP_A, BTP_B, BTP_C, BTP_D$.

From the point of view of the plant operational practice, it would be advisable that the maximum value of temperature is reached for all the segments in correspondence to approximately the same distance from the strand end, as this implies that the sintering process is quite homogeneous in all the portions of the strand itself.

A preparatory statistical analysis conducted on historical data coming from the plant has put into evidence that the transversal alignment of $BTP_A, \ldots, BTP_D$ is related to a specific configuration of $K_A, \ldots, K_D$. Thus, the idea behind the strategies is to control the charging gates in order to obtain the suitable permeability configuration in the macro-area $\mathcal{B}$.

Indeed, as the permeability configuration is supposed to be symmetric, the difference between the external permeabilities (K_A, K_D) and the internal ones (K_B, K_C) can leads to crucial information about the suitable configuration to be obtained.

Thus, in order to pursue such investigation, the following 4 permeability ratios ($\mathbf{K}_r$) have been defined:

$$(r_1, r_2, r_3, r_4)^T = \left(\frac{K_A}{K_D}, \frac{K_B}{K_C}, \frac{K_A}{K_B}, \frac{K_D}{K_C} \right)^T . \tag{1}$$

3.2 Description of the Strategies

Two control strategies have been designed, which both aim at maximizing the
increase of the average permeability. The first strategy ('a') takes into account only
this target; the second one ('b') takes also into account the stress on the actuator
and tries to minimize the movement of each gate, namely, if more actions involving
different gates are equally physically feasible, this latter strategy suggests the action
that produces the minimum displacement of the gates from their current position.

Variables and symbols for both strategies are summarized in Table 1 and a description of the first strategy ('a') is the following:

1. **Gates and Infrared Indices**: the gate and the infrared values are normalized
   using its operative limits (G_{min}, G_{max}, IR_{min}, IR_{max});
2. **Control Coefficients**: four control coefficients that express the contribution of
   each couple of gates (i.e. 1,2 - 2,3 - 4,5 - 5,6) on each segment of the bed are
   computed using the above indices;
3. **Action Indices**: four "action-indices" that contain the information about the operations to perform on each couple of gates (e.g. a value less than zero indicates
   that the gates of a couple need to be closed) are computed and limited between
   suitable thresholds (c_{min}, c_{max});
4. **Targets Computation**: four different target permeability profiles, namely 4 different vectors of 4 entries each, are obtained using the permeability ratios $\mathbf{K}_r$; the
   ith element of the ith vector is obtained by imposing $\mathbf{K}_{K_i}^{target}(i) = \mathbf{K}_t(i)$ while the
   other elements are calculated using $\mathbf{K}_r$; the Eq. (2) show the computations for the
   first ($i = 1$) profile:

$$\mathbf{K}_{K_1}^{target}(1) \stackrel{def}{=} K_A^{target} = \mathbf{K}_t(1) \equiv K_A \tag{2a}$$

$$\mathbf{K}_{K_1}^{target}(2) \stackrel{def}{=} K_B^{target} = K_A^{target}/r_3 \tag{2b}$$

$$\mathbf{K}_{K_1}^{target}(3) \stackrel{def}{=} K_C^{target} = K_B^{target}/r_2 \tag{2c}$$

$$\mathbf{K}_{K_1}^{target}(4) \stackrel{def}{=} K_D^{target} = K_A^{target}/r_1 \tag{2d}$$

where the relation (2a) concern the imposed elements ($i = 1$), while the formulas
(2b–2d) concern the computed elements of the profile;
5. **Gaps between Current and Target**: for each target profile the gaps between the
   current profile and the target is evaluated, obtaining 4 different vectors of 4 entries
   each that are computed according to the following equations, where $i = 1, \ldots, 4$:

$$\mathbf{K}_{K_i}^{gaps}(1) - K_A - \mathbf{K}_{K_i}^{target}(1) \tag{3a}$$

$$\mathbf{K}_{K_i}^{gaps}(2) = K_B - \mathbf{K}_{K_i}^{target}(2) \tag{3b}$$

Table 1 Symbols used within the control strategies

Symbol	Meaning[†]
$\mathbf{G}_t$	Opening percentage of the gates
$\mathbf{IR}_t$	Measure of the height of the sinter bed along the transversal direction
$\mathbf{K}_t$	Permeability values i.e. $\mathbf{K}_t = (K_1, \ldots, K_4)^T \equiv (K_A, \ldots, K_D)^T$
$\mathbf{K}_r$	Permeability ratios i.e. $\mathbf{K}_r \equiv (r_1, r_2, r_3, r_4)^T$
$\mathbf{K}_{K_i}^{\text{target}}$	Permeability values computed using $\mathbf{K}_r$ by imposing $\mathbf{K}_{K_i}^{\text{target}}(i) = \mathbf{K}_t(i), \ i = 1, \ldots, 4$
$G_{\min}, G_{\max}$	Working range limits of the gates
$IR_{\min}, IR_{\max}$	Minimum and maximum allowed height of the sinter bed
$c_{\min}, c_{\max}$	Suitable thresholds
$\mathbf{Y}$	Selected control actions

[†] where $(\cdot)_t$ means that the values are taken at the current time instant i.e. they are on-line values

$$\mathbf{K}_{K_i}^{\text{gaps}}(3) = K_C - \mathbf{K}_{K_i}^{\text{target}}(3) \tag{3c}$$

$$\mathbf{K}_{K_i}^{\text{gaps}}(4) = K_D - \mathbf{K}_{K_i}^{\text{target}}(4) \tag{3d}$$

6. **Control Amount for each Couple of Gates**: for each vector of gaps the control amount values to be applied on each couple of gates is calculated by obtaining four vectors with four elements each;
7. **Calculate the Feasibility**: for each target profile a feasibility coefficient is computed informing if the related action on the couple of gates are physically feasible or not (e.g. it is required to close a gate that it is already completely closed);
8. **Calculate the Increase of the Average Permeability**: for each target profile the gains of the average values are evaluated in order to use these values as performance indicators; each of them is related to a vector of control amount values so that the better the control on the gates the *higher* will be the index.
9. **Select the Actions**: among some target profile all equally physically feasible, it is selected the one which optimizes the performance indicator gaining the four control amount values denoted by u_{12}, u_{23}, u_{45} and u_{56}.

In Fig. 3 the conceptual diagram of control strategy 'a' is reported.

The description of strategy 'b' is similar to the one of the strategy 'a' where the point 8 is modified as follows:

8. **Calculate the Stress on the Actuator**: for each target profile, the stress produced on the actuator is evaluated by summing the overall gaps of the profile[1] in order to use these values (one for each target profile) as performance indicators; each of them is related to a vector of control amount values so that the better the control on the gates the *lower* will be the index.

---

[1]In fact, the gaps are related to the distance between the current and the desired position of the gates.

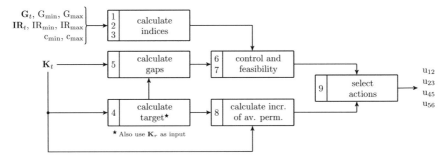

Fig. 3 Conceptual scheme of the first control strategy

Thus, the conceptual diagram of the second control strategy is analogous to the one of the strategy 'a' except for the block number 8.

4 Advisory System

The new value of the gates, required to reach the target permeability profile, can be computed through the following relations:

$$\mathbf{G}^{new}(1) = \mathbf{G}_t(1) + 0.5\,u_{12} \ , \tag{4a}$$

$$\mathbf{G}^{new}(2) = \mathbf{G}_t(2) + 0.5\,u_{12} + 0.5\,u_{23} \ , \tag{4b}$$

$$\mathbf{G}^{new}(3) = \mathbf{G}_t(3) + 0.5\,u_{23} \ , \tag{4c}$$

$$\mathbf{G}^{new}(4) = \mathbf{G}_t(4) + 0.5\,u_{45} \ , \tag{4d}$$

$$\mathbf{G}^{new}(5) = \mathbf{G}_t(5) + 0.5\,u_{45} + 0.5\,u_{56} \ , \tag{4e}$$

$$\mathbf{G}^{new}(6) = \mathbf{G}_t(6) + 0.5\,u_{56} \tag{4f}$$

where the coefficient of each term has been set heuristically using the knowledge of the technicians' expertise.

Comparing the new values of the gates with the current values, the advisory system is able to provide information about the actions to be performed on *all* of the charging gates. The actions are also related to a rank number in order to inform about the most promising of them.

The advisory system is an expert system (ES) founded on multiple-valued logic with a rule base that reproduce the knowledge of the plant operators. Thus the system belongs to the larger family of the Fuzzy Rule-Based Expert Systems (FRBES) [31].

The system is designed using the zero-order Takagi-Sugeno-Kang (TSK) model [32] where the *j*th fuzzy rule (R_j) of the form "IF ⟨*premise*⟩ THEN ⟨*conclusion*⟩" is given by:

$$R_j: \text{ IF } (x \text{ is } A_i) \text{ AND/OR } (y \text{ is } B_i) \text{ THEN } z_j = c_j \tag{5}$$

where x and y are the inputs, A_i and B_i are fuzzy sets and c_j is a crisp adjustable parameter.

The system evaluates each rule (*implication*) collecting together the results (*aggregation*) in order to produce a unique output fuzzy set. The crisp value extracted from this fuzzy set (*defuzzification*) represents the output of the entire inference process.

The aggregation and the defuzzification tasks can be merged in a unique operation in the TSK model as follows:

$$z = \frac{\sum\limits_{j=1}^{N} w_j c_j}{\sum\limits_{j=1}^{N} w_j}, \text{ with } w_j = F(\mu_i, v_i) \tag{6}$$

where N is the number of rules, w_j is the firing strength of the jth rule (i.e. the "degree of truth" of the $\langle premise \rangle$), F is the method that implements the AND operator (F is a t-norm) or the OR operator (F is a t-conorm), μ_i is the membership degree of x to A_i and v_i is the membership degree of y to B_i.

The system is composed of 6 specialized fuzzy inference systems (FISs), one for each gate (FIS$_i$, $i = 1, \ldots, 6$), whose inputs are the differences $\Delta G_i = \mathbf{G}^{\text{new}}(i) - \mathbf{G}_t(i)$ (after a proper normalization stage forcing any crisp input to lie in the range $[-1, 1]$).

Moreover, ΔG_2 is in input to both the FIS$_1$ and the FIS$_3$ as additional input as well as ΔG_5 is in input to both the FIS$_4$ and the FIS$_6$. These additional gaps will be denoted by the DGN term within the rules.

The FIS$_2$ has ΔG_1 and ΔG_3 as additional inputs as well as the FIS$_5$ receives ΔG_4 and ΔG_6. These additional gaps will be denoted, respectively, by the DGDx and DGSx terms within the rules. Table 2 summarizes the inputs of each FIS.

The inputs of the FIS$_{1,3,4,6}$ are 2 linguistic variables, while the inputs of the FIS$_{2,5}$ are 3 linguistic variables. For each input variable, 3 linguistic terms (i.e. fuzzy sets) are defined: 'Negative' (NEG), 'Null' (NUL) and 'Positive' (POS). The membership functions of all the fuzzy sets are bell-shaped functions whose parameters (μ, σ) have been heuristically set as described in Table 3.

Table 2 Inputs of each fuzzy inference system

	ΔG_1	ΔG_2	ΔG_3	ΔG_4	ΔG_5	ΔG_6
FIS$_1$	•	•				
FIS$_2$	•	•	•			
FIS$_3$		•	•			
FIS$_4$				•	•	
FIS$_5$				•	•	•
FIS$_6$					•	•

Table 3 Characteristic parameters μ and σ of the linguistic terms

| | $FIS_{1,3,4,6}$ | | $FIS_{2,5}$ | | |
	DG	DGN	DG	DGSx	DGDx
NEG	$-1, 0.34$	$-1, 0.34$	$-1, 0.34$	$-1, 0.34$	$-1, 0.34$
NUL	$0, 0.1$	$0, 0.1$	$0, 0.16$	$0, 0.14$	$0, 0.13$
POS	$1, 0.34$	$1, 0.34$	$1, 0.34$	$1, 0.34$	$1, 0.34$

Table 4 Rules for the $FIS_{1,3,4,6}$

	DGN is NEG	DGN is POS	DGN is NUL
DG is NEG	-1	0	-1
DG is POS	0	1	1
DG is NUL	0	0	0

The rule bases for the systems have been obtained after fruitful discussions with the plant operators and considering also the operative practices of the operators too.

All the rules for the $FIS_{1,3,4,6}$ are shown in Table 4 while in Tables 5, 6 and 7 are shown the rules for the $FIS_{2,5}$.

In all the rule bases the numerical value of the output means that the gate must be closed (-1) or opened (1) or, finally, that no operation must be performed on the gate (0). The $\prod(x, y) = x * y$ operator is the t-norm that implements the AND connection of each rule and the relation (6) is used to defuzzify the inferred output fuzzy set.

Table 5 Rules for the $FIS_{2,5}$ when DGSx is POS

| | DGSx is POS | | |
	DGDx is NEG	DGDx is POS	DGDx is NUL
DG is NEG	0	0	-1
DG is POS	0	1	0
DG is NUL	0	0	0

Table 6 Rules for the $FIS_{2,5}$ when DGSx is NUL

| | DGSx is NUL | | |
	DGDx is NEG	DGDx is POS	DGDx is NUL
DG is NEG	0	-1	-1
DG is POS	1	0	1
DG is NUL	0	0	0

Table 7 Rules for the $FIS_{2,5}$ when DGSx is NEG

	DGSx is NEG		
	DGDx is NEG	DGDx is POS	DGDx is NUL
DG is NEG	−1	0	0
DG is POS	0	0	1
DG is NUL	0	0	0

5 Experimental Results

Two different scenarios have been considered for the plant tests. The first one ("short period scenario") takes into account a period of 4 h using the strategy 'a' for the first 2 h and the strategy 'b' during the last 2 h.

Some characteristic conditions can be highlighted during the tests within this scenario: (a) no stoppage of the strand occurred; (b) the percentages of lime and limestone within the mix were fixed; (c) the ratio between the speed of the drum feeder and the speed of the strand was constant; (d) the moisture of the mix was kept as constant as possible.

The qualitative results of the first session showed that the strategy 'a' leads to an actual increment of the average permeability and thus to a better yield of the plant: this can be assessed considering the lower Internal Return Fines (IRFs) consumption and thus a lower wet coke consumption.

During the tests, the plant operators confirmed that: (i) the actions proposed by the system have been always fully safe despite the change of the operational conditions in which each decision has been taken; (ii) the operations have been always coherent with the personnel's expertise. Thus, in the short period the advisory system proved to behave according to the best practice of the operators. Therefore, it is expected that, in the long period, the automatic implementation of such system will lead to improvement of the process performances with a reduction of the operators' effort.

The second scenario ("long period scenario") considers a period of 84 h and compares the system behavior during intervals when one of the developed control strategies was applied to intervals when none of them was used and the plant was controlled by exploiting only the expertise of the plant technicians. The detailed description of each period of time is summarized in Table 8.

The long period of eighty-four hours is the best one that can be obtained minimizing the external influences such as, for example, a different chemical composition in the mix caused by a different Blend Iron Ore (BIO) in the mix. In Taranto, in fact, the typical amount of BIO is of 160000–180000 tonnes and they are used within the mix during a typical period of five days. Thus, the tests have been performed after a suitable stabilization period after the change of the BIO. Moreover, during the first day of the long period test a significant stoppage of the strand occurred. In order to take this fact into account, the results have been computed by using only the data deriving from a stable condition of the process (i.e. about 90 min after the restart).

Table 8 Time intervals of the long period tests

Test number	Duration (h)	Time interval	Control strategy
1	12	Day 1 (from 00:00 to 12:00)	Technicians expertise
2	12	Day 1 (from 12:00 to 00:00)	'a'
3	12	Day 2 (from 00:00 to 12:00)	Technicians expertise
4	12	Day 2 (from 12:00 to 00:00)	'b'
5	12	Day 3 (from 00:00 to 12:00)	Technicians expertise
6	12	Day 3 (from 12:00 to 00:00)	'a'
7	12	Day 4 (from 00:00 to 12:00)	'b'

Let ΔS be the amount of the produced sinter (in tonnes), σ_C the specific coke consumption (i.e. the amount of the coke consumption—measured in Kg—per tonne of produced sinter) and σ_{IRF} the specific IRF production (i.e. the amount of produced IRFs production—measured in Kg—per tonne of produced sinter); 3 significant Key Performance Indicators (KPIs) can be defined as follows (where $(\cdot)/h$ means "per hour"):

$$\text{KPI}_1 \text{ i.e. Sinter Production} = \frac{\Delta S}{h} \, , \tag{7a}$$

$$\text{KPI}_2 \text{ i.e. Wet Coke Consumption} = \frac{\sigma_C}{h} \, , \tag{7b}$$

$$\text{KPI}_3 \text{ i.e. IRF Production} = \frac{\sigma_{IRF}}{h} \, . \tag{7c}$$

In order to evaluate the goodness of the results the trends of the KPIs have been evaluated. In particular, the IRFs production that can be related directly to the yield of the plant through the following equation:

$$\text{Yield}(\%) = 100 - 0.1 \, \sigma_{IRF} \, . \tag{8}$$

The results of the long period tests are numerically described in Table 9: using the strategy 'a' an average increment of the produced sinter and an average reduction of the coke consumption as well as of the IRFs production have been gained. The strategy 'b', on the other hand, led to comparable results regarding the coke consumption and the IRFs production, but caused an average decrement of the productivity.

During both the short and the long period tests the opinions of the technicians have been taken into account in order to evaluate the practical goodness of the strategies as well as the KPI's variations.

The behaviour most frequently used by the plant experts was very similar to that of the strategy 'b' and in contrast with the strategy 'a'. Indeed, the strategy 'a' takes into account only the average permeability and whenever a variation of the gates position is required in order to improve the permeability, the variation is reported to the gates. This leads to frequent variation in the gates positions and sometimes in abrupt

Table 9 Experimental results

Control strategy	KPI_1	KPI_2	KPI_3
Expertise	4.25	−0.019	6.60
Strategy 'a'	2.38	−0.236	−6.75
Expertise	−6.25	0.208	4.66
Strategy 'b'	1.45	−0.062	−2.50
Expertise	−1.18	0.048	2.17
Strategy 'a'	3.92	−0.015	−0.43
Strategy 'b'	−5.76	−0.231	−0.93
Expertise (mean)	−1.06	0.08	4.48
Strategy 'a' (mean)	3.15	−0.13	−3.59
Strategy 'b' (mean)	−2.16	−0.15	−1.72

changes. The technicians, on the other hand, use a more conservative approach that tends to perform slight modification in the gates position and rarely abrupt changes.

Summing up the results of the tests of the two control strategies: strategy 'a' gives better automatic results, but it is less coherent to the technicians' standard operating practice, thus they can experience higher efforts in order to follow the plant behavior when this control strategy is applied. Strategy 'b' leads to fairly good results, but it is more coherent to the standard operating practice.

Finally, it can be noticeable that the advisory system has been designed to be improved through its use, as the overall software system supports data collection and analysis. After a longer period of use it will be possible to refine the performances of the proposed system using the same statistical parameters that supplied the system.

6 'Hybrid' Control Strategy: A Feasibility Study

Considering the results obtained with the two above described control strategies a third one that combines the strong points of the two developed ones has been investigated. This 'hybrid' approach, exploiting the advantages of each developed strategy, should lead to better results during all the production phases.

The 'hybrid' strategy is able to detect the operating conditions in which each of the two developed ones ('a' or 'b') is mostly suitable and performs an automatic switching between them.

The former point has been discussed in depth with the technicians and they agreed to use of strategy 'a'/'b' when the plant is characterized by low/high productivity. These conditions can be detected by monitoring the time trend of the KPI_1 and the ratio between the speed of the drum feeder and the speed of the strand. In fact, after a stoppage, the plant is characterized by a very low speed of the strand while the trend

Fig. 4 Transition graph of
the 'hybrid' strategy

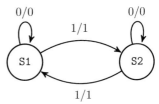

of the KPI_1 is not yet affected by the stoppage. Thus, monitoring only the KPI_1 can be not sufficient to properly detect the correct operating conditions.

The automatic switching can be performed by means of a finite-state machine, characterized by two states S1 and S2, one input variable in and one output variable out with the following meaning:

- the state S1 and S2 are, respectively, the state where the strategy 'a' or 'b' is used;
- the variable in claims if the switching conditions have been properly detected (in=1) or not (in=0);
- the variable out is more properly the control action on the switch, i.e. it commands (out=1) or not (out=0) the switching between the strategies.

The initial state of the automaton can be S1 as well as S2, depending on the particular condition of the plant when the 'hybrid' strategy starts to operate on the plant. Whichever the initial state is, the transition graph is depicted in Fig. 4 (where the notation $\langle in \rangle / \langle out \rangle$ stands for the couple of values of the variable in and out and the relation between to each other).

7 Conclusion

A new approach based on fuzzy rule-based expert systems and a new advisory system to control the charging gates of a sinter plant is presented. Two new control strategies have been developed and tested on the field.

Strategy 'a' is more invasive within the process operational conditions, as it aims at maximizing the productivity without any kind of trade-off. It can be used when the plant is characterized by lower productivity (e.g. the plant is restarted after a stoppage): in these cases the greater the control amount, the shorter the time elapsed before reaching fair operating conditions.

On the other hand, strategy 'b' is more conservative, as it aims at maximizing the productivity but considers also the stress on the gates' actuators and produces less perturbation in the operating conditions with respect to the other strategy. It can be used when the plant is characterized by higher productivity and the machine shows a higher sensitivity to the changes on the gates.

Real-time tests are still ongoing at the sinter plant of ILVA S.p.A. (Taranto Works, Italy) and satisfactory results confirm the goodness of the automatic control system.

Future work will deal with the plant test of the 'hybrid' strategy, that has been developed in the feasibility study.

Acknowledgments The work described in the present paper was developed within the project entitled "New measurement and control techniques for total control in iron ore sinter plants" ToSiCo (Contract No. RFSR-CT-2009-00001) that has received funding from the Research Fund for Coal and Steel of the European Union. The sole responsibility of the issues treated in the present paper lies with the authors; the Commission is not responsible for any use that may be made of the information contained therein.

References

1. Kanjilal, P.P., Rose, E.: Application of adaptive prediction and control method for improved operation of the sintering process. Ironmak. Steelmak. **13**, 289–293 (1986)
2. Hu, J.Q., Rose, E.: Predictive fuzzy control applied to the sinter strand process. Control Eng. Pract. **5**, 247–252 (1997)
3. Arbeithuber, C., Jörgl, H.P., Raml, H.: Fuzzy control of an iron ore sintering plant. Control Eng. Pract. **3**, 1669–1674 (1995)
4. Abe, S., Lan, M.S.: Fuzzy rules extraction directly from numerical data for function approximation. IEEE Trans. Syst., Man Cybern. **25**, 119–129 (1995)
5. Jahromi, M.Z., Taheri, M.: A proposed method for learning rule weights in fuzzy rule-based classification systems. Fuzzy Sets Syst. **159**, 449–459 (2008)
6. Mansoori, E.G., Zolghadri, M.J., Katebi, S.D.: A weighting function for improving fuzzy classification systems performance. Fuzzy Sets Syst. **158**, 583–591 (2007)
7. Sugeno, M., Yasukawa, T.: A fuzzy-logic-based approach to qualitative modeling. IEEE Trans. Fuzzy Syst. **1**, 7–31 (1993)
8. Wang, L.X., Mendel, J.: Generating fuzzy rules by learning from examples. IEEE Trans. Syst., Man Cybern. **22**, 1414–1427 (1992)
9. Chen, C.Y., Li, S.A., Liu, T.K., Chen, K.Y., Wong, C.C.: A clustering-based algorithm to extracting fuzzy rules for system modeling. Int. J. Adv. Comput. Technol. (IJACT) **3**, 394–401 (2011)
10. Martens, D., Huysmans, J., Setiono, R., Vanthienen, J., Baesens, B.: Rule extraction from support vector machines: an overview of issues and application in credit scoring. In: Diederich, J. (ed.) Rule Extraction from Support Vector Machines. Studies in Computational Intelligence, vol. 80, pp. 33–63. Springer, Berlin (2008)
11. Nelles, O., Fischer, M., Muller, B.: Fuzzy rule extraction by a genetic algorithm and constrained nonlinear optimization of membership functions. In: Proceedings of the Fifth IEEE International Conference on Fuzzy Systems, vol. 1, pp. 213–219 (1996)
12. Chaves, A., Vellasco, M., Tanscheit, R.: Fuzzy rule extraction from support vector machines. In: Fifth International Conference on Hybrid Intelligent Systems. HIS '05 (2005)
13. del Jesus, M., Gonzalez, P., Herrera, F.: Multiobjective genetic algorithm for extracting subgroup discovery fuzzy rules. In: IEEE Symposium on Computational Intelligence in Multicriteria Decision Making, pp. 50–57 (2007)
14. Nawa, N., Furuhashi, T.: Fuzzy system parameters discovery by bacterial evolutionary algorithm. IEEE Trans. Fuzzy Syst. **7**, 608–616 (1999)
15. Botzheim, J., Cabrita, C., Kczy, L.T., Ruano, A.E.: Fuzzy rule extraction by bacterial memetic algorithms. Int. J. Intell. Syst. **24**, 312–339 (2009)
16. Gal, L., Botzheim, J., Koczy, L.: Improvements to the bacterial memetic algorithm used for fuzzy rule base extraction. In: 2008 IEEE International Conference on Computational Intelligence for Measurement Systems and Applications. CIMSA 2008, pp. 38–43 (2008)

17. Liu, Z.Q., Zhang, Y.J.: Rule extraction using a neuro-fuzzy learning algorithm. In: The 12th IEEE International Conference on Fuzzy Systems. FUZZ '03, vol. 2, pp. 1401–1405 (2003)
18. Dehzangi, O., Zolghadri, M., Taheri, S., Fakhrahmad, S.: Efficient fuzzy rule generation: a new approach using data mining principles and rule weighting. In: Fourth International Conference on Fuzzy Systems and Knowledge Discovery. FSKD 2007, vol. 2, pp. 134–139 (2007)
19. Allotta, B., Colla, V., Malvezzi, M.: Train position and speed estimation using wheel velocity measurements. Proceedings of the Institution of Mechanical Engineers. Part F, J. Rail Rapid Transit **216**, pp. 207–225 (2002)
20. Borselli, A., Colla, V., Vannucci, M., Veroli, M.: A fuzzy inference system applied to defect detection in flat steel production. In: IEEE World Congress on Computational Intelligence, WCCI 2010, pp. 1–6 (2010)
21. Jang, J.S.: ANFIS: adaptive-network-based fuzzy inference system. IEEE Trans. Syst., Man Cybern. **23**, 665–685 (1993)
22. Young, R.W.: Dynamic mathematical model of a sintering process. Ironmak. Steelmak. **4**, 321–328 (1977)
23. Augustin, M., Arbeithuber, C., Jörgl, H.P.: Modeling and simulation of an iron ore sinterstrand. In: EUROSIM, pp. 873–878 (1995)
24. de Castro, J.A., Sazaki, Y., Yagi, J.: Three dimensional mathematical model of the iron ore sintering process based on multiphase theory. Mater. Res. **15**, 848–858 (2012)
25. Zadeh, L.A.: Fuzzy sets. Inf. Control **8**, 338–353 (1965)
26. Lee, C.C.: Fuzzy logic in control systems: fuzzy logic controller—parts 1 and 2. IEEE Trans. Syst., Man Cybern. **20**, 404–435 (1990)
27. Pedrycz, W.: Fuzzy Control and Fuzzy Systems. Research Studies Press Ltd (1989)
28. Passino, K.M., Yurkovich, S.: Fuzzy Control. Addison Wesley (1998)
29. Jackson, P.: Introduction to Expert Systems, 3rd edn. Addison Wesley (1998)
30. Durkin, J.: Expert Systems: Design and Development. Macmillan Publishing Company (1998)
31. Geyer-Schulz, A.: Fuzzy Rule-Based Expert Systems and Genetic Machine Learning. Physica-Verlag (1995)
32. Takagi, T., Sugeno, M.: Fuzzy identification of systems and its applications to modelling and control. IEEE Trans. Syst., Man Cybern. **15**, 116–132 (1985)

Handling Uncertainty Degrees in the Evaluation of Relevant Opinions Within a Large Group

Ana Tapia-Rosero, Robin De Mol and Guy De Tré

Abstract Social media makes it possible to involve a large group of people to express their opinions, but not all of these opinions are considered to be relevant from a decision-maker's point of view. Our approach splits a large group of opinions into clusters—here, a cluster represents a group of similar opinions over a criterion specification. Then, these clusters are categorized as more or less relevant taking into account the decision maker's point of view over some characteristics—like the level of togetherness (or cohesion) among opinions, and their representativeness. However, these characteristics might include some level of uncertainty. Thus, the aim of this paper is to evaluate relevant opinions taking into account any associated uncertainty. What follows is to produce a meaningful selection of clusters based on their evaluation and their uncertainty degree. Finally, this proposal includes the steps describing the process through an illustrative example.

1 Introduction

Nowadays, a company might involve a large amount of people to share their opinions with regard to their products—e.g., clients of a company expressing their product preferences by means of social media. Thereby, in a decision making context, a company could make decisions in favor of a product that best suits the preferences of their clients. In this way, a company could gather opinions given by their clients with regard to a feature like size, weight, color, or usefulness of a product. Although,

A. Tapia-Rosero (✉)
ESPOL University, Campus Gustavo Galindo, Guayaquil, Ecuador
e-mail: atapia@espol.edu.ec

A. Tapia-Rosero · R. De Mol · G. De Tré
Ghent University, St-Pieternieuwstraat 41, B-9000 Ghent, Belgium
e-mail: Ana.Tapia@UGent.be

R. De Mol
e-mail: Robin.DeMol@UGent.be

G. De Tré
e-mail: Guy.DeTre@UGent.be

© Springer International Publishing Switzerland 2016
J.J. Merelo et al. (eds.), *Computational Intelligence*,
Studies in Computational Intelligence 620, DOI 10.1007/978-3-319-26393-9_17

all of these opinions could be considered given by experts, not all of them might be considered relevant. Hence, this paper intends to identify and evaluate relevant opinions (from the decision-maker's point of view) where different points of view (given by the clients) are present, and some of these opinions are more representative than others. But, *how do we evaluate relevant opinions when uncertainty can stem from different sources including the accuracy or expertise level of our experts?* Here, the purpose of this paper is twofold. First, our aim is to evaluate relevance through the wisdom of the crowd in a simplified fashion—i.e., avoiding to be overwhelmed with a huge amount of opinions. Second, our purpose is to incorporate the level of present uncertainty as a quality measure. Afterwards, the decision maker will be able to select a meaningful group of opinions that could lead to make a higher quality decision (compared to the decision when uncertainty is not taken into account).

Let us consider that a company wants to know the "level of usefulness" (criterion) of a product (e.g., a smart phone with augmented reality, a wristband with a pedometer or a heart rate monitor, among others). In this way, it is possible to gather this information from the company's social media (e.g., the fan page of a company's product) where opinions are given by different levels of knowledge (students, non-experts and professionals), areas of expertise (engineering, marketing, economics, etc.), and personal profiles (single, married, parents, etc.). Bearing in mind these profile differences, it is desired to differentiate opinions that are considered more important (or worthy of notice) than others (e.g., the opinion of some specific professionals versus the opinion of some regular users of a product). Although, incorporating the number of noticeable opinions in a model might improve the evaluation of relevant opinions, it is possible that representing the level of expertise of an expert becomes a subjective value and hence a source of uncertainty.

Using soft computing techniques a person could express his/her preferences through membership functions setting his/her level of preference over a criterion (e.g., level of usefulness) by providing some values. These values will be used to define the attribute criterion in a membership function [7]. For example, an expert may use expressions like "the usefulness level is above 50 %", or "the usefulness level is around 30 %" to express his/her preference. Therefore, it is not necessary that all the experts have preknowledge on soft computing techniques to represent their preference $P(x)$ as a matter of degree—i.e., $0 \leq P(x) \leq 1$ where 0 denotes a non preferred value, and 1 denotes the highest preference level.

When all membership functions have been gathered, we use a shape-similarity method [13] to simplify the number of membership functions (representing expert opinions). In a similar way that an executive summary reduces a longer report, this method allows a decision-maker to visualize the opinion's trends instead of all individual opinions. That is, by grouping membership functions reflecting similar expert opinions into clusters that could be differentiated based on their attributes. Some cluster attributes are inherent like the number of membership functions and its boundaries—approximation to represent a group of expert opinions through its upper and lower bounds–, among other characteristics with regard to their information source (e.g., number of noticeable opinions based on the social media profile of the experts). Meanwhile, new cluster attributes could be obtained by computing inherent attributes. For instance, the cluster boundaries allows us to compute a *cohesion measure* among the contained membership functions, where a higher value denotes more togetherness and hence expresses a group of more confident opinions [14]. It is worth to notice that any cluster with a single membership function will obtain the highest cohesion value. Therefore, additional attributes like the *representativeness* of a cluster—obtained by combining the number of membership functions and the number of noticeable opinions—was considered in the model to evaluate relevant opinions presented in [15]. In this paper, considering that uncertainty is an attribute of information [16], the model to evaluate relevant opinions has been extended taking into account the *uncertainty over the attributes of the clusters* (e.g., the number of noticeable opinions with regard to their level of expertise).

Herein, we use the *logic scoring of preference (LSP)* method [4, 8] to obtain the overall evaluation value for each cluster based on a combination of the cluster attributes. The LSP method is based on the verbalized approach of the Generalized Conjunction/Disjunction (GCD) [6], which makes possible to derive GCD aggregators using a verbal specification of the overall importance of the involved cluster characteristics. Additionally, to deal with the uncertainty between cluster characteristics we use an extension of the GCD proposed by the authors in [1] that reflects the *overall evaluation* and the *overall degree of uncertainty* through a vector. In this paper, the vector components are used—by the decision-maker—to select the relevant clusters. Here, it is possible to select one cluster with the best combination of evaluation and uncertainty degree (i.e., the highest overall evaluation value and the lowest overall degree of uncertainty), or to select a group of the top clusters.

This proposal identifies relevant opinions from the decision-maker's point of view where it is a challenge trying to reflect someone's point of view accurately. Then, we combine the uncertainties, associated to the evaluation model inputs (cluster attributes), to obtain an overall uncertainty degree. Finally, the process produces a meaningful selection of opinions based on their evaluation and their uncertainty level. By using soft computing techniques we provide a method to model and handle relevant opinions under uncertainty, and we studied how a large group of opinions is reduced to some of them considered to be relevant (by the decision-maker) under uncertainty as well. An advantage within the scope of this proposal is that it evaluates different points of view, given by a large group of people (as preferences), through social media. Furthermore, the evaluation allows a decision-maker to select the group of opinions that best suits his/her choice based on the combination of some cluster attributes under uncertainty.

The remainder of this paper is structured as follows. Section 2 gives some preliminary concepts for representing expert opinions and clustering similar opinions. Section 3 presents the generalized conjunction/disjunction (GCD) aggregators and some extensions. Section 4 describes the evaluation of relevant opinions under uncertainty with LSP, based on the GCD aggregators and its extensions. Section 5 presents an example that illustrates our proposal in a decision-making context. Section 6 concludes the paper and presents some opportunities for future work.

2 Preliminaries

This section defines preliminary concepts to properly understand the remaining sections. These include concepts on fuzzy sets for representing expert opinions and some definitions to cluster similarly shaped membership functions.

2.1 Representing Expert Opinions

A membership function $f_A(x)$, from the preference point of view, represents a set of more or less preferred values of a decision variable x in a fuzzy set A. Hereby, $f_A(x)$ represents the intensity of preference or preference level in favor of value x [2].

In this paper, trapezoidal membership functions are used to represent the expert preferences over criteria [3]. An advantage of using trapezoidal membership functions is that they could be built with a few input values, i.e., parameters a, b, c, and d (Eq. 1). Dividing points between the segments, denoted by the aforementioned parameters, satisfy the relation $a \leq b \leq c \leq d$ among them. Triangular membership functions are treated as a particular case of trapezium considering that b and c have equal values.

$$f_A(x) = \begin{cases} 0 & , x \leq a \\ \frac{x-a}{b-a} & , a < x < b \\ 1 & , b \leq x \leq c \\ \frac{d-x}{d-c} & , c < x < d \\ 0 & , x \geq d \end{cases} \tag{1}$$

In this way, trapezoidal membership functions allow experts to denote their preferences using percentages [11, 12]. For instance, experts denoting the *usefulness level* of a product, could use expressions like "the usefulness level is above 50 %" (Fig. 1a) hereby $b = 50\%$, "it is below 50 %" (Fig. 1b) hereby $c = 50\%$, or "it is between 25 and 50 %" (Fig. 1c) hereby $b = 25\%$ and $c = 50\%$. These are cases where $f_A(x) = 1$ denote the highest level of preference. In a similar fashion, other expressions (given by experts) will lead us to denote the lowest level of preference $f_A(x) = 0$ on the criterion. Triangular membership functions could be used by experts through expressions "around x" where x denotes the highest level of agreement on the criteria ($x = b = c$), and the spread of less preferred values (i.e., between a and d) may vary among experts (Fig. 1d).

2.2 Clustering Similar Opinions

Considering that we use a shape-similarity method, within this subsection some definitions proposed in [13] are included to make this paper self-contained. The

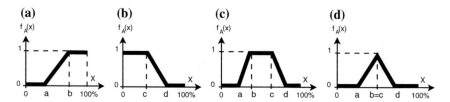

Fig. 1 Trapezoidal membership functions expressing expert preferences through percentages

shape-similarity method assumes that similar opinions are reflected by similarly shaped membership functions. It uses as inputs several membership functions, representing the opinion of experts over a specific criterion, and builds clusters of similar opinions. The shape-similarity method has three phases that could be summarized as follows:

1. A *shape-symbolic notation* for each normalized membership function is built, which depicts a membership function through a sequence of symbolic-characters (See Definition 1).
2. A *similarity measure* in the unit interval among shape-symbolic notations is obtained, where 0 denotes no similarity and 1 denotes full similarity between them.
3. A *clustering step* is performed, based on the aforementioned similarity measure between notations, where it is possible to adjust the clusters from the highest similarity with the smallest number of membership functions (i.e., where each cluster contains a single membership function) to a lower similarity containing the highest possible number of membership functions (i.e., one cluster containing all the membership functions).

Let $S^{category} = \{+, -, 0, 1, L, I, H\}$ be the set that is used to represent the category of a segment in a membership function, and S^{length} a linguistic term set used to represent its relative length on the X-axis compared to the sum of all segments. Using the aforementioned sets, a *symbolic-character* is defined as follows:

Definition 1 A **symbolic-character** is a representation of a segment in a membership function as a pair $\langle t, r \rangle$ with $t \in S^{category}$ and $r \in S^{length}$; where t represents the category of the segment and r depicts its relative length by means of a linguistic term.

In this way, each segment of the membership function uses a sign $\{+,-\}$ to represent its slope, a value $\{0, 1\}$ to represent its preference level on the criterion (i.e., the lowest level or the highest level of preference respectively) and a letter $\{L, I, H\}$ to denote a *low, intermediate* or *high* point (e.g., a peak in a triangular membership function corresponds to a high point annotated as H). The linguistic term set S^{length} expresses the relative length of the segment on the X-axis by means of labels (Eq. 2).

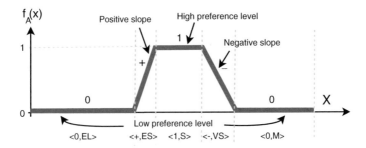

Fig. 2 Segments of a trapezium and their corresponding shape-symbolic characters

$$S^{length} = \begin{cases} \text{ES} = \text{"extremely short"}, \text{VS} = \text{"very short"}, \text{S} = \text{"short"}, \\ \text{M} = \text{"medium"}, \text{L} = \text{"long"}, \text{VL} = \text{"very long"}, \\ \text{EL} = \text{"extremely long"} \end{cases} \quad (2)$$

Figure 2 shows a trapezoidal membership function with five segments, each of them represented by a shape-symbolic character. Therefore, the *shape-symbolic notation* for this figure could be expressed as $\langle 0, EL \rangle \langle +, ES \rangle \langle 1, S \rangle \langle -, VS \rangle \langle 0, M \rangle$.

Hereafter we will consider that a set of k different clusters $C = \{C_1, \ldots, C_k\}$ containing similarly shaped membership functions were obtained. Moreover, each cluster C_j will be represented through an array of n attributes.

3 Generalized Conjunction/Disjunction and Some Extensions

The generalized conjunction disjunction (GCD) operator is a continuous logic function that integrates conjunctive and disjunctive properties in a single function [9], denoted as $y = x_1 \Diamond \cdots \Diamond x_n$, $x_i \in [0, 1]$, $i = 1, \ldots, n$, and $y \in [0, 1]$. GCD includes two parameters: the *andness* and the *orness*. The *andness* α, denotes simultaneity and expresses the conjunction degree. Meanwhile, the *orness* ω, denotes replaceability and expresses the disjunction degree [10]. These parameters are complementary, i.e., $\alpha + \omega = 1$. Although GCD can be implemented in several ways [5], within this paper we will only consider an implementation based on the weighted power means (WPM) as follows:

$$x_1 \Diamond \cdots \Diamond x_n = (W_1 x_1^r + \cdots + W_n x_n^r)^{\frac{1}{r}}, \quad (3)$$

here W_i denotes the weight assigned to the parameter x_i and the parameter r can be computed as a function of andness α using a suitable numerical approximation [4].

Table 1 Aggregation operators for 17 levels of GCD implemented by WPM

Symbol	Orness (ω)	Andness (α)	Exponent r
D	1	0	$+\infty$
D++	0.9375	0.0625	20.63
D+	0.8750	0.1250	9.521
D+−	0.8125	0.1875	5.802
DA	0.7500	0.2500	3.929
D−+	0.6875	0.3125	2.792
D−	0.6250	0.3750	2.018
D−−	0.5625	0.4375	1.449
A	0.5	0.5	1
C−−	0.4375	0.5625	0.619
C−	0.3750	0.6250	0.261
C−+	0.3125	0.6875	−0.148
CA	0.2500	0.7500	−0.72
C+−	0.1875	0.8125	−1.655
C+	0.1250	0.8750	−3.510
C++	0.0625	0.9375	−9.06
C	0	1	$-\infty$

Reprinted from International Journal of Approximate Reasoning, 41(1), Dujmović, J. and Nagashima, H., LSP method and its use for evaluation of Java IDEs, pp. 3–22, Copyright (2006), with permission from Elsevier

Table 1 includes, as a reference, the corresponding orness, andness and exponent r for 17 levels of GCD implemented using WPM. Notice that symbols D and C correspond to full disjunction ($\omega = 1$), and full conjunction ($\alpha = 1$) respectively.

3.1 GCD Verbalized Approach

The GCD verbalized approach [6] facilitates the selection of aggregators by means of a multi-level overall importance scale. In this approach, the decision-maker specifies the overall degree of importance for each attribute using a scale with L levels from "lowest" to "highest" (Table 2). Besides, the overall importance the decision-maker should provide the selection of simultaneity S or replaceability R.

In the case of n attributes of overall importance, $(S_1, \ldots, S_n)$ for simultaneity, the andness α is defined as the mean normalized overall importance:

$$\alpha = \frac{(S_1 + \cdots + S_n)}{nL}, \ S_i \in [0, L] \tag{4}$$

Table 2 Overall importance scale with $L = 16$ levels [6]

Level	Overall importance
16	Highest
15	*Slightly below highest*
14	Very high
13	*Slightly above high*
12	High
11	*Slightly below high*
10	Medium-high
9	*Slightly above medium*
8	Medium
...	...
0	Lowest

Analogously, in the case of replaceability $(R_1, \ldots, R_n)$ the orness ω is defined as:

$$\omega = \frac{(R_1 + \cdots + R_n)}{nL}, \quad R_i \in [0, L] \tag{5}$$

Once the level of andness/orness have been given by the decision-maker, it is necessary to map these into normalized weights $W_1 + \cdots + W_n = 1$. Although it is possible to use the GCD verbalized approach to compute the weights [6], within this paper we consider that these will be given by the decision-maker as well when obtaining the appropriate aggregator.

3.2 Extended Generalized Conjunction/Disjunction

A GCD aggregator is usually defined as a function of the form $F : [0, 1]^n \to [0, 1]$, but in this paper we use an extension proposed by the authors in [1]. The extension of GCD, denoted hereafter as EGCD, is defined as a vector function. The $U : ([0, 1], [0, 1])^n \to ([0, 1], [0, 1])$, where the abscissa of an input maps to the original preference parameter of GCD, and the ordinate corresponds to a degree of uncertainty. In this way, the output of an EGCD function is a vector where the first component corresponds to the overall preference evaluation and the second component corresponds to the aggregated overall uncertainty.

The output of the EGCD with regard to the degree of uncertainty, depends on the selected aggregator operator as follows:

- In the case of full conjunction ($\alpha = 1$), the uncertainty of the lowest input is propagated. Meanwhile in the case of full disjunction ($\omega = 1$), the uncertainty of the highest input is propagated.

- In the case of the *neutrality function* ($\alpha = \omega = 0.5$), which is implemented as the arithmetic mean, the overall uncertainty also produces the average of the elementary uncertainties.
- In the case of the *partial conjunction* ($\alpha > 0.5 > \omega$) and the *partial disjunction* ($\alpha < 0.5 < \omega$), we observe a gradient proportional to the global preference, i.e., the overall uncertainty leans to the elementary uncertainty of the preferred inputs.

The implementation of EGCD for aggregators with two inputs is based on the weighted average function, and it has shown to produce an intuitive output when dealing with uncertain data. The implementation for aggregators with more than two inputs needs further research.

In this paper we use this extension considering that sometimes incorporating inputs in the model (to evaluate relevant opinions) might improve the evaluation results, but these inputs may also become a source of uncertainty (i.e., considering that uncertainty is an attribute of information [16]).

4 Evaluating Relevant Opinions Under Uncertainty

With the purpose of evaluating relevant opinions within a large group, our model uses the LSP method to reflect the decision-maker's point of view expressing what he/she considers as relevant opinions. Besides evaluating relevant opinions, we consider that it is important to incorporate the level of present uncertainty as a quality measure. Thus, the main contribution of this proposal is to provide a method that allows a decision-maker to make a higher quality selection of relevant opinions compared to the selection when uncertainty is not taken into account. Therefore, in this section we describe how to evaluate relevant opinions with LSP, and how to incorporate the uncertainty by means of an extension of the generalized conjunction/disjunction (EGCD) aggregator.

On the assumption that similar opinions have been clustered by shape-similarity as described in Sect. 2.2, a set of k clusters is available. Hence, each cluster C_j is represented through an array of n attributes $a_{i,j}$ where i is the attribute identifier and j is the cluster identifier. In this way, each attribute $a_{i,j}$ is a pair $\langle v, u \rangle$, where v represents its value and u represents its associated uncertainty. Hereafter, we will refer to $a_{i,j}.v$ as the value and we will refer to $a_{i,j}.u$ as the uncertainty of the attribute i in cluster j. Figure 3 depicts the general architecture of our method that has the following steps:

1. *Creation of a System Attribute Tree*

    This step takes into account the decision-maker's point of view and it is based on the set of available attributes of the cluster. The attributes are hierarchically organized in a tree, and its leaves correspond to the elementary attributes of a cluster. These are not further decomposed, they have been previously measured, and their levels of uncertainty have been associated. The root of the tree will

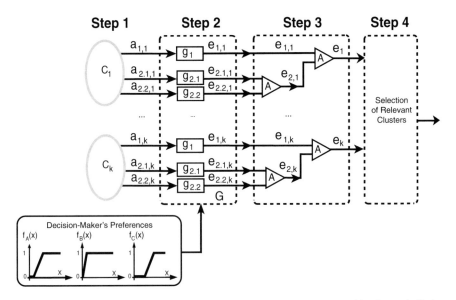

Fig. 3 Evaluation of relevant opinions under uncertainty from clusters grouped by shape-similarity

lead us to e_j denoting the evaluation of cluster j, which is a pair $\langle v, u \rangle$ where $e_j.v$ corresponds to the overall evaluation value and $e_j.u$ corresponds to the overall uncertainty level.

When intermediate nodes are present, this indicates that the attribute has been decomposed in more elementary attributes. For example, *representativeness* has been decomposed in *number of membership functions* and *number of notice-able opinions*. Figure 4 corresponds to the system attribute tree representing the decision-maker's point of view for evaluating relevant opinions under uncertainty.

For readability purposes, Fig. 4 includes the attribute identifiers in parentheses. Thus, the identifiers for cohesion, representativeness, number of membership functions and number of noticeable opinions are 1, 2, 2.1 and 2.2 respectively.

2. *Definition of Elementary Criteria*

   Here through functions G_i the decision-maker reflects the acceptable and unacceptable values for each elementary attribute i. Within this step, a membership function $f(x)$ is used to represent the decision-maker's preference for each

Fig. 4 System attribute tree for evaluating relevant opinions under uncertainty

Fig. 5 Example of the decision-maker's preference for elementary attribute "number of membership functions"

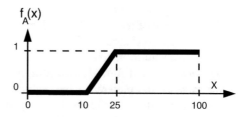

Fig. 5 Example of the decision-maker's preference for elementary attribute "number of membership functions"

elementary attribute. Figure 5 depicts an example where the decision-maker (by means of membership function $f_A(x)$) accepts clusters with "at least 10 opinions", but he/she prefers "more than 25". Moreover, clusters with "less than 10 opinions" are not acceptable.

Once all the elementary criteria G have been defined, it is possible to evaluate all the cluster attributes that are present in the system attribute tree. For instance, $e_{i,j} = g_i(e_{i,j}.v)$ corresponds to the evaluation of attribute i in cluster j.

3. *Creation of an Aggregation Structure*

This step establishes the proper aggregation operators and weights reflecting the decision-maker's point of view. In this paper, the aggregation operators are selected using the GCD verbalized approach described in Sect. 3.1, which allows the decision-maker to use an overall importance scale to represent each attribute. Additionally, the decision-maker should provide the selection of simultaneity or replaceability, and the weights for each attribute. Then, the aggregation structure uses EGCD which gives us a vector $e_j = (v, u)$ for cluster j. The vector components v and u correspond to the overall evaluation and the overall uncertainty respectively. For example, in order to obtain the evaluation of cluster C_j with regard to its *representativeness*, it is necessary to take into account its components (i.e., attributes $a_{2.1,j}$ and $a_{2.2,j}$) and the level of simultaneity or replaceability among them. Figure 6 shows the aforementioned evaluation of *representativeness* annotated as $e_{2,j}$. In a similar way, we will obtain the overall evaluation of *relevant opinions* for cluster C_j, here we need to aggregate its evaluation components $e_{1,j}$ and $e_{2,j}$ to obtain e_j. The level of simultaneity or replaceability will be given by the proper selection of the aggregation operators represented as A in this figure. It is important to notice that this proposal allows to change the input parameters, given by the decision-maker, to accurately represent his/her point of view. For example, a different result is expected when changing the overall degree of importance among attributes (i.e., which results in a different selection of aggregators) or changing their weights.

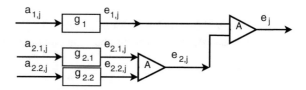

Fig. 6 Example of the aggregation structure for elementary attributes 1, 2.1 and 2.2 of cluster C_j

4. *Selection of Relevant Clusters Under Uncertainty*

The selection of relevant clusters is made by the decision-maker, where his/her decision will be based on the vector that describes each cluster in terms of the overall evaluation and the overall uncertainty. This proposal keeps these two values separated, and allows the decision-maker to differentiate among clusters with the same overall preference, but with different overall uncertainties. Hence, the selection of relevant opinions (by means of clusters of similar opinions) will give the decision-maker the possibility to make a selection based on the overall evaluation, the overall uncertainty, or a combination of both using LSP (i.e., through steps 1–3).

Next section describes the aforementioned process, to evaluate relevant opinions within a large group incorporating an uncertainty degree, by means of an illustrative example.

5 Illustrative Example

A company wants to know the perceived "level of usefulness" of adding a pedometer (new feature) in a previously well positioned wrist-clock model (product). The company has a large amount of followers in social media, and the product manager of the company considers that the opinion given by these followers is important due to their diversity (i.e., different profiles). Therefore, the company gathered the preferences over the "level of usefulness" using the product's fan page. In this example, the product manager will be considered as the decision-maker, and he considers that all the opinions are important, but those given by athletes are noticeable.

The aforementioned preferences were clustered by the shape-similarity method presented in Sect. 2.2. Hereby, a set of $k = 50$ clusters $C = \{C_1, \dots, C_{50}\}$ have been obtained representing a total of $t = 120$ opinions. For each cluster C_j a set of n attributes, where each attribute $a_{i,j}$ is given by a pair $\langle v, u \rangle$ denoting its value v and its uncertainty u, has been obtained as well.

Example of a Cluster and its Attributes. Figure 7 depicts a cluster C_j and its attributes. Herein, the attribute i in cluster j is denoted as $a_{i,j} = \langle v, u \rangle$, where $a_{i,j}.v$ stands for its value and $a_{i,j}.u$ stands for its uncertainty. This figure illustrates cluster C_{50} and the *number of membership functions* attribute has identifier $i = 2.1$, its value is given by the number of membership functions contained in the cluster $a_{2.1,50}.v = 5$ and considering that the number of represented opinions is well-known there is no uncertainty associated to this attribute $a_{2.1,50}.u = 0$. The value and the uncertainty of the *number of noticeable opinions* attribute are both 0. These are inherent cluster attributes.

Additionally, the *cohesion* attribute has identifier $i = 1$ and its value and its uncertainty has been obtained by additional computations. That is, the attribute value

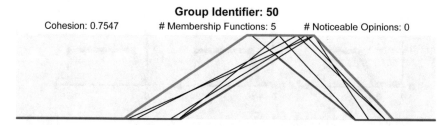

Fig. 7 Example of a cluster C_j and its attributes $a_{i,j}$. The cluster C_{50} has the following attributes $a_{i,50}$: cohesion ($i = 1$), number of membership functions ($i = 2.1$), and number of noticeable opinions ($i = 2.2$)

$a_{1,50}.v = 0.7547$ has been obtained as the area between the upper and lower bounds, and its uncertainty $a_{1,50}.u = 0.3$ has been computed based on the length of the upper-bound support.

The process presented in Sect. 4, to evaluate relevant opinions within a large group incorporating an uncertainty degree, is illustrated below through the following phases:

Step 1. Creation of a System Attribute Tree. The decision-maker considers that the *representativeness* of a cluster of similar opinions is given by a combination of the number of membership functions and the number of noticeable opinions. Additionally, he considers that the *evaluation of relevant opinions* is given by the cohesion and the representativeness of the cluster. Therefore, the decision-maker's point of view is reflected in the system attribute tree shown in Fig. 8.

Step 2. Definition of Elementary Criteria. The decision-maker's preferences reflecting his acceptable and unacceptable values were given through trapezoidal membership functions $f(x)$ for each elementary attribute. Thus, $f_A(x)$ reflects his preferences for elementary attribute *cohesion*, in a similar way $f_B(x)$ stands for the *number of membership functions* attribute, and $f_C(x)$ stands for the *number of noticeable opinions* attribute. These membership functions are shown in Fig. 9 as a reference.

In this step, the evaluation of the elementary attributes is based on the decision-maker's level of preference. Thus, through functions G_i it is possible to determine the elementary preference score reflecting the acceptable and unacceptable values for attributes i. *Within this step, all the attributes of each cluster will be evaluated using their corresponding function to reflect the decision-maker's preferences.* Thus, for

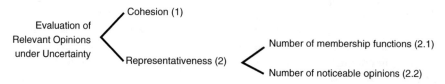

Fig. 8 System attribute tree for evaluating relevant opinions within a large group under uncertainty

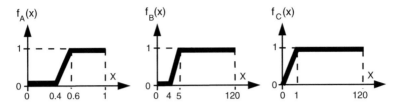

Fig. 9 Decision-maker's preferences for elementary attributes cohesion $f_A(x)$, number of membership functions $f_B(x)$ and number of noticeable opinions $f_C(x)$

cluster C_{50} we obtain for the cohesion attribute $e_{1,50} = g_1(a1, 50.v) = g_1(0.7547) = 1$. In a similar way, we obtain $e_{2.1,50} = 1$, and $e_{2.2,50} = 0$.

It is important to notice that if the attribute value of a cluster lies in a slope of its corresponding membership function, the evaluation of this attribute could be obtained using a linear approximation. For instance, let us consider cluster C_{29} with cohesion value 0.4455. In this case to evaluate the cohesion $e_{1,29}$ from the decision-maker's point of view, we need to interpolate its value using $g_1(x) = (x - 0.4)/(0.6 - 0.4)$ as follows:

$$e_{1,29} = g_1(0.4455) = \frac{0.4455 - 0.4}{0.6 - 0.4} = 0.2275$$

Step 3. Creation of an Aggregation Structure. In this example, the decision maker considers that clusters with a large number of opinions are important, but smaller clusters with a single opinion given by an athlete (i.e., noticeable opinion) might be also relevant. Thus, the "representativeness" of a cluster indicates *replaceability* (denoted as R) among the number of opinions (i.e., *number of membership functions* attribute) and those considered noticeable (i.e., *number of noticeable opinions* attribute). Moreover, he considers that the "representativeness" is given by the *number of membership functions* ($R_1 = 14$ considered as *very high* and the *number of noticeable opinions* ($R_2 = 12$ considered as *high*). Thus, the aggregator is obtained using Eq. 5, which is used to obtain the *orness* or disjunction degree as follows:

$$\omega = \frac{(R_1 + R_2)}{nL} = \frac{14 + 12}{2(16)} = 0.8125.$$

Then using Table 1, on column *Orness* (ω) we look up for the previously obtained value to obtain the aggregation operator. In this example, the aggregator operator corresponds to a partial disjunction annotated by the $D + -$ symbol. Finally, the weights for the elementary attributes of the *representativeness* should be given. In the case that the decision-maker considers that the *number of membership functions* is equally important as the *number of noticeable opinions* then both weights are 0.5 (i.e., $W_{2.1} = W_{2.2} = 0.5$).

In a similar way, the aggregator to combine the *cohesion* and *representativeness* attributes (See Fig. 8) must be obtained. In this example, the decision-maker

Fig. 10 Aggregation
structure based on the
decision-maker's point of
view

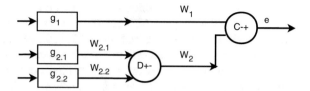

considers that these attributes should be simultaneously satisfied. Then, using Eq. 4
we obtain the *andness* α. For instance, with *cohesion* ($S_1 = 12$ considered "high")
and *representativeness* ($S_2 = 10$ considered "medium high") the $\alpha = 0.6875$ corresponding to the aggregator annotated by the $C - +$ symbol (partial conjunction).
Additionally, the *cohesion* is considered two times more important than the representativeness hence their corresponding weights are $W_1 = 0.67$ and $W_2 = 0.33$.
Figure 10 illustrates the aggregation structure reflecting the decision-maker's point
of view.

*Handling Uncertainty Degrees in the Evaluation of Relevant Opinions within a
Large Group.* Using the previously obtained aggregation structure, using EGCD, we
obtain a vector e_j representing the overall evaluation of relevant opinions e_i and the
overall uncertainty e_u for each cluster j. For illustration purposes, let us compute
the overall evaluation of relevant opinions for the cluster given as an example at
the beginning of this section. First, let us evaluate its *representativeness* given the
selected aggregator $D + -$. Notice that Table 1 could be used as a reference to obtain
the r exponent used in the implementation of the weighted power means (Eq. 3).

$$e_{2,50}.v = (0.5(e_{2.1,50}.v)^r + 0.5(e_{2.2,50}.v)^r)^{\frac{1}{r}}$$
$$e_{2,50}.v = (0.5(1)^{5.802} + 0.5(0)^{5.802})^{\frac{1}{5.802}}$$
$$e_{2,50}.v = 0.887393$$

The uncertainty present in the *representativeness* attribute is $a_{2,50}.u = 0$, due to
the uncertainties given by its elementary attributes which are $a_{2.1,50}.u = a_{2.2,50} = 0$.

Then, in a similar way, using aggregator $C - +$ we will compute the overall evaluation of relevant opinions under uncertainty as follows:

$$e_{50}.v = (0.67(e_{1,50}.v)^r + 0.33(e_{2,50}.v)^r)^{\frac{1}{r}}$$
$$e_{50}.v = (0.67(1)^{-0.148} + 0.33(0.8874)^{-0.148})^{\frac{1}{-0.148}}$$
$$e_{50}.v = 0.9611$$

Although the *cohesion* attribute has uncertainty $e_{1,50}.u = 0.3$, due to the nature
of the aggregator $C - +$ the uncertainty is propagated as $e_{50}.u = 0.1964$ based on
the weighted average as mentioned in Sect. 3.2. Thus, the evaluation of cluster C_{50}
is given by e_{50} as a vector $(0.9611, 0.1964)$ based on the previously obtained values.

Notice, that using this approach it is possible to easily change the input parameters, given by the decision-maker, in order to accurately represent his/her point

of view. For example, if the decision-maker would have changed the given weights (i.e., 33 % for cohesion and 67 % for representativeness) in the aggregation structure, the evaluation e_j would have been $(0.9229, 0.0945)$. In this case, due to a higher weight on the representativeness attribute, it becomes a more dominant input for the partial conjunction $C - +$ operator. What follows is that the uncertainty associated to the representativeness will get more weight as well, hence reducing the overall uncertainty.

Step 4. Selection of Relevant Clusters. This step reflects the cluster that best suits the decision-maker's evaluation. Here it is possible to select the cluster with the highest evaluation value, the cluster with the lowest uncertainty value, or the cluster that satisfies the decision-maker's point of view as a combination of these two (e.g., using LSP through steps 1–3).

Based on the set of clusters used within this example, we obtained several clusters that perfectly satisfied the decision-maker's preferences based on the overall evaluation value. However, taking into account the uncertainty the selection of a cluster representing relevant opinions allowed us to select the one with the lowest overall uncertainty.

6 Conclusions and Further Work

This paper proposed to handle uncertainty degrees in the evaluation of relevant opinions within a large number of expert opinions. These opinions, that might be gathered through social media, are expressed by means of membership functions setting their level of preference over a criterion specification (e.g., level of usefulness). We use a shape-similarity method to cluster similar preferences in order to reduce the number of evaluations. The evaluation results from selecting the best combination of *cohesion* and *representativeness* within the available clusters from the decision-maker's point of view. The cohesion is a measure obtained from computing the area among the upper and lower bounds of the cluster compared to the total available area, while its representativeness is given by aggregating the number of membership functions and the number of noticeable opinions. In order to properly reflect the decision-maker's point of view this proposal uses the LSP method that builds a precise representative model of logic aggregation of preferences.

The main advantage of this proposal is that it can handle a large group of opinions gathered through social media, where the preferences initially given are not modified and subject to uncertainty degrees. Furthermore, it permits the decision-maker to select the group of opinions that best suits his/her choice (i.e., given as preferences in the LSP method) based on the combination of some cluster attributes taking into account any present uncertainty. Within this paper the cluster attributes are a cohesion measure, the number of membership functions and the number of noticeable opinions, each of them represented by its value ant its uncertainty.

Further work will focus in evaluating opinions in organizations, where more than one decision maker is present. Additionally, evaluating relevant opinions with different strategies, in order to compare them with the presented approach, are subject to further study.

Acknowledgments This research is supported by Escuela Superior Politécnica del Litoral (ESPOL) and SENESCYT under Ph.D. studies 2015.

References

1. De Mol, R., Tapia-Rosero, A., De Tré, G.: An approach for uncertainty aggregation using generalised conjunction/disjunction aggregators. In: Advances in Intelligent Systems Research. vol. 89, pp. 1499–1506. Atlantis Press (2015)
2. Dubois, D., Prade, H.: The three semantics of fuzzy sets. Fuzzy Sets Syst. **90**, 141–150 (1997)
3. Dubois, D., Prade, H.: Fundamentals of Fuzzy Sets (The Handbooks of Fuzzy Sets, vol. 7). Springer (2000)
4. Dujmović, J.: Continuous preference logic for system evaluation. IEEE Trans. Fuzzy Syst. **15**(6), 1082–1099 (2007)
5. Dujmović, J.: Characteristic forms of generalized conjunction/disjunction. In: 2008 IEEE International Conference on Fuzzy Systems (IEEE World Congress on Computational Intelligence), pp. 1075–1080 (2008)
6. Dujmović, J.: Andness and orness as a mean of overall importance. In: 2012 IEEE International Conference on Fuzzy Systems (FUZZ-IEEE), IEEE, pp. 1–6 (2012)
7. Dujmović, J., De Tré, G.: Multicriteria methods and logic aggregation in suitability maps. Int. J. Intell. Syst. **26**(10), 971–1001 (2011)
8. Dujmović, J., De Tré, G., Van De Weghe, N.: LSP suitability maps. Soft Comput. **14**(5), 421–434 (2010)
9. Dujmović, J., Larsen, H.L.: Generalized conjunction/disjunction. Int. J. Approx. Reason. **46**(3), 423–446 (2007)
10. Dujmović, J., Nagashima, H.: LSP method and its use for evaluation of Java IDEs. Int. J. Approx. Reason. **41**(1), 3–22 (2006)
11. Eshragh, F.t., Mamdani, E.: A general approach to linguistic approximation. Int. J. Man-Mach. Stud. **11**(4), 501–519 (1979)
12. Pedrycz, W.: Granular Computing: Analysis and Design of Intelligent Systems. CRC Press (2013)
13. Tapia-Rosero, A., Bronselaer, A., De Tré, G.: A method based on shape-similarity for detecting similar opinions in group decision-making. Inf. Sci. **258**, 291–311 (2014)
14. Tapia-Rosero, A., De Tré, G.: A cohesion measure for expert preferences in group decision-making. In: Atanassov, K., Baczynski, M., Drewniak, J., Kacprzyk, J., Krawczak, M., Szmidt, E., Wygralak, M., Zadrozny, S. (eds.) Modern Approaches in Fuzzy Sets, Intuitionistic Fuzzy Sets, Generalized Nets and Related Topics. Volume II: Applications, vol. 2, pp. 125–142. SRI-PAS (2014)
15. Tapia-Rosero, A., De Tré, G.: Evaluating relevant opinions within a large group. In: Dourado, A., Cadenas, J., Filipe, J. (eds.) 6th International Conference on Fuzzy Computation Theory and Applications, Proceedings, pp. 76–86. SciTePress (2014)
16. Zadeh, L.A.: Generalized theory of uncertainty (GTU)principal concepts and ideas. Comput. Stat. Data Anal. **51**(1), 15–46 (2006)

Part III
Neural Computation Theory
and Applications

Opening the Black Box: Analysing MLP Functionality Using Walsh Functions

Kevin Swingler

Abstract The Multilayer Perceptron (MLP) is a neural network architecture that is widely used for regression, classification and time series forecasting. One often cited disadvantage of the MLP, however, is the difficulty associated with human understanding of a particular MLP's function. This so called black box limitation is due to the fact that the weights of the network reveal little about structure of the function they implement. This paper proposes a method for understanding the structure of the function learned by MLPs that model functions of the class $f : \{-1, 1\}^n \rightarrow \mathbb{R}^m$. This includes regression and classification models. A Walsh decomposition of the function implemented by a trained MLP is performed and the coefficients analysed. The advantage of a Walsh decomposition is that it explicitly separates the contribution to the function made by each subset of input neurons. It also allows networks to be compared in terms of their structure and complexity. The method is demonstrated on some small toy functions and on the larger problem of the MNIST handwritten digit classification data set.

Keywords Multilayer perceptrons · Walsh functions · Network function analysis

1 Introduction

The multilayer perceptron (MLP) [1] is a widely used neural network architecture. It has been applied to regression, classification and novelty detection problems and has been extended in various ways to process time varying data, e.g. [2]. In the field of data mining MLPs are a common choice amongst other candidates such as classification trees, support vector machines and multiple regression. Due to the wide variety of tasks for which they are suited, and their ability as universal approximators, MLPs have become very popular. However, there is one aspect of the MLP that restricts and complicates its application, and that is the role of the hidden neurons. A common criticism of the MLP is that its knowledge is not represented in a human

K. Swingler (✉)
Computing Science and Maths, University of Stirling, Stirling FK9 4LA, Scotland, UK
e-mail: kms@cs.stir.ac.uk

© Springer International Publishing Switzerland 2016　　　　　　　　　　　　303
J.J. Merelo et al. (eds.), *Computational Intelligence*,
Studies in Computational Intelligence 620, DOI 10.1007/978-3-319-26393-9_18

readable form. The comparison that is often made is with classification or regression trees, which represent partitions in the input space explicitly in their structure. This makes human understanding of the underlying function and the reasons behind any given output quite easy. Given a picture of a classification tree, a human may apply it to an input pattern without even needing a computer to run the algorithm. This is far from simple with an MLP.

The hidden units in an MLP act as feature detectors, combining inputs from below into higher order features that are, in turn, combined by higher layers still. The common learning algorithms such as back propagation of error [1] have no explicit means of ensuring that the features are optimally arranged. Different neurons can share the same feature, or have overlapping representations. In networks where each layer is fully connected to the one above, every hidden neuron in a layer shares the same receptive field, so their roles often overlap. This makes analysis even more difficult as hidden neurons do not have independent roles. The inclusion of additional layers of hidden neurons compounds the problem further.

Some work has been carried out on the analysis of hidden neurons in MLPs. For example, [3] used an entropy based analysis to identify important hidden units (known as principal hidden units) in a network for the purpose of pruning an oversize hidden layer. [4] proposed a method of contribution analysis based on the products of hidden unit activations and weights and [5] presented a specific analysis of the hidden units of a network trained to classify sonar targets.

The question of how to extract rules from multilayer perceptrons has received more attention and is still a very active field of research. [6] propose a fuzzy rule extraction method for neural networks, which they call Fuzzy DIFACONN. [7] propose a clustering based approach to MLP rule extraction that uses genetic algorithm based clustering to identify clusters of hidden unit activations which are then used to generate classification rules. [8] use an inversion method to generate rules in the form of hyperplanes. Inverting an MLP (i.e. finding the inputs that lead to a desired output) is done by gradient descent and using an evolutionary algorithm. Both [9, 10] present recent comparative studies of neural network rule extraction, distinguishing between methods that are decompositional, pedagogical and eclectic. A decompositional approach extracts rules from the weights and activations of the neural network itself. The pedagogical approach, which is taken in this paper, treats the neural network as a black box and generates rules based on the outputs generated by the network in response to a set of input patterns. Eclectic rule extraction combines both of the aforementioned approaches.

More work has concentrated on choosing the right number of hidden units for a specific data set. [11] bound the number of weights by the target error size, [12] bounded the number of hidden units by the number of patterns to be learned, [13] chose a bound based on the number of output units in the network, and [14] pointed out that the amount of noise in the training data has an impact on the number of units used. Some have taken a dynamic approach to network structure discovery, for example [15] used an information theoretic approach to add or remove hidden neurons during training. The problem with this approach to training an MLP is that the existing weights are found in an attempt to minimise error for that number of hidden neurons. Adding a new one may mean the existing weights are starting in a configuration that is unsuitable for a network with more neurons. Other search methods have also been applied to finding the right structure in an MLP. [16, 17] used genetic algorithms to search the space of network structures, for example.

When using MLPs (and other machine learning techniques), it is common practice to produce several models to be used in an ensemble [18]. Due to the random start point of the weight values and the differences in architecture across the networks in an ensemble, it is not easy to know whether or not different networks are functionally different. It is possible to train a number of different MLPs that all implement the same function (perhaps with differing quality of fit across the weights) with very different configurations of weight values. For example, one could re-order the hidden units of any trained network (along with their weights) and produce many different looking networks, all with identical functionality. One way to compare MLPs is to compare their outputs, but a structural comparison might also be desirable, and that is what we present here.

Note the distinction between the structure of an MLP, which is defined by the neurons and connecting weights, and the structure of the function it implements, which can be viewed in a number of other ways. This paper views the underlying function implemented by an MLP in terms of the contribution of subsets of input variables. The number of variables in a subset is called its order, and there are $\binom{n}{k}$ subsets of order k in a network of n inputs. The first order subsets are the single input variables alone. The second order subsets are each of the possible pairs of variables, and so on. There is a single order n set, which is the entire set of inputs. Any function can be represented as a weighted sum of the values in each of these subsets. The weights (known as coefficients in the chosen analysis) are independent (unlike the weights in an MLP, whose values are determined to an extent by other weights in the network) and specific to their variable subset. The first order coefficients describe the effect of each variable in isolation, the second order coefficients describe the contribution of variable pairs, and so on. The method for decomposing a neural network function into separate components described in this paper is the Walsh transform. When the phrase "network functionality" is used in this paper, it means the form the function takes in terms of how the interactions between different subsets of input variables affect each output variable.

Section 2 describes the Walsh transform in some detail. This is followed by a description of the method for producing Walsh coefficients from a neural network in Sect. 3. Section 4 introduces some functions that will be used in experiments described in following sections. Section 5 demonstrates how the method can be used to track the complexity of MLPs during training and Sect. 6 demonstrates how a partial transform on a small sample from a larger network can provide useful insights. The Walsh method is compared to other methods of understanding network structure in Sect. 7. Finally, Sects. 8 and 9 offer some conclusions and ideas for further work.

2 Walsh Functions

Walsh functions [19, 20] form a basis for real valued functions of binary vectors. Any function $f : \{-1, 1\}^n \to \mathbb{R}$ can be represented as a weighted linear sum of Walsh functions. The Walsh functions take the form of a sequence of bit strings over $\{-1, 1\}^{2^n}$ where n is the number of variables in the function input. n is known as the Walsh function order. There are 2^n Walsh functions of order n, each 2^n bits long. Figure 1 shows a representation of the order 3 Walsh functions. Each Walsh function has an index from 1 to 2^n, with the jth function being ψ_j and bit number x of the jth Walsh function is $\psi_j(x)$. As Fig. 1 shows, the Walsh functions can be viewed as a matrix of values from $\{-1, 1\}$ with rows representing each Walsh function and columns representing each bit.

A Walsh representation of a function $f(\mathbf{x})$ is defined by a vector of parameters, the Walsh coefficients, $\omega = \omega_0 \dots \omega_{2^n-1}$. Each ω_j is associated with the Walsh function ψ_j, that is a row in the Walsh matrix. Each possible input, $\mathbf{x}$ is given an index, x, which is calculated by replacing any -1 in $\mathbf{x}$ with 0 and converting the result to base 10. For example if $\mathbf{x} = (1, -1, 1)$, then $x = 5$. Each column of the Walsh matrix corresponds to a value of x.

The Walsh representation of $f(\mathbf{x})$ is constructed as a sum over all ω_j. In the sum, each ω_j is either added to or subtracted from the total, depending on the value of the bit corresponding to x (i.e. column x in the Walsh matrix), which gives the function for the Walsh sum:

$$f(\mathbf{x}) = \sum_{j=0}^{2^n} \omega_j \psi_j(x) \qquad (1)$$

Fig. 1 A pictorial representation of an order 3 Walsh matrix with *black squares* representing 1 and *white squares* −1. A Walsh sum is calculated by summing the product of the Walsh coefficient associated with each row by the values in the column indexed by the function input

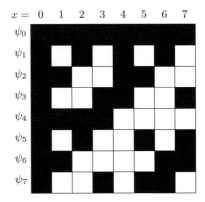

2.1 Constructing the Walsh Functions

The value of a single cell in the Walsh matrix, $\psi_j(x)$ is calculated from the binary representation of the coordinates (j, x), of $\mathbf{j}$ and $\mathbf{x}$, and returns $+1$ or -1 depending on the parity of the number of 1 bits in shared positions. Using logical notation, a Walsh function is derived from the result of an XOR (parity count) of an AND (agreement of bits with a value of 1):

$$\psi_j(x) = \oplus_{i=1}^{n} (x_i \wedge j_i) \tag{2}$$

where $\oplus$ is a parity operator, which returns 1 if the argument list contains an even number of 1 s and -1 otherwise.

2.2 Calculating the Coefficients—the Walsh Transform

The Walsh transform of an n-bit function, $f(\mathbf{x})$, produces 2^n Walsh coefficients, ω_x, indexed by the 2^n combinations across $f(\mathbf{x})$. Each Walsh coefficient, ω_x is calculated by

$$\omega_x = \frac{1}{2^n} \sum_{j=0}^{2^n-1} f(j)\psi_j(x) \tag{3}$$

Each of the resulting Walsh coefficients has an index, which defines the set of input variables over which it operates. Converting the index to a binary representation over n bits produces a representation of the variables associated with the coefficient where a 1 in position i indicates that x_i contributes to the effect of that coefficient. For example, over 4 bits, the coefficient ω_9 produces a binary word 1001, which tells us that x_1 and x_4 contribute to the effect of ω_9. The order of a coefficient is defined as the number of bits it contains that are set to 1. For example, ω_2 and ω_8 are first order as they have one bit set to 1, and ω_9 is second order. The magnitude of a coefficient indicates its importance in contributing to the output of the function on average across all possible input patterns.

A function of n inputs produces 2^n Walsh coefficients, so it is not always possible to consider the value of each coefficient individually. In this work we look at individual coefficients and also define some simple aggregate measures for summarising the results of a Walsh transform. They are the number of non-zero coefficients, which is taken as a crude measure of overall complexity, and the average magnitude of coefficients at each order, which produces a set of values that measure the contribution to the function's output made on average by interactions of each possible order.

3 Method

In this context, the Walsh transform is not used to understand the training data, but
to understand a neural network that was trained on that data. The analysis is in terms
of the inputs to and the outputs from the network, not its weights or activations,
making this a pedagogical approach. The black box of the neural function is assessed
in terms of its Walsh decomposition. Walsh functions map a vector of binary valued
inputs onto a real valued output, so any function with this structure is amenable to
the analysis. As shown below, multiple output neurons and classification networks
may also be analysed with this approach, so the outputs can be nominal, discrete or
continuous.

As neural networks can generalise and produce an output for any given input pattern, we can generate an exhaustive or randomly sampled data set from which to perform the Walsh transform. A full Walsh decomposition, as defined in Eq. 3 requires
an exhaustive sample of the input space. In all but the smallest of networks, this is
unfeasible in an acceptable time period, so the coefficients must be calculated from a
sample. In either case, the sample used to calculate the coefficients is generated from
the whole input space, not just the training data. The significant coefficients (those
that are significantly far from zero) can be very informative about the underlying
structure of the function (in this case, the MLP). The procedure is similar to that of
pedagogical rule discovery in that it treats the MLP as a black box and performs an
analysis on the output values that the network produces in response to input patterns.
The method proceeds as follows:

1. Build a single MLP using your chosen method of design and weight learning;
2. Generate input patterns (either exhaustively or at random) and allow the MLP to
   generate its associated output, thus producing $(\mathbf{x}, f(\mathbf{x}))$ pairs;
3. Use the resulting $(\mathbf{x}, f(\mathbf{x}))$ pairs to perform a Walsh transform using Eq. 3;
4. Analyse the significant coefficient values, ω_x.

The method can also be used for MLPs designed for classification rather than
regression. In such cases, there is normally a single output neuron for each class,
with a target output value of one when the input belongs to the neuron's designated
class and zero otherwise. Properly trained, each neuron represents the probability
of a new pattern belonging to its designated class. Such a network is effectively a
number of related functions (one for each class) with a continuous output between
zero and one. Each output neuron can be analysed in turn using the same procedure.

Step 4, the analysis of the ω_x values can take many forms. This paper discriminates
between analysis during training (Sect. 5) where the goal is to gain an insight into
the level of complexity a network achieves as learning progresses, and post training
analysis, designed to provide insights into the function of the trained network. The
example of such an analysis in Sect. 6.1 shows how the generalisation ability of a
network may be investigated from the results of the Walsh analysis. The goal of the
analysis is not to generate rules, so this is not another rule extraction method, rather
it is designed to give human insights into the hidden life of the MLP.

4 Experiments

A set of functions of increasing complexity[1] were chosen to generate data to test this analysis. They are:

OneMax, which simply counts the number of values set to one across the inputs. This is a first order function as each variable contributes to the output independently of any others. The OneMax function is calculated as

$$f(\mathbf{x}) = \sum_{i=1}^{n} x_i \tag{4}$$

Vertical symmetry, which arranges the bits in the input pattern in a square and measures symmetry across the vertical centre line. This is a second order function and is calculated as

$$f(\mathbf{x}) = \sum_{i=1}^{n} \sum_{j=1}^{n} \delta_{ij} s_{ij} \tag{5}$$

where δ_{ij} is the Kronecker delta between x_i and x_j, and s_{ij} is 1 when i and j are in symmetrical positions and 0 otherwise.

K-bit trap functions are defined by the number of inputs with a value of one. The output is highest when all the inputs are set to one, but when at least one input has a value of zero, the output is equal to one less than the number of inputs with a value of zero. For example, in three bits, $f(111) = 3$ is the function's maximum, $f(000) = 2$ produces the next highest output, and $f(011) = 0$ is a global minimum. A k-bit trap function over n inputs, where k is a factor of n is defined by concatenating subsets of k inputs n/k times. Let $\mathbf{b} \in \mathbf{x}$ be one such subset and $c_0(\mathbf{b})$ be the number of bits in $\mathbf{b}$ set to zero.

$$f(\mathbf{x}) = \sum_{\mathbf{b} \in \mathbf{x}} f(\mathbf{b}) \tag{6}$$

where

$$f(\mathbf{b}) = \begin{cases} c_0(\mathbf{b}) - 1, & \text{if } c_0(\mathbf{b}) > 0 \\ k, & \text{if } c_0(\mathbf{b}) = 0 \end{cases} \tag{7}$$

The first case in Eq. 7, which applies to all but 1 in 2^k patterns, could be modelled with a first order network (a linear perceptron, for example), which is a local minimum in the error space. The 'trap' part of the function is caused by the second case in Eq. 7, which requires the output to be high when all of the inputs have a value of one. This requires a higher order function, including components at orders from 1 to k, but only a small proportion of the data (1 in 2^k of them) contains any clue to this.

---

[1]Complexity has a specific meaning in this context. It describes the number and order of the interactions between inputs that produce a function's output.

5 Analysis During Training

Experiments were conducted to investigate the structure of the function represented by an MLP as it learns. The MLP used in these experiments had a single hidden layer and one linear output neuron. The functions described above were used to generate training data, which was used to train a standard MLP using the error back propagation algorithm. At the end of each epoch (a single full pass through the training data), a Walsh transform was performed on the predictions made by the network in its current state.

Summary statistics designed to reflect the complexity of the function the network has implemented and the level of contribution from each order of interaction were calculated from the Walsh coefficients. The complexity of the function was calculated as the number of significant non-zero Walsh coefficients. The size of the contribution from an order of interaction, o was calculated as the average of the absolute value of the coefficients of order o. Experiment 1 trained networks on the simple OneMax function (Eq. 4). Figure 2 shows the training error and network complexity of an MLP trained on the OneMax function. During learning, the network initially becomes over complex and then, as the error drops, the network complexity also drops to the correct level.

In experiment 2, an MLP was trained on the symmetry function of Eq. 5, which contains only second order features. Figure 3 shows the results of analysing the Walsh coefficients of the network function during learning. Three lines are shown. The solid line shows the network prediction error over time and the broken lines show the contribution of the first and second order coefficients in the Walsh analysis of the network function. Note the point in the error plot where the error falls quickly corresponds to the point in the Walsh analysis where the second order coefficients grow past those of first order. Compare this chart to that in Fig. 4, where the same problem is given to another MLP with the same structure, but which becomes trapped at a local error minimum, which is a first order dominated approximation to the

Fig. 2 Comparing training error with network complexity during learning of the OneMax function with an MLP with one hidden unit. Note that complexity falls almost 1000 epochs after the training error has settled at its minimum

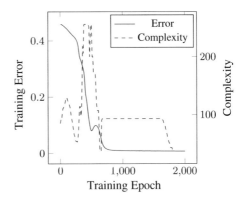

Fig. 3 Network error and contribution of first and second order Walsh coefficients during training of an MLP on a second order function. Note the fall in the error rate when the second order coefficients overtake the first

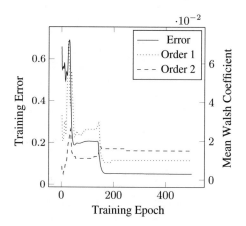

Fig. 4 Contribution of first and second order Walsh coefficients during training of an MLP on a second order function, stuck in a local minimum

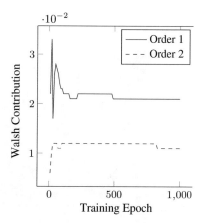

function. The plot suggests that the higher order components cannot increase their contribution and that this network is unlikely to improve.

Figure 5 shows the error of an MLP decrease as it learns the 4 bit trap problem described by Eq. 6. The contribution of the first, second, third and fourth order Walsh coefficients are each summed and plotted separately. The final, correct configuration can be seen in the right hand part of the plot, with the first order coefficients having the strongest contribution, but with the second, third and fourth also required to escape the 'trap' of the order below. The plot shows the first order coefficients growing first (as they did in Fig. 3), causing the average error to rise due to the higher order trap part of the function. The first order components are suppressed by the high error they cause, but the error doesn't settle to its lowest point until the first order coefficients recover the correct level of contribution.

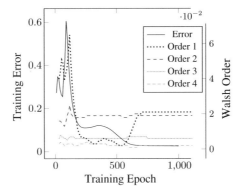

Fig. 5 The contribution of first, second, third and fourth order Walsh coefficients during training of an MLP on a the 4-bit trap function plotted with the average error per pass through the data set (*solid line*)

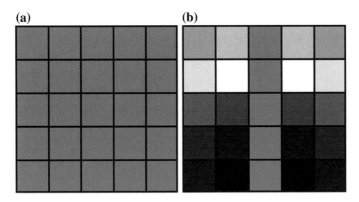

Fig. 6 First and second order coefficients of a symmetry counting function. In **a** the coefficients are all zero. In **b** the shade of *gray* indicates a non zero second order coefficient across the two pixels with shared gray level. **a** First order, **b** second order

5.1 A Second Order Function

In the following experiment, a second order function is investigated. The function is a measure of pattern symmetry, as defined in Eq. 5. Figure 6a shows the first order coefficients of a network trained to measure the symmetry of an image. Unsurprisingly, it shows no first order coefficients of importance. Mid grey indicates values close to zero, which suggests either that the variable that corresponds to the coefficient is unimportant or that variables are involved at higher orders. The higher order coefficient values tell us which of these possibilities is true.

Figure 6b shows the second order coefficients of a Walsh transform of the symmetry predicting MLP. The plot is produced by finding pairs of inputs that share a non-zero second order coefficient and setting them both to the same, unique shade of gray. Note that the centre column inputs share no second order relationships and are shaded mid-gray. The others are shaded so that their gray level matches that of the inputs with which they share a non-zero second order coefficient. The depth of shade does not indicate the size of the parameter, just that a connection exists. The shading is to discriminate between input pairs.

It is clear from Fig. 6b that each input is important to the calculation of the function output, so the interpretation of the zero valued first order coefficients is that the inputs' contributions are important, but only at orders above one.

The next experiment described in this paper makes use of partial samples from both the coefficients and the input space to gain an insight into the structure of an MLP trained on a pattern recognition task. It also illustrates the way error rates can be compared to determine how much of the network's functionality has been explained by the computed coefficients.

6 Partial Walsh Analysis

For even moderately large numbers of inputs, calculating every Walsh coefficient can take an impractically long amount of time. In such cases, a partial Walsh analysis can still be useful. A partial analysis calculates the values of only a small subset of the Walsh coefficients. An obvious choice for the subset of coefficients to calculate are those of the lower orders. ω_0 is the average output of the function (in this case, the MLP) across the sampled data. The first order coefficients, $\omega_1, \omega_2, \omega_4, \ldots$ represent the average contribution of each input in isolation. In general, order k coefficients represent the additional contribution of each subset of inputs of size k to the function output. The number of coefficients of order k from a set of inputs of size n is $\binom{n}{k}$, a figure which rises exponentially with k up to $k = n/2$ and then falls exponentially after that, to the point where there is just one order n coefficient. In general, one might expect a function to have significant interactions at the lower orders rather than the higher ones, so the number of coefficients of interest can be said to rise exponentially with their order.

It is also possible to estimate the Walsh coefficients from a sample of random input patterns and their associated predicted outputs from the network, rather than analysing every input pattern exhaustively. As with the calculation of any statistic from a sample, the values gained are estimates, but they can still provide useful insights into the functioning of a neural network. The number of samples required to estimate coefficients accurately grows exponentially with their order, so the low order coefficients can be estimated with smaller samples than the higher order coefficients require.

The values that are found as a result of sampling a small proportion of the Walsh coefficients can be used to reconstruct an estimate of the function implemented by the MLP that produced them. This reconstructed function can be used to generate predictions on the test data. The accuracy of this model will almost always be worse than that of the MLP but by comparing the respective error rates, the proportion of MLP'a ability that is captured by the Walsh coefficients can be measured.

6.1 Measuring Generalisation

Generalisation is the ability of an MLP to produce correct outputs for patterns that were not in its training data. As the weights of the network are difficult to analyse, the performance of the learned function in areas of input space that are outside those covered by the training data can be difficult to assess. Test and validation sets perform this task to a degree, but this paper proposes a new method based on a Walsh analysis where the network is analysed with random input patterns. The use of random inputs (i.e. patterns where each input takes an independent, uniformly distributed random value) allows a trained network to be tested on potentially massive test sets. Of course, these random patterns do not have associated target outputs, but as the Walsh analysis makes use of only the predicted output from the network, the test patterns do not need a target (or desired) output. This allows the analysis to explore a far greater variety of the input/output space of a trained neural network.

The Walsh coefficients of an MLP function are generated by randomly sampling from the whole input space, not just the part of it covered by the training or test data. The coefficients give a picture of the general shape of the function, not just its behaviour on the training or test data. The experiments described in this section demonstrate the use of a Walsh decomposition of an MLP trained on the MNIST [21] handwritten digit data set. The goal is not to produce a better classification rate than those already reported in the literature. The goal is to train some different networks and use a Walsh analysis to gain an insight into their structure.

6.1.1 Learning the MNIST Data

The MNIST images are made up of 28×28 pixels, making 784 inputs, each with a value from 0 to 255, indicating a grey level in an anti-aliased image. In this work, input values were passed through a threshold to create binary patterns rather than the grey level images of the raw MNIST data. A neural network with 784 inputs, 20 hidden units and 10 outputs (1 for each of the digits from 0 to 9) was trained on the standard MNIST training set, where the images are centred on their centre of mass. The resulting network implements 10 different functions, each mapping the input pattern to a continuous output variable that reflects how well the input pattern matches digits that correspond to its class (i.e. the identity of the digit). These functions are not independent as they share the weights between the input and hidden layer. Across a well trained network, the outputs should sum to one. The network

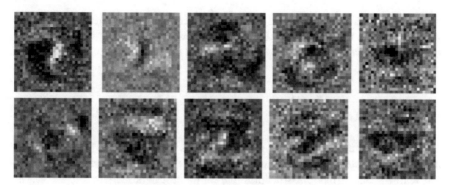

Fig. 7 First order Walsh coefficients from a network trained on the MNIST data with no added noise. *Grey squares* indicate no contribution to classification from a first order component. Greater depth of *black* or *white* indicates stronger contribution (positive or negative)

achieved a correct classification rate of 89 %, which is poor compared to any serious attempt, but useful for illustration purposes here.

An individual Walsh decomposition for each output neuron based on 50,000 random input samples and their associated network output was performed after training had completed. This produced ten Walsh decompositions. Initially, only the first order coefficients were calculated. The first order coefficients were plotted on a grid where the pixel locations from the inputs correspond to the first order coefficients of the Walsh decomposition, as shown in Fig. 7, in which it is clear that the network does not even have a very general first order model of the patterns that make each digit. Take the coefficients for the digit "1", for example. Very few of the pixels are used—four or five central pixels have large positive coefficients (making a positive contribution the output neuron value for class "1") and there are a small number of negative (shown in white) pixels to the top left and bottom right which cause pixels in their respective locations to diminish the output for the class "1" output neuron. This shows the network to have a reliance (one might argue and over-reliance) on specific inputs for making a classification. This would manifest itself as a poor ability to cope with noise in any test data where the specific pixels were altered.

When used to reconstruct a first order approximation to the network's function, the first order coefficients alone for this model achieved a root mean squared error (RMSE) of 0.17, indicating that the first order coefficients have captured a large proportion of the network's functionality. The RMSE of a decomposition with respect to the MLP it was derived from is calculated by using Eq. 1 to produce an output for a number of random samples from the input space, which is compared with the output from the MLP given the same input. A very low RMSE for a partial decomposition indicates that the remaining coefficients (those excluded from the partial decomposition) make very little additional contribution to the function output.

Note that all of the images in this paper are produced by normalising the coefficients being plotted to a range that causes the colours to vary from white to black. This leads to some distortion when the range is very small, as there may be a small distance between the highest and lowest coefficient. If all coefficients are close to zero, the highest will still appear black and the lowest white, though in reality they are all similarly small. This means that you cannot compare one plot with another in terms of absolute values.

6.1.2 Adding Training Noise

A common method of improving generalisation is to add noise to the input values in the training data. By randomly flipping 10 % of the input bits each time a pattern was learned, the correct classification rate on the test data increased to 93 %. The impact on the first order coefficients can be seen in Fig. 8 where it is clear that some (but not all) of the digits are now quite clearly identified across more of the input variables. The coefficients reveal the degree to which some classes have a clearly defined shape in the network and others do not. Returning to the example of the digit "1", Fig. 8 shows how a larger number of central pixels have a positive effect on the output neuron for class "1". The white inhibitory pixels are also more clear and widespread in these figures.

When used to reconstruct a first order approximation to the network's function, the first order coefficients alone for this model achieved a root mean squared error of 0.32, showing the first order coefficients to be responsible for less of the MLP's functionality, even though more of them are used. The remaining network functionality is of a higher order, leading to the conclusion that this network, which has better generalisation ability than the first, is more complex in the sense that it relies on more higher order interactions between the input variables to make its classifications. Discovering the higher order coefficients of interest is not trivial as significant coefficients are sparse among all possible coefficients.

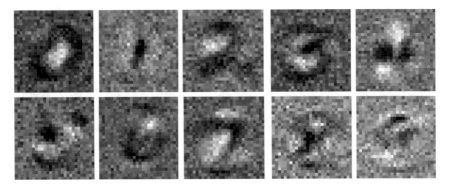

Fig. 8 First order Walsh coefficients from a network trained on the MNIST data with 10 % added noise. The noise ensures that no individual input can be relied upon to produce a correct classification, and so produces a model that covers more of the input space, and so is better at generalisation

6.1.3   Adding Training Jitter

Another method for attempting to improve the generalisation ability of a handwritten digit classifier is to jitter the training data, which means to shift each training image a small random number of pixels in a random direction before each is presented to the network for learning. Each training pattern is learned several times, each time with a new random shift to its location. This has the effect of blurring the first order coefficients across the input space, making them less useful for classification. Imagine an extreme case where the digit for "1" is moved to any location in the input field. No single pixel would be set to 1 (i.e. black) more often than any other when the input pattern represents a 1, so there would be no useful first order information in the data. Second order contributions between pixel pairs would be required for a good classification model. Pairs of pixels that are both set to 1 or both set to 0 when the input pattern represents "1" would have a positive coefficient and those that differed would have a negative coefficient. The strongest effect we would expect in the example of the digit "1" would be positive weights between pixels that were above and below each other in the image field.

The hypothesis is that an MLP trained on jittered inputs would produce weaker first order and stronger second order coefficients. To illustrate this point, take a simpler example than the MNIST data using hand designed digits over 25 pixels arranged in a 5×5 grid. Concentrating on the example for digit "1" and allowing that digit to be represented by any pattern of three or more black pixels above each other on an otherwise white background. Figure 9 shows some example "1" digit images. The other digits (0 and 2–9) were also hand designed and a data set was created containing equal numbers of examples of each. The other digits were fixed in their location, but the "1" digits were placed at random in the input field and given random lengths of 3, 4, or 5 pixels. After training a neural network to distinguish between the examples of "1" digits and designs for the other digits from 0 to 9, a full first and second order Walsh decomposition was performed by sampling random input patterns and the associated output from the neuron corresponding the class "1".

Figure 10 shows the first order coefficients for the output for "1". Note that there is very little variation in the values as the value of any individual pixel makes no consistent contribution to the output. Second order coefficients are not as straight forward to plot and view as those of first order, as there is one coefficient for every pair of input variables. To visually represent some of the second order coefficients, those with the highest absolute values were chosen and plotted as pairs joined with a

Fig. 9 Three examples of small training patterns for the digit "1", varied by location and length, but maintaining the vertical quality

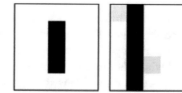

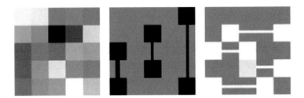

Fig. 10 First order coefficients (*left*) of a network trained on the randomly placed "1" digits of Fig. 9, and some examples of large positive (*middle*) and negative (*right*) second order coefficients from the same network

line, as shown in Fig. 10, in which black dots indicate a positive coefficient between the two inputs and white dots indicate a negative coefficient. Each figure shows a small number of coefficients (three positive and four negative). It is clear that pairs of pixels that are in the same vertical line share a positive coefficient and pairs that are in different columns share a negative coefficient. Not every second order coefficient has a significant value. As with the first order coefficients, a small number of them are sufficient to allow correct classifications to be made, so there is no pressure during training for further weight changes to produce a function where every coefficient has the expected value.

On a toy example, this is easy to see. On the larger MNIST data, the process is not as straight forward. This is partly because there are many more second order coefficients to sample and partly because not all of them need to take a value. As pointed out above, and as seen in the first order examples before noise is added, a sparse subset of coefficients are actually needed to reproduce the functionality of the network and once the error is sufficiently low, there is no pressure to change the weights further. However, the process was repeated for the MNIST data. An MLP was trained on the MNIST data and during training, each input pattern was moved by up to 4 pixels in one of the eight possible directions.

Figure 11 shows the first order coefficients, with the result of the shifted input patterns clearer to see in some classes than others. The root mean squared error between the first order Walsh decomposition of this MLP and the output of the MLP was 0.10, indicating that, contrary to expectations, the network in which the patterns were shifted around is better able to rely on the pixel values in isolation, rather than needing higher order coefficients.

6.1.4 Choosing Higher Order Coefficients

The number of coefficients that might be calculated grows exponentially with the number of variables and the order of the coefficients so it is crucial to choose the coefficients of to sample carefully. This section discusses some possible heuristics that might be employed when choosing which higher order coefficients to sample. The coefficients can be used to approximate the function that the neural network

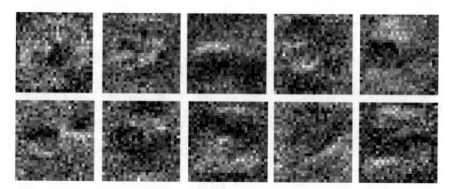

Fig. 11 First order coefficients calculated from the output neurons corresponding to each class of hand written digit from training data subjected to random jitter

has implemented, using Eq. 1 and assuming a zero value for all of the un-calculated coefficients. The mean difference between the network output and that predicted by the Walsh decomposition indicates the mean size of the missing coefficients. When that is sufficiently small, there can be few remaining coefficients to discover.

The best heuristic in most circumstances is to start with the lowest order coefficients first. In the case of the MNIST data, the first order coefficients accounted for most of the network's ability, and so gave a reasonable picture of its functional shape. The identity of the significant first order coefficients may be used to drive the choice of second order coefficients. The choice is between an assumption that variables that are useful at the first order level will also be useful in combination and the opposite assumption that variables that were found to have no first order contribution might play their role in higher orders. With human knowledge of the training process, one might be able to make informed choices as to which pixels might interact.

Another option with image data such as the MNIST set is to calculate coefficients that join neighbouring pixels. For example, each of the 784 pixels might share a second order coefficient with any of the remaining 783, but each pixel has only eight immediate neighbours. By hypothesising that written digits are made by a continuous stroke and so neighbouring pixels are more likely to interact to influence classification output, a small field around each pixel can be chosen, greatly reducing the number of calculations to be made. Second order Walsh coefficients in a small neighbourhood act as edge detectors, giving negative coefficients at points where neighbouring pixels that disagree contribute positively to the function output. This approach shares a great deal with the use of Markov Random Fields for image processing such as segmentation and edge detection, (see [22] for example). The key difference in this context is that we are not using the local field to process an image, but to analyse an MLP that was trained to classify a set of images. If the MLP has not captured certain features, then the analysis cannot reveal them, so a failure to find features that might have been expected is not a reflection on the method, but on the particular MLP's underlying functional shape.

7 Comparison with Other Methods

Recent published work in this field, such as the papers mentioned in the introduction, has concentrated on rule discovery, though what constitutes a rule is quite flexible. [23], for example build a binary truth table to represent the function of the MLP. The Walsh method is a pedagogical approach, according to the definitions in [10] as it treats the MLP as a black box. One of the advantages of the pedagogical approach is that the rules that are produced are easy to interpret.

The Walsh decomposition approach certainly aids interpretability, but it cannot be considered a rule extraction algorithm as it does not generate rules. Instead, it provides insight into the complexity of an MLP, highlighting both the level of complexity, and the variables involved. For example, in the k-bit trap function, it is clear from an examination of the coefficients that inputs are organised into subsets which interact within the traps, but that they are independent across traps.

One advantage of the Walsh method is that the coefficients may be easily visualised. Figure 12 shows the coefficients generated from an MLP that has learned a 5-bit trap function over 30 inputs. The figure is generated by discarding non-significant coefficients and then sorting the remaining coefficients into combinatorial sequence so that low order coefficients are at the top of the figure. Each row of the figure represents a single coefficient as the binary equivalent of its index. For example ω_5 is a second order coefficient with binary representation 101, meaning that the coefficient measures the interaction between inputs 1 and 3. Dark pixels represent connected inputs in the figure.

Another advantage of the pedagogical rule extraction approach is that it is portable across network architectures as it treats the network as a black box. The Walsh method shares this advantage. A common feature of rule extraction methods is that they accept a reduction in accuracy in return for a simpler set of rules. The rule set can be evaluated on the same test data as the MLP that generated the rules and the

Fig. 12 5 bit trap Walsh coefficients

Order   Weights

1

2

3

4

5

trade-off between accuracy and size of the rule set needs to be managed. To reproduce the functionality of the network perfectly with a rule set can require a great many rules and a large number of exceptions (or rules that apply to a very small area of input space). The Walsh method shares this limitation, but for different reasons. As the Walsh functions are a basis set, there is no function that they cannot represent, so there is no network whose behaviour cannot be perfectly reproduced. Any network with binary inputs can have its behaviour perfectly reproduced by a Walsh decomposition, but only by a full decomposition from an exhaustive sample of input,output pairs. This is possible for small networks, but infeasible for networks with large numbers of inputs. A sample of coefficients must then be calculated from a sample of data points, which will lead to an approximate representation of the MLP function.

Classification rules are generally local in that they partition a data set into subspaces that share the same output. This works well when the inputs are numeric as the conditional part of the rule can specify a range. When the inputs are discrete, as in the binary case studied here, the rules cannot partition the input space across a range. In such cases, a rule set may not be the best way to understand a function. Take the character recognition task for example, we can learn more by visualising the coefficients (even just those of low order) than by studying a long list of rules. Walsh coefficients are global as they describe the contribution of an input or group of inputs across the entire input space. This means that it is not possible to partition the input space and so derive simple rules. Every coefficient plays a part in calculating the output from every input pattern. General statements can still be made, however, but they are of the form "When variable $x = 1$, the output increases" or "When variables a and b are equal, the output decreases". These statements can be generated directly from the coefficients.

8 Conclusions

An MLP trained on binary input data with either numeric or categorical output neurons can be analysed using Walsh functions. Such an analysis can reveal the relative complexity of different networks, give an insight into the way the function represented by an MLP evolves during learning and shed light on which areas of input space a network has utilised in learning that function. This understanding can help in understanding how well a network will generalise to new data and where its likely points of failure may be. An exhaustive Walsh decomposition is only possible for small networks, but a partial decomposition based on a random sample from the network's input space can still be used to gain valuable insights into the specific function learned by an MLP.

9 Further Work

This work has used Walsh functions as its method of complexity analysis, but other basis functions–particularly those suitable for real valued inputs–are also worthy of investigation. As the analysis is not designed to reconstruct the function, merely to shed light on its structure in a human readable form, it should be possible to use an information theoretic measure of interaction such as mutual entropy.

The method provides a useful measure of network complexity that is not based on the number of weights in the network. Training methods that favour simple models over more complex ones often use parameter counts (in the case of MLP, the weights) as a measure of complexity. For example, minimum description length (MDL) methods are often based on parameter counts, but might usefully be adapted to account for other types of complexity such as that described here. The Walsh analysis reveals that two networks of equal size do not necessarily share an equal complexity. The relationship between network complexity and network size is an interesting field of study in its own right. Of course, this analysis is not restricted to use with MLPs. Any regression function may be used, but it is well applied to MLPs as they are difficult to analyse in terms of the structure of their weights alone.

The number of Walsh coefficients to consider grows exponentially with the number of inputs to the network, so it is not possible to exhaustively calculate every possible one in a large network. For networks that contain key interactions at a number of different higher orders, the task of finding the significant coefficients becomes a great problem. Work on heuristics for finding the significant high order coefficients in a sparse coefficient space is ongoing. One approach is to build a probabilistic model of the importance of different neurons and connection orders and sample coefficients from that model. As more coefficients are found, the quality of the model improves and allows the faster discovery of others.

References

1. Rumelhart, D.E., Hinton, G.E., Williams, R.J.: Parallel Distributed Processing: Explorations in the Microstructure of Cognition, vol. 1. MIT Press, Cambridge, MA, USA (1986)
2. Elman, J.L.: Finding structure in time. Cognitive Sci. **14**, 179–211 (1990)
3. Kamimura, R.: Principal hidden unit analysis with minimum entropy method. In: Gielen, S., Kappen, B. (eds.) ICANN 1993, pp. 760–763. Springer, London (1993)
4. Sanger, D.: Contribution analysis: a technique for assigning responsibilities to hidden units in connectionist networks. Connection Sci. **1**, 115–138 (1989)
5. Gorman, R.P., Sejnowski, T.J.: Analysis of hidden units in a layered network trained to classify sonar targets. Neural Netw. **1**, 75–89 (1988)
6. Kulluk, S., Özbakir, L., Baykasoğglu, A.: Fuzzy difaconn-miner: A novel approach for fuzzy rule extraction from neural networks. Expert Systems with Applications 40, : 938–946 FUZZYSS11: 2nd International Fuzzy Systems Symposium 17–18 November 2011. Ankara, Turkey (2013)
7. Hruschka, E. R., Ebecken, N. F.: Extracting rules from multilayer perceptrons in classification problems: A clustering-based approach. Neurocomputing 70 (2006) 384–397

Neural Networks Selected Papers from the 7th Brazilian Symposium on Neural Networks (SBRN '04) 7th Brazilian Symposium on Neural Networks

8. Saad, E., Wunsch II, D.: Neural network explanation using inversion. Neural Networks 20 (2007) 78–93 cited By (since 1996)22

9. Augasta, M., Kathirvalavakumar, T.: Rule Extraction from Neural Networks - A Comparative Study, pp. 404–408 (2012). cited By (since 1996)

10. Jivani, K., Ambasana, J., Kanani, S.: A survey on rule extraction approaches based techniques for data classification using neural network. International Journal of Futuristic Trends in Engineering and Technology 1 (2014)

11. Baum, E.B., Haussler, D.: What size net gives valid generalization? Neural Comput. **1**, 151–160 (1989)

12. Uphadyaya, B., Eryurek, E.: Application of neural networks for sensor validation and plant monitoring. Neural Technology (1992) 170–176

13. Widrow, B., Lehr, M.: 30 years of adaptive neural networks: perceptron, madaline, and back-propagation. Proc. IEEE **78**, 1415–1442 (1990)

14. Weigend, A. S., Huberman, B. A., Rumelhart, D. E.: Predicting Sunspots and Exchange Rates with Connectionist Networks. In Casdagli, M., Eubank, S., eds.: Nonlinear modeling and forecasting. Addison-Wesley (1992) 395–432

15. Bartlett, E.B.: Dynamic node architecture learning: an information theoretic approach. Neural Netw. **7**, 129–140 (1994)

16. Castillo, P. A., Carpio, J., Merelo, J., Prieto, A., Rivas, V., Romero, G.: Evolving multilayer perceptrons (2000)

17. Yao, X.: Evolving artificial neural networks. Proc. IEEE **87**, 1423–1447 (1999)

18. Krogh, A., Vedelsby, J.: Neural network ensembles, cross validation, and active learning. In: NIPS. (1994) 231–238

19. Walsh, J.: A closed set of normal orthogonal functions. Amer. J. Math **45**, 5–24 (1923)

20. Beauchamp, K.: Applications of Walsh and Related Functions. Academic Press, London (1984)

21. Lecun, Y., Cortes, C.: (The MNIST database of handwritten digits). <!– Missing/Wrong Year –>

22. Li, S. Z.: Markov random field modeling in computer vision. Springer-Verlag New York, Inc. (1995)

23. Jian-guo, W., Jian-hong, Y., Wen-xing, Z., Jin-wu, X.: Rule extraction from artificial neural network with optimized activation functions. In: Intelligent System and Knowledge Engineering, 2008. ISKE 2008. 3rd International Conference on. Volume 1, IEEE (2008) 873–879

Towards an Objective Tool for Evaluating the Surgical Skill

Giovanni Costantini, Giovanni Saggio, Laura Sbernini,
Nicola Di Lorenzo and Daniele Casali

Abstract In this paper we present a system for the evaluation of the skill of a physician or physician student by means of the analysis of the movements of the hand. By comparing these movements to the ones of a set of subjects known to be skilled, we could tell if they are correct. We consider the execution of a typical surgical task: the suture. For the data acquisition we used the HiTEg sensory glove, then, we extract a set of features from data analysis and classify it by means of different kind of classifiers. We compared results from an RBF neural network and a Bayesian classifier. The system has been tested on a set of 18 subjects. We found that accuracy depends on the feature set that is used, and it can reach 94 % when we consider a set of 20 features: 9 of them are taken from data of bending sensor, 10 from accelerometers and gyroscopes, and one feature is the length of the gesture.

Keywords Neural networks · Data glove · Hand-gesture · Classification · Surgery

1 Introduction

It is indubitable for surgeons the need to acquire a certain skill in operating and to maintain their ability, so to deliver acceptable results. How the skill level can affect clinical outcomes has been already evidenced [1]. However, a great number of failure during operations have been time-by-time recorded [2], and tens of operations have been demonstrated to be necessary before reaching an aimed plateau in the learning curve [3]. Unavoidably, this results in more or less traumatic consequences

G. Costantini · G. Saggio · D. Casali (✉)
Departemet of Electronic a Engineering, University of Rome "Tor Vergata", Rome, Italy
e-mail: daniele.casali@uniroma2.it

L. Sbernini · N. Di Lorenzo
Department of Experimental Medicine and Surgery, Tor Vergata University, Rome, Italy

© Springer International Publishing Switzerland 2016
J.J. Merelo et al. (eds.), *Computational Intelligence*,
Studies in Computational Intelligence 620, DOI 10.1007/978-3-319-26393-9_19

for patients. Therefore, a key point is to understand if there is the possibility to lower the number of failures, and to more rapidly acquire the necessary skill.

But, in what the skill relies on, how to evaluate it, and how to reduce the time to obtaining it?

A skillfully performed operation is based on decision making mixed to dexterity [4], the amount of their percentages depending on the specialties of surgery.

The evaluation of the skillfulness is mainly due to direct supervision of senior surgeons who observe both decision making mixed to dexterity of novices and provides verbal feedbacks [5]. But, this evaluation is prone to human errors and can potentially rely on bad or biased judgments, hence the need of freeing from human sensitivity, developing objective criteria. To this aim, objective measurements of technical skill have been searched, and some answers have been found. In particular, psychomotor skills have been evaluated considering the reduction in the operation time [6–10], the corresponding increasing in operation speed [3, 11], the precision [11], the accordance with a task-specific checklist [12], the force/torque "signatures" [13–15], and the economy of motion (the number of movements made [16], the trajectory [15] and the total distance travelled [17]).

Although these psychomotor skills effectively change during the learning curve, it remains difficult to interpret their real meaning, in absolute values. Consequently, up to now we have to rely on criteria levels based on expert performance, i.e. comparing, and revealing the discrepancies, between the gestures performed by two well defined groups consisting of novice and expert surgeons respectively, as in [8, 15]. To this aim, we developed and adopted a sensory glove capable to measure the hand dexterity.

In order to compare different sets of data coming from the different groups, different mathematical tools can be adopted, such as the Mann–Whitney U test [18], or the ANOVA [11].

Here, we propone the adoption of a sensory glove and a classifier, with the aim of recognizing patterns related to skilled surgeon and novices. We do a comparison between different classifiers and feature sets. The paper is structured as follows: in the second section we describe the data glove that we used to measure the hand movements. In the third section we describe the classification system. Finally in the fourth section we describe the experiment and discuss results.

2   The Sensory Glove

On the basis of our previous experience on bending and inertial sensors [19], we designed and realized a sensory glove (Fig. 1) suitable for the experiments. The glove was capable to measure flex/extension of each finger joint by means of bending sensors (by Flexpoint Inc., South Draper, UT, USA), and the movements of the wrist by means of inertial units including 3D type accelerometer and gyroscope (by SparkFun Electronics, Niwot, CO, USA) [20, 21]. Data from the sensory

Fig. 1 The sensory glove
dressed by the user during the
test phase

glove were acquired and sent to a personal computer, via USB port, by means of an
ad-hoc realized electronic circuitry.

For sake of completeness and to furnish a visual feedback to the user, the hand
movements were reproduced on a computer screen via 3D avatar.

Table 1 identifies the sensors supported by the glove. In particular, two flex
sensors were for the thumb (1–2), and three for the other fingers (3–14), plus 3D
accelerometer (15–17) and 3D gyroscope (18–20) were for the wrist.

3 The Classification System

We asked to each subject to repeat a specific gesture for a given number of times.
After a pre-processing, where data is filtered with a moving average filter, we cut
initial and final parts of data because this data do not describe any movement. Data
are then re-sampled in order to have the same number of samples for every subject.
Every repetition is normalized to the length of 1000 samples, so the whole gesture
is 1000 n samples length, where n is the number of repetitions for the gesture.

We separately took into account information regarding the actual duration of the
gesture.

Data obtained from the 20 sensors are splitted into non-overlapped windows of
W samples. We considered two option for the length of W: the first one is $W = 10$,
obtaining 100 windows in total, the second one is $W = 20$, obtaining 50 windows.
Every window is a representation of the state of the system in a specific interval of
time. For example, window 1 represents the beginning of the gesture, from its start
to $1/W$ of its length. For every window we calculate the mean value of its samples;
every repetition is considered as a different instance. With 20 sensors and 100 (or
50) time-series values, we have a total of 2000 values that can be considered as
features for classification when we use $W = 10$, or 1000 when we use $W = 20$. In
addition, we also consider the median value of the time length of the gesture.
Medians of the duration time of the repetitions are shown in Fig. 2. For every one of

Table 1 List of sensors

#	Sensor
1	1PIPJ
2	1DIPJ
3	2MCPJ
4	2PIPJ
5	2DIPJ
6	3MCPJ
7	3PIPJ
8	3DIPJ
9	4MCPJ
10	4MCPJ
11	4PIPJ
12	5DIPJ
13	5PIPJ
14	5DIPJ
15	Accelerometer, x axis
16	Accelerometer, y axis
17	Accelerometer, z axis
18	Gyroscope, x axis
19	Gyroscope, y axis
20	Gyroscope, z axis

"1PIPJ" stands for thumb Proximal Interphalangeal Junction angle, "1 DIPJ" stands for thumb Distal Interphalangeal Junction angle, "2MCPJ" stands for first finger Met-acarpo Phalangeal angle, and so on

the 18 subjects, the first box represents the median value for the first session, and the second box the median value for the second session. The first 9 subjects are expert, while the second 9 are non-expert. This feature can be useful for the classification due to the fact that the duration for non-experts is often longer.

For the classification, we used a Radial Basis Function (RBF) neural network and Bayesian Networks. We considered three feature sets: first we consider all bending sensors, gyroscope and accelerometer data for every window, plus time length, for a total of 2001 features for $W = 10$ or 1001 for $W = 20$. We also made another test with a subset of the first one, taking only some time windows of some sensors, for a total of 20 features. We based our selection by adapting feature that showed to be optimal in our previous work. Finally we considered a third feature set, which only considers data from accelerometers and gyroscopes, which have shown to be the most important, to see if and how much performance degrades when we avoid bending sensors.. All features are normalized as required by neural networks.

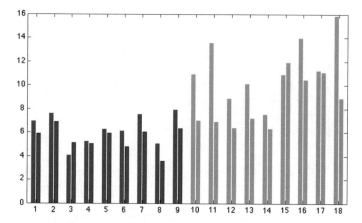

Fig. 2 Medians of the duration time of the repetitions. The first nine subjects are expert, while the second nine are non-expert

4 Experiments and Results

4.1 Experimental Procedure

For the experiment, we involved 18 subjects: 9 of them were skilled surgeons and 9 were novices on their starting learning curve. All of them were asked to perform the same task: a surgical suture simulated on a plastic material designed to have the same characteristics of human skin. The gestures always started and finished on the same rest position. Every subject repeated the gesture 10 times for every session.

Two sessions were recorded for every subject, on two different days, and every session included 10 repetitions, so we totally collected a total of 360 instances: 180 from skilled and 180 from unskilled subjects.

The duration times of every gesture for every subject were calculated.

4.2 Feature Extraction

Data analysis clearly shows differences between skilled surgeon and novices. For example, Fig. 3 reports data from sensor 20 (gyroscope, axis z), windowed with $W = 20$, in a box-plot. In the axis x we reported the time window (1–50), in axis y the values from of expert subjects. On each box, the central mark is the median, the edges of the box are the 25th and 75th percentiles, the whiskers extend to the most extreme data points not considered outliers, and outliers are plotted individually.

From the graph, it appears that the trajectory followed by the experts are very similar: almost all of them behave starting with a value around 1.24, slightly

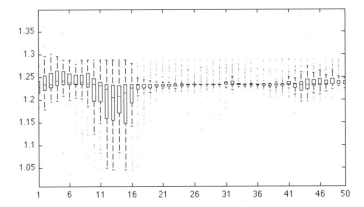

Fig. 3 Box-plot of sensor 20, for all experts, with $W = 20$ from time window 1 to 50 (begin to end of every repetition)

increasing, reducing to 1.16, then rising up again to 1.23, at half of the total duration. Figure 4 reports data from novice subjects: dispersion is higher, and there is no recognizable standard sequence.

4.3 Feature Selection

The box-plots of the experts report that value dispersion is not identical in time: in some time instants (for example in the central position of the graph in Fig. 3) it is very low, while it is higher elsewhere. Moreover, this can change with the sensor. For example, in Fig. 5 we show the values relative to sensor 1, which is the proximal interphalangeal junction angle of the thumb. Dispersion of this value is

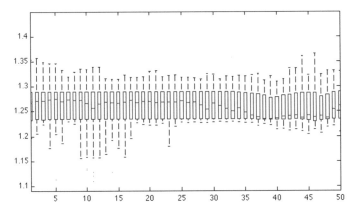

Fig. 4 Box-plot of sensor 20, with $W = 20$, for all non-experts

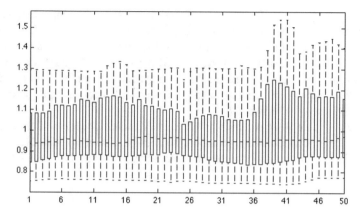

Fig. 5 Box-plot of sensor 1, for all experts

quite high among the experts too. This could mean that the position of the thumb can vary, and is not a discriminant factor for the recognition of the skill.

In our previous work [22] we discovered a subset of features that gives good results, by means of the CFS feature selection algorithm [23]. So, we adapted these features to our conditions. Selected features are described in Table 2.

We randomly selected 12 subjects, with a total number of 240 instances and used them as training set. Then we tested the classifier with data from the remaining 6 subjects (120 instances). We repeated the test 5 times, with different random choices for training and test sets, taking care that instances from the same subject always fell on the same set.

As a comparison we also tested the system with a Bayesian Network classifier. Results are summarized the following tables. In Tables 3 and 4 we show the confusion matrix using $W = 10$ and $W = 20$, analyzing these data we find that true positive (TP) rate for experts is 0.87 and 0.88 respectively, while false positive

Table 2 Selected features for the reduced dataset

Sensor	Time window ($W = 10$)	Time window ($W = 20$)
4 (2PIPJ)	1	1
7 (3PIPJ)	30	15
9 (4MCPJ)	84	42
10 (4PIPJ)	36, 100	18, 50
11 (4DIPJ)	30	15
13 (5PIPJ)	48, 62, 64	24, 31, 32
16 (acc.Y)	8, 48, 92, 94	4, 24, 46, 47
17 (acc. Z)	10	5
18 (gyr. X)	42	21
19 (gyr. Y)	90	45
20 (gyr. Z)	26, 28, 34	13, 14, 17

(FP) rate is 0.17 and 0.19. In Tables 5 and 6 we show the confusion matrices obtained using the reduced dataset with the 21 features reported in Table 2, we can see that in this case TP rate for experts increases to 0.91 and 0.98 and FP rate decreases to 0.30 and 0.16. In Tables 7 and 8 we show the confusion matrices obtained using only gyroscopes and accelerometers data, we can see that in this case TP rate for experts decreases to 0.77 and 0.80 and FP rate grows to an inacceptable rate of 0.53 and 0.68.

Finally, in Table 9 we summarize mean and standard deviation of the accuracy (number of correctly classified instances/number of instances) with all combinations of window lengths and feature sets, comparing results of RBF with a Bayesian network.

Table 3 Confusion matrix of RBF with $W = 10$, using all data

RBF $W = 10$, all data	Classified as expert	Classified as novice
Expert	52.4	7.6
Novice	10.2	49.8

Values are the averaged accuracy on the test set of 5 trials (120 instances for every trial: 60 experts and 60 novices)

Table 4 Confusion matrix of RBF with $W = 20$, using all data

RBF $W = 20$, all data	Classified as expert	Classified as novice
Expert	52.8	7.2
Novice	11.2	48.8

Values are the averaged results of 5 trials (120 instances for every trial: 60 experts and 60 novices)

Table 5 Confusion matrix of RBF with $W = 10$, using the reduced data set

RBF $W = 10$, all data	Classified as expert	Classified as novice
Expert	54.4	5.6
Novice	18.2	41,8

Values are the averaged accuracy on the test set of 5 trials (120 instances for every trial: 60 experts and 60 novices)

Table 6 Confusion matrix of RBF with $W = 20$, using the reduced data set

RBF $W = 20$, all data	Classified as expert	Classified as novice
Expert	58,8	1.2
Novice	9.6	50.4

Values are the averaged results of 5 trials (120 instances for every trial: 60 experts and 60 novice)

Table 7 Confusion matrix of RBF with $W = 10$, using only data from gyroscopes and accelerometers

RBF $W = 10$, Gyr. and Acc.	Classified as expert	Classified as novice
Expert	46.2	13.8
Novice	32	28

Values are the averaged results of 5 trials (120 instances for every trial: 60 experts and 60 novices)

Table 8 Confusion matrix of RBF with $W = 20$, using only data from gyroscopes and accelerometers

RBF $W = 20$, Gyr. and Acc.	Classified as expert	Classified as novice
Expert	47.8	12.2
Novice	41	19

Values are the averaged results of 5 trials (120 instances for every trial: 60 experts and 60 novices)

Table 9 Mean and standard deviation of accuracy with Bayesian Network and RBF network, considering all channels, and the two reduced datasets

Window length	All features (%)	20 features (%)	6 features (only data of gyroscopes and accelerometers) (%)
RBF			
$W = 10$	85.16 ± 4.5	$80,16 \pm 6.01$	61.83 ± 3.97
$W = 20$	86.67 ± 5.36	91.01 ± 8.38	55.67 ± 4.06
Bayesian Network			
$W = 10$	78.67 ± 8.29	$94,06 \pm 2.95$	68.46 ± 9.47
$W = 20$	78.34 ± 5.60	$93,66 \pm 3.6$	65.67 ± 7.49

5 Conclusions

We designed and developed a system for the evaluation of the skill of a surgeon while performing a suture. We compared different classifiers, different feature sets and different parameters for feature extraction. The system makes use of a sensory glove to obtain the exact position of the hand and movements of the fingers. Features were extracted by averaging the values of the 20 sensors in windows of W samples, with $W = 50$ or $W = 100$. Finally, the median of the duration of the gesture was added to the feature set. The dataset was classified by means of an RBF neural network and a Bayesian network. As far as window length is concerned, we found no significant difference between $W = 10$ and $W = 20$, being always around 85 %. A reduced feature set that includes 20 selected features improves the performance up to more than 90 %, but if the dataset is further reduced to only include data of gyroscope and accelerometers the performance notably decreases to about 60 %. Neural network performs better Bayesian network when using all features and worse when using 20 features.

References

1. Cox, M., Irby, D.M., Reznick, R.K., MacRae, H.: Teaching surgical skills—changes in the wind. N. Engl. J. Med. **355**, 2664–2669 (2006)
2. Poloniecki, J., Valencia, O., Littlejohns, P.: Cumulative risk adjusted mortality chart for detecting changes in death rate: observational study of heart surgery. BMJ **316**, 1697–1700 (1998)
3. Furuya, S., Furuya, R., Ogura, H., Araki, T., Arita, T.: A study of 4,031 patients of transurethral resection of the prostate performed by one surgeon: learning curve, surgical results and postoperative complications. Hinyokika kiyo Acta urologica Japonica **52**, 609–614 (2006)
4. Spencer, F.: Teaching and measuring surgical techniques: the technical evaluation of competence. Bull. Am. Coll. Surg. **63**, 9–12 (1978)
5. Reiley, C.E., Lin, H.C., Yuh, D.D., Hager, G.D.: Review of methods for objective surgical skill evaluation. Surg. Endosc. **25**, 356–366 (2011)
6. Van Rij, A., McDonald, J., Pettigrew, R., Putterill, M., Reddy, C., Wright, J.: Cusum as an aid to early assessment of the surgical trainee. Br. J. Surg. **82**, 1500–1503 (1995)
7. Hanna, G.B., Shimi, S.M., Cuschieri, A.: Randomised study of influence of two-dimensional versus three-dimensional imaging on performance of laparoscopic cholecystectomy. Lancet **351**, 248–251 (1998)
8. Santosuosso, G.L., Saggio, G., Sora, F., Sbernini, L., Di Lorenzo, N.: Advanced algorithms for surgical gesture classification. In: IEEE International Conference on Acoustics, Speech and Signal Processing (ICASSP), pp. 3596–3600. IEEE 2014
9. Costantini, G., Saggio, G., Sbernini, L., Di Lorenzo, N., Di Paolo, F., Casali, D.: Surgical skill evaluation by means of a sensory glove and a neural network. In: 6th International Joint Conference on Computational Intelligence, 22–24 October 2014, Rome, Italy (IJCCI 2014)–Session: 6th International Conference on Fuzzy Computation Theory and Applications (FCTA 2014), SCITEPRESS–Science and Technology Publications 2014, pp. 105–10 (2014)
10. Traxer, O., Gettman, M.T., Napper, C.A., Scott, D.J., Jones, D.B., Roehrborn, C.G., et al.: The impact of intense laparoscopic skills training on the operative performance of urology residents. J. Urol. **166**, 1658–1661 (2001)
11. Derossis, A.M., Fried, G.M., Abrahamowicz, M., Sigman, H.H., Barkun, J.S., Meakins, J.L.: Development of a model for training and evaluation of laparoscopic skills. Am. J. Surg. **175**, 482–487 (1998)
12. Goff, B.A., Lentz, G.M., Lee, D., Houmard, B., Mandel, L.S.: Development of an objective structured assessment of technical skills for obstetric and gynecology residents. Obstet. Gynecol. **96**, 146–150 (2000)
13. Rosen, J., Solazzo, M., Hannaford, B., Sinanan, M.: Objective laparoscopic skills assessments of surgical residents using Hidden Markov Models based on haptic information and tool/tissue interactions. Stud. Health Technol. Inform. 417–23 (2001)
14. Yamauchi, Y., Yamashita, J., Morikawa, O., Hashimoto, R., Mochimaru, M, Fukui, Y., et al.: Surgical skill evaluation by force data for endoscopic sinus surgery training system. In: Medical Image Computing and Computer-Assisted Intervention—MICCAI 2002, pp. 44–51. Springer, 2002
15. Tashiro, Y., Miura, H., Nakanishi, Y., Okazaki, K., Iwamoto, Y.: Evaluation of skills in arthroscopic training based on trajectory and force data. Clin. Orthop. Relat. Res. **467**, 546–552 (2009)
16. Taffinder, N., Smith, S., Jansen, J., Ardehali, B., Russell, R., Darzi, A.: Objective measurement of surgical dexterity-validation of the Imperial College Surgical Assessment Device (ICSAD). Minim. Invasive Ther. Allied Tech. **7**, 11 (1998)
17. Judkins, T.N., Oleynikov, D., Stergiou, N.: Objective evaluation of expert and novice performance during robotic surgical training tasks. Surg. Endosc. **23**, 590–597 (2009)

18. Fraser, S., Klassen, D., Feldman, L., Ghitulescu, G., Stanbridge, D., Fried, G.: Evaluating laparoscopic skills. Surg. Endosc. **17**, 964–967 (2003)
19. Saggio, G.: Mechanical model of flex sensors used to sense finger movements. Sens. Actuators A **185**, 53–58 (2012)
20. Saggio, G., Bocchetti, S., Pinto, G.A., Orengo, G., Giannini, F.: A novel application method for wearable bend sensors. In: ISABEL2009, 2nd International Symposium on Applied Sciences in Biomedical and Communication Technologies, Bratislava, Slovak Republic, 24–27 Nov 2009
21. Saggio, G., De Sanctis, M., Cianca, E., Latessa, G., De Santis, F., Giannini, F: Long term measurement of human joint movements for health care and rehabilitation purposes". In: Wireless Vitae09—Wireless Communications, Vehicular Technology, Information Theory and Aerospace and Electronic Systems Technology, Aalborg (Denmark), pp. 674–678, 17–20 May 2009
22. Costantini, G., Saggio, G., Sbernini, L., Di Lorenzo, N., Di Paolo, F., Casali, D.: Surgical skill evaluation by means of a sensory glove and a neural network. In: Proceedings of the International Conference on Neural Computation Theory and Applications (NCTA-2014), pp 105–110 (2014)
23. Hall, M.A.: Correlation-based Feature Subset Selection for Machine Learning Hamilton. New Zealand (1998)

Neurons with Non-standard Behaviors Can Be Computationally Relevant

Stylianos Kampakis

Abstract Neurons can exhibit many different kinds of behaviors, such as bursting, oscillating or rebound spiking. However, research in spiking neural networks has largely focused on the neuron type known as "integrator". Recent researches have suggested that using neural networks equipped with neurons other than the integrator, might carry computational advantages. However, there still lacks an experimental validation of this idea. This study used a spiking neural network with a biologically realistic neuron model in order to provide experimental evidence on this hypothesis. The study contains two experiments. In the first experiment the optimization of the network is conducted by setting the weights to random values and then adjusting the parameters of the neurons in order to adapt the neural behaviors. In the second experiment, the parameter optimization is used in order to improve the network's performance after the weights have been trained. The results illustrate that neurons with non-standard behaviors can provide computational advantages for a network. Further implications of this study and suggestions for future research are discussed.

1 Introduction

Spiking neural networks have been called the third generation of neural networks [17]. They have been tested on a variety of machine learning tasks such as unsupervised [3, 19], supervised [2, 7, 8] and reinforcement learning [21]. In many studies, the neuron model being used is usually an integrate-and-fire neuron or some of its variants or generalizations, like the leaky integrate-and-fire model and the spike response model. This is for example the case for the aforementioned studies.

S. Kampakis (✉)
Department of Computer Science, University College London, London, UK
e-mail: stylianos.kampakis@gmail.com

© Springer International Publishing Switzerland 2016
J.J. Merelo et al. (eds.), *Computational Intelligence*,
Studies in Computational Intelligence 620, DOI 10.1007/978-3-319-26393-9_20

However, realistic neuron models can exhibit different behaviors, which the neural models used in these studies cannot replicate. Izhikevich [10] presents 20 different neural behaviors (Fig. 1) that real neurons can exhibit, while also developing a model that can support all of these behaviors [9]. According to this classification, integrate-and-fire neurons are fall into the "integrator" category.

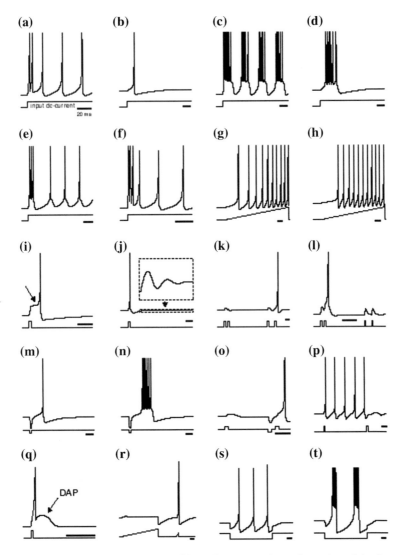

Fig. 1 Different neural behaviors exhibited by real neurons. Electronic version of the figure and reproduction permissions are freely available at www.izhikevich.com

Previous researches, such as those mentioned in the first paragraph usually just try to optimize the weights of the network, and in some cases, some other parameters, such as synaptic delays.

However, they do not optimize the behaviors of each individual neuron. This is, of course, difficult to do for a model of limited realism. In fact, the integrate-and-fire model, and its variants, offer limited flexibility with respect to the set of neural behaviors they can exhibit. Therefore, it is difficult to obtain computational advantages for specific tasks (e.g. supervised learning) by trying to adapt the neurons' behavior.

However, it could be the case that the flexibility of the model could help it adapt better to the task at hand. In fact, this model was used by Kampakis [12] in the context of spiking neural network for a supervised learning task. That research provided some evidence that different neural behaviors, other than the ones assumed by the integrate-and-fire neuron, could be useful. More specifically, it was demonstrated that a spiking neural network could learn the XOR logic gate with three neurons by using rebound spiking. This is something which could not be achieved with simple integrators.

Kampakis [13, 14] looked into the issue of the computational power of these neurons from a theoretical perspective. This study investigated the advantages that some specific non-standard behaviors can offer. The study focused on oscillators, bursting neurons and rebound spiking neurons and demonstrated how the use of these neurons for some particular tasks can reduce the number of synapses or neurons used in a network. However, a practical investigation was not pursued in that study. So, it remained unclear how non-standard behaviors could actually be used in a real setting and whether they would be useful.

Some similar ideas have emerged from other researchers as well. Maul [18] discussed the idea of a "Neural Diversity Machine". A neural diversity machine is an artificial neural network whose neurons can have different types of weights and activation functions. Nodes with different activation functions in an artificial neural network can be thought as equivalent to neurons with different behaviors in a spiking neural network. The inspiration behind this idea is similar to the inspiration behind the investigation of neurons with non-standard behaviors. However, Neural Diversity Machines have not been studied in the context of more realistic neuron models.

There is further justification in the literature to support this idea. First of all, neural diversity exists in the brain [20] and it has also been suggested that it can be computationally relevant for neural processing [16]. Secondly, there is evidence that artificial neural networks whose neurons use different activation functions can have more power [18]. Finally, Buzsaki et al. [4] have shown that biological neuronal diversification leads, both to a reduction of the number of neurons used by a network and to their wiring length.

The idea behind neural diversification is also justified from the perspective of inductive bias. This was something that was discussed by Kampakis [13, 14] through the theory of "rational neural processing". Diverse neural behaviors possess different inductive biases. This can make some neural behaviors more suitable for some tasks. Research conducted by Maul [18] and Cohen and Intrator [5] has proven that this can be true for artificial neural networks, as well.

However, while it became clear in theory [13, 14] that neural diversity provides greater flexibility in spiking neurons, which can lead to improved performance, no practical evidence of that has been provided yet. In the goal of the research outlined in this study is to provide this practical evidence. The experiments use a spiking neural network in order to test whether diverse neural behaviors can be computationally relevant. The network is trained on a supervised learning task by optimizing the parameters that control the neurons' behavior. Experiment 1 compares a network with optimized neuron parameters against an unoptimized network. Experiment 2 combines the parameter optimization with weight training in order to identify whether parameter optimization can provide any improvements in performance beyond weight optimization.

The purpose of this study is to provide some first experimental evidence that neural diversity can be computationally relevant for spiking neural networks, while also connecting this evidence with some of the recent research in the field.

2   Theoretical Motivation

This section will describe, in brief, the theoretical justification behind this research. Kampakis [13, 14] outlined a theoretical framework called "Theory of Rational Neural Processing". This theory is based on three main points:

(1) The neural circuits in the brain that lead to cognitive processes over specific tasks are characterized by specific biases, either in learning or in optimizing.
(2) These biases have been developed in order to optimize over the specific characteristics of the optimization problems that a neural circuit faces on a systematic basis.
(3) These biases are caused in the brain by the way that neurons are either connected or by the specifics of their behavior.

According to this theory the various neural behaviors that can be exhibited by neurons exist because they have been developed to provide computational advantages to different regions in the brain, according to the tasks they perform.

If this theory is correct, then it would be possible to simulate some of these computational advantages provided by the neural behaviors in an artificial spiking neural network. However, an additional problem is distinguishing between neural behaviors that are *computationally relevant* and those that are *biologically relevant*.

The term computationally relevant refers to behaviors that can be advantageous for a task, irrespective of whether they are in the brain or in an artificial neural network. The term biologically relevant refers to neural behaviors that have been developed in order to address biological limitations that are not relevant for an artificial system.

This study takes the position that using non-standard neural behaviors in computational tasks, such as supervised learning, can indicate whether the advantages are computationally relevant or biologically relevant.

3 Methods and Data

3.1 Neuron Model

This research used the Izhikevich neuron model. The neuron model of Izhikevich [9] is described by the following set of equations:

$$v' = 0.04v^2 + 5v + 140 - u + I \qquad (1)$$

$$u' = a(bv - u) \qquad (2)$$

The following condition ensures that the membrane voltage is reset after a spike:

$$if\ v \geq 30\,\mathrm{mV}, then \begin{cases} v \leftarrow c \\ u \leftarrow u + d \end{cases} \qquad (3)$$

The letters a, b, and d are dimensionless parameters of the model. I is the input, v is the voltage of the neuron's membrane, and u is the recovery variable. The parameter c is voltage in mV.

Wang [28] proposed an improvement over the original model, which prohibits the membrane voltage from reaching unrealistically high values. This improvement was implemented in this research as well. So, the condition from 3 changed to:

$$if\ v \geq 30\,\mathrm{mV}, then \begin{cases} v \leftarrow c \\ u \leftarrow u + d \end{cases} \\ if\ v \geq 30\ then\ v = 30 \\ if\ v = 30\ then\ v = c \qquad (4)$$

Figure 2 shows examples of different neural behaviors for different values of the parameters a, b, c and d.

3.2 Neural Architecture

The neural architecture used in this study is the same one as the one used by Kampakis [12] for the iris classification task and it is shown in Fig. 3.

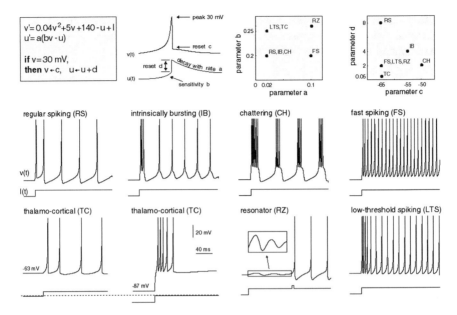

Fig. 2 Examples of various neural behaviors for different values of *a*, *b*, *c* and *d*. Reproduced with permission from www.izhikevich.com

Fig. 3 Architecture of the network used for the supervised learning task as it was presented in [12]

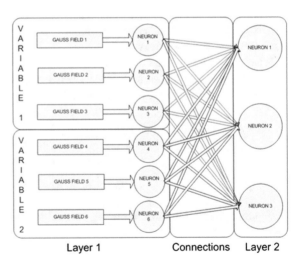

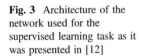

The architecture consists of two layers. The first layer consists of pairs of receptive fields with Izhikevich neurons. The receptive fields are comprised of Gaussian radial basis functions (hence the name "Gauss field" in Fig. 2). The equation for a receptive field is defined in (5).

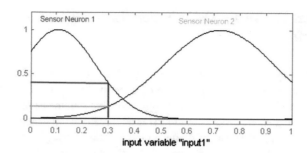

Fig. 4 An input is coded with two different Gaussian receptive fields. (Reproduced from Kampakis [12])

$$f(x; \sigma, c) = e^{\frac{-(x-c)^2}{2\sigma^2}} \tag{5}$$

In Eq. (5), σ is the standard deviation of the function, and c is the center. The centers of the receptive fields are uncovered by k-means clustering. The receptive fields receive the input in the form of a real number. The output of each radial basis function is fed into its respective input neuron. The output of the receptive field becomes the variable I in Eq. (1).

Figure 4 shows an example of an input encoded by two receptive fields termed "Sensor Neuron 1" and "Sensor Neuron 2". It can be seen that the same input, causes different responses to the two sensor neurons. This allows the receptive fields to encode different regions of the input space.

The input layer is fully connected to the output layer. The output is encoded by using a "winner-takes-all" strategy where the first output neuron to fire signifies the classification result.

3.3 Supervised Learning Task and Dataset

The chosen supervised learning task is the correct classification of the iris flowers in Fisher's iris dataset [6]. There are three iris types: Iris setosa, Iris virginica and Iris versicolor. Each type is represented in the dataset by 50 instances, for a total of 150 instances.

Each instance contains four attributes: sepal length, sepal width, petal length and petal width. Only sepal length and sepal width were used, like in [12] because the rest of the attributes are noisy.

3.4   Neural Parameter Optimization Through Genetic Algorithms

The parameters of the network were optimized through the use of a genetic algorithm. Parameter search is a standard use of genetic algorithms [22]. For example, recently, Wu et al. [29] used a genetic algorithm for parameter optimization in a support vector regression task. Tutkun [24] used a real-valued genetic algorithm for parameter search in mathematical models. Optimization through genetic algorithms has also been used successfully for optimizing the parameters of neuron models to experimental data [1, 15, 23].

There are other choices for optimizing neuron model parameters. A comprehensive review is provided by Van Geit et al. [26]. Some other choices besides meta-heuristic optimization include hand tuning, brute force and gradient descent methods. In practice, hand tuning is infeasible for this case, due to the large number of tests required for our purpose. Brute force is infeasible as well, due to the large computational demands required.

Gradient descent methods would require us to specify a differentiable error function. However, in practice, this seemed to be very difficult. On the other hand, genetic algorithms make no assumptions about the problem, and provide a very nice balance between exploitation of found solutions and exploration of new ones.

4   Experimental Setup

4.1   Experiments

This study consisted of two experiments. For the first experiment, two networks are created. The networks' weights are initialized by assigning a random set of weights sampled from the standard normal distribution.

Then, one network is trained by using a genetic algorithm in order to affect the parameters of each neuron in the network individually. Affecting the parameters changes the behavior and the response of the neurons. The experimental hypothesis was whether this can lead to improvements of accuracy, therefore demonstrating that diverse neural behaviors can be computationally relevant.

The objective function being optimized was the training accuracy on the supervised learning task. The training and testing is done by using 10-fold cross-validation. The training accuracy is recorded as the percentage of correct classifications across the data that were used for training, and the testing accuracy is recorded as the percentage of correct classifications for the fold that was not used in the training.

In order to identify whether affecting the parameters of the neurons can lead to improvements over the accuracy, the network was tested for 25 rounds of 10-fold cross validation against the network with random (unoptimized) weights.

In the second experiment the network is trained through a two-step optimization procedure. The network's weights are first trained using a genetic algorithm. Then, the neurons' parameters are optimized by using the genetic algorithm from the first experiment in order to improve the accuracy even further. The second experiment is used in order to examine whether parameter optimization can offer advantages in addition to standard training over the weights of the network.

It could be the case that any potential improvements in accuracy in the first experiment might not be significant when compared to standard training that optimizes the weights only. Therefore, this experiment was devised in order to explore whether parameter optimization can be computationally relevant when used in conjunction with standard weight training, or whether any advantages vanish. For that case, as for the first experiment, the procedure was repeated for 25 rounds of 10-fold cross validation and the objective function was the training accuracy.

For the second experiment the weights are optimized through the use of a genetic algorithm. The genetic algorithm used for optimizing the weights had the exact same configuration as the one in [12]: two populations that ranged from 50 to 100 members each, with crossover ratio 0.6 and 1 elite. The algorithm terminated after 150 generations had passed.

4.2 Parameter Optimization

After the weight optimization, a genetic algorithm is used in both experiments in order to optimize over the parameters of every neuron in the network, without affecting the weights.

The algorithm optimizes all the parameters of each neuron (a, b, c and d). The size of each individual in the population was 36 (this is equal to the sum of the number of neurons times 4). The manipulation of these parameters allows the neurons to exhibit many different behaviors, which were shown in Fig. 1.

The tweaking of these parameters not only affects the general behavior, but can also affect details within each behavior, such as the frequency of bursting, or the threshold of a neuron [11].

The genetic algorithms were executed by using the genetic algorithm toolbox of Matlab version 2011Rb. The default settings were used except for the following parameters: The population was set to 75 and the crossover rate was changed from the default of 0.8–0.6. This allowed a greater exploration of the parameter space, which seemed to improve the results in the pilot runs. The number of generations was set to 50. The upper and lower bounds of the variables were set to the interval [−100; 100]. The type of mutation was Gaussian and the selection method was stochastic uniform.

The optimization stopped as soon as the genetic algorithm reached the upper limit of generations. The parameters of the neurons and the architecture were the same as the ones used by Kampakis [12].

5   Results

Table 1 presents the results of the first experiment. The first and the third columns present the results for the optimized network. The third and fourth columns present the results from the unoptimized network. The first row presents the mean across all runs.

Table 2 demonstrates the results of the second experiment. The table shows the comparison between simple weight training (first two columns) and the two-step optimization procedure (last two columns). A Wilcoxon signed rank test for the null hypothesis that the medians of the two populations are unequal has a p-value of 0.0252.

Table 3 shows a comparison with other algorithms published in the literature. The comparisons include SpikeProp [2], SWAT [27] and MuSpiNN [7]. The reported scores are all mean averages of the accuracy for the iris classification task.

Table 1 Results of the optimization procedure and the unoptimized neural network

	Training accuracy	Test accuracy (%)	Random training (%)	Random test (%)
Mean	67.8	60.5	52.3	52.28
Std	15.2 %	3.6	11.8	10.9
Max	98.5 %	97.0	88.9	78.4
Min	55.2 %	46.7	22.2	7.8

Table 2 Comparison between weight training and the two step optimization procedure

	Weight (train) (%)	Weight (test) (%)	Two-step (train) (%)	Two-step (test) (%)
Mean	97.0	96	97.7	97.3
Std	1.2	1.9	0.6	2.3

Table 3 Comparison with other algorithms

Algorithm	Neurons	Training accuracy	Test accuracy (%)
SpikeProp	63	97.4 %	96.1
SWAT	208	97.3 %	94.7
MuSpiNN	17	Not reported	94.5
Two-step	9	97.7 %	97.3

6 Discussion

From the first experiment it is clear that the parameter optimization leads to improvements in performance over a randomly initialized network. The difference between the trained network in both the training accuracy and the test accuracy and the random network is quite prominent. The optimized network manages to generalize, obtaining a performance that is clearly better than what would be expected at random.

From the comparison between simple weight training and the two-step optimization procedure it seems that the parameter optimization can lead to further improvements in the accuracy after the network's weights have been trained. Furthermore, the two-step optimization's accuracy is comparable to other results reported in the literature, but it uses fewer neurons.

Therefore, it seems that optimizing the neural behavior of each neuron individually can provide improvements in accuracy that might not be achievable by using weight optimization alone.

7 Conclusion

This study provided evidence that parameter optimization for a biologically plausible neuron model is a feasible strategy to improve the performance of a supervised learning task. This was done in alignment with recent research that has promoted the idea that neural networks (biological, spiking and artificial) with heterogeneous neurons and non-standard behaviors might possess increased computational power. This study provided additional evidence for this idea by showing that it holds true for spiking neural networks, as well.

This study provides evidence that biologically simple neuron models, such as the integrate-and-fire model, might offer limited computational capabilities compared to more biologically realistic neuron models. Furthermore, the study provided evidence that biologically realistic features in neuron models can be computationally relevant and that they might provide feasible targets for an optimization procedure when considering specific tasks, such as supervised learning.

A question worth investigating is whether additional improvements in performance could be gained by adding more components that are biologically relevant. A possible choice, for example, could be to implement more realistic synaptic dynamics. Furthermore, future research could try to test other coding schemes for this network and indicate whether different coding schemes could provide advantages for different tasks.

Also, many other issues remain open, such as, how to connect the results for the spiking neural networks with artificial neural networks and biological networks. Some further issues include the creation of a mathematical theory that can link these different results in neural networks and, also, an investigation of how these results could be applied in a real setting. Finally, future research could focus on developing a training algorithm that takes into account neural diversity.

References

1. Achard, P., Schutter, E.D.: Complex parameter landscape for a complex neuron model. PLoS Comput. Biol. **2**(7) (2006)
2. Bohte, S., Kok, J., Poutre, H.L.: Error backpropagationin temporally encoded networks of spiking neurons. Neurocomputing **48**, 17–37 (2002)
3. Bohte, S.M., Poutre, H.L., Kok, J.N.: Unsupervised clustering with spiking neurons by sparse temporal coding and multi-layer RBF networks. IEEE Trans. Neural Netw. **XX** (2001)
4. Buzsaki, G., Geisler, C., Henze, D.A., Wang, X.J.: Interneuron diversity series: circuit complexity and axon wiring economy of cortical interneurons. Trends Neurosci. **27**(4), 186–193 (2004)
5. Cohen, S., Intrator, N.: A hybrid projection-based and radial basis function architecture: initial values and global optimisation. Pattern Anal. Appl. **5**(2), 113–120 (2002)
6. Fisher, R.A.: The use of multiple measurements in taxonomic problems. Ann. Eugenics 179–188 (1936)
7. Ghosh-Dastidar, S., Adeli, H.: A new supervised learning algorithm for multiple spiking neural networks with application in epilepsy and seizure detection. Neural Netw. **22** (2009)
8. Ianella, N., Back, A.D.: A spiking neural network architecture for nonlinear function approximation. Neural Netw. **14**(2001), 933–939 (2001)
9. Izhikevich, E.: Simple model of spiking neurons. IEEE Trans. Neural Netw. **14**(6) (2003)
10. Izhikevich, E.M.: Which model to use for cortical spiking neurons? IEEE Trans. Neural Netw. **15**(5), 1063–1070 (2004)
11. Izhikevich, E.M.: Dynamical Systems in Neuroscience: The Geometry of Excitability and Bursting. MIT Press, s.l. (2006)
12. Kampakis, S.: Improved Izhikevich neurons for spiking neural networks. J. Soft. Comput. (2011)
13. Kampakis, S.: (under review). ReSpiN: a supervised training algorithm for rebound spiking neurons. J. Soft Comput. (2013)
14. Kampakis, S.: Investigating the computational power of spiking neurons with non-standard behaviors. Neural Netw. **43**, 41–54 (2013)
15. Keren, N., Peled, N., Korngreen, A.: Constraining compartmental models using multiple voltage recordings and genetic algorithms. J. Neurophysiol. 3730–3742 (2006)
16. Klausberger, T., Somogyi, P.: Neuronal diversity and temporal dynamics: the unity of hippocampal circuit operations. Science **321**(5885), 53–57 (2008)
17. Maass, W.: Networks of spiking neurons: the third generation of spiking neural networks. Neural Netw. **10**(9), 1659–1671 (1997)
18. Maul, T.: (in press, accepted manuscript). Early experiments with neural diversity machines (2013)
19. Meftah, B., Lezoray, O., Benyettou, A.: Segmentation and edge detection based on spiking neural network model. Neural Process. Lett. **32**(2), 131–146 (2010)
20. Moore, C.I., Carlen, M., Knoblich, U., Cardin, J.A.: Neocortical interneurons: from diversity, strength. Cell **142**(2), 189–193 (2010)
21. Potjans, W., Morrison, A., Diesmann, M.. A spiking neural network model of an actor-critic learning agent. Neural Comput. **21**(2), 301–339 (2009)
22. Rawlins, G.J.E. (ed.): Foundations of Genetic Algorithms (FOGA 1). Morgan Kaufmann, s.l. (1991)

23. Taylor, A.M., Enoka, R.M.: Optimization of input patterns and neuronal properties to evoke motor neuron synchronization. J. Comput. Neurosci. **16**(2), 139–157 (2004)
24. Tutkun, N.: Parameter estimation in mathematical models using the real coded genetic algorithms. Expert Syst. Appl. **36**(2), 3342–3345 (2009)
25. Valko, M., Marques, N.C., Castellani, M.: Evolutionary Feature Selection for Spiking Neural Network Pattern Classifiers, pp. 181–187. IEEE, Covilha (2005)
26. Van Geit, W., De Schutter, E., Achard, P.: Automated neuron model optimization techniques: a review. Biol. Cybern. **99**, 241–251 (2008)
27. Wade, J.J., McDaid, L.J., Santos, J.A., Sayers, H.M.: SWAT: An Unsupervised SNN Training Algorithm for Classification Problems, pp. 2648–2655. IEEE, Hong Kong (2008)
28. Wang, H.: Improvement of Izhikevich's neuronal and neural network model. IEEE, Wuhan, China (2009)
29. Wu, C.-H., Tzeng, G.-H., Lin, R.-H.: A novel hybrid genetic algorithm for kernel function and parameter optimization in support vector regression. Expert Syst. Appl. **36**(3), 4725–4735 (2009)

Single-Hidden Layer Feedforward Neual Network Training Using Class Geometric Information

Alexandros Iosifidis, Anastasios Tefas and Ioannis Pitas

Abstract Single-hidden Layer Feedforward (SLFN) networks have been proven to be effective in many pattern classification problems. In this chapter, we provide an overview of a relatively new approach for SLFN network training that is based on Extreme Learning. Subsequently, extended versions of the Extreme Learning Machine algorithm that exploit local class data geometric information in the optimization process followed for the calculation of the network output weights are discussed. An experimental study comparing the two approaches on facial image classification problems concludes this chapter.

1 Introduction

Single-hidden Layer Feedforward (SLFN) networks have been proven to be effective in many pattern classification problems, since they are able to approximate any continuous function arbitrary well [1]. Extreme Learning Machine is a relatively new algorithm for Single-hidden Layer Feedforward Neural (SLFN) networks training [2] that leads to fast network training requiring low human supervision. Conventional SLFN network training algorithms require the input weights and the hidden layer biases to be adjusted using a parameter optimization approach, like gradient descend. However, gradient descend-based learning techniques are generally slow and may decrease the network's generalization ability, since they may lead to local minima. Unlike the popular thinking that the network's parameters need to be tuned, in ELM the input weights and the hidden layer biases are randomly assigned. The network output weights are, subsequently, analytically calculated. ELM not only tends to reach the smallest training error, but also the smallest norm of output weights. For

A. Iosifidis · A. Tefas (✉) · I. Pitas
Department of Informatics, Aristotle University of Thessaloniki, Thessaloniki, Greece
e-mail: aiosif@aiia.csd.auth.gr

A. Tefas
e-mail: tefas@aiia.csd.auth.gr

I. Pitas
e-mail: pitas@aiia.csd.auth.gr

© Springer International Publishing Switzerland 2016
J.J. Merelo et al. (eds.), *Computational Intelligence*,
Studies in Computational Intelligence 620, DOI 10.1007/978-3-319-26393-9_21

feedforward networks reaching a small training error, the smaller the norm of weights is, the better generalization performance the networks tend to have [3]. Despite the fact that the determination of the network hidden layer output is a result of randomly assigned weights, it has been shown that SLFN networks trained by using the ELM algorithm have the properties of global approximators [4]. Due to its effectiveness and its fast learning process, the ELM network has been widely adopted in many classification problems, including facial image classification [5–14].

Despite its success in many classification problems, the ability of the original ELM algorithm to calculate the output weights is limited due to the fact that the network hidden layer output matrix is, usually, singular. In order to address this issue, the Effective ELM (EELM) algorithm has been proposed in [15], where the strictly diagonally dominant criterion for nonsingular matrices is exploited, in order to choose proper network input weights and bias values. However, the EELM algorithm has been designed only for a special case of SLFN networks employing Gaussian Radial Basis Functions (RBF) for the input layer neurons. In [9], an optimization-based regularized version of the ELM algorithm (RELM) aiming at both overcoming the full rank assumption for the network hidden layer output matrix and at enhancing the generalization properties of the ELM algorithm has been proposed. RELM has been evaluated on a large number of classification problems providing very satisfactory classification performance.

By using a sufficiently large number of hidden layer neurons, the ELM classification scheme, when approached from a Subspace Learning point of view, can be considered as a learning process formed by two processing steps [16]. The first step corresponds to a mapping process of the input space to a high-dimensional feature space preserving some properties of interest for the training data. In the second step, an optimization scheme is employed for the determination of a linear projection of the high-dimensional data to a low-dimensional feature space determined by the network target vectors, where classification is performed by a linear classifier. Based on this observation, the RELM algorithm has been extended in order to exploit subspace learning criteria in its optimization process [16, 17]. Specifically, it has been shown that the incorporation of the within-class and total scatter of the training data (represented in the feature space determined by the network hidden layer outputs) in the optimization process followed for the calculation of the network output weights enhances the network classification performance.

In this Chapter, we provide an overview of the ELM algorithm for SLFN network training [2, 9]. Extensions of the ELM algorithm exploiting subspace learning criteria on its optimization process are also described. Subsequently, an extension of the ELM algorithm which exploits local class information in the ELM optimization problem is described in detail. The so-called Local Class Variance ELM (LCVELM) algorithm aims at minimizing both the network output weights norm and the within class variance of the training data in the ELM space, expressed by employing locality constraints. An experimental study comparing the performance of ELM [2], RELM [9], MCVELM [16] and LCVELM [18] networks is facial image classification problems is finally provided.

The chapter is structured as follows. In Sect. 2 we briefly describe the ELM algorithms. RELM is described in Sect. 2. In Sect. 4 ELM algorithms exploiting dispersion information in their optimization problem are described. In Sect. 5, we describe the LCVELM algorithm exploiting intrinsic graph structures for SLFN network training. Section 7 presents an experimental study evaluating the performance of ELM-based classification in facial image classification problems. Finally, conclusions are drawn in Sect. 8.

2 Extreme Learning Machine

ELM network has been proposed for SLFN network-based classification [2]. Let us denote by $\{\mathbf{x}_i, c_i\}$, $i = 1, \dots, N$ a set of N vectors $\mathbf{x}_i \in \mathbb{R}^D$ and the corresponding class labels $c_i \in \{1, \dots, C\}$. We employ $\{\mathbf{x}_i, c_i\}$, $i = 1, \dots, N$ in order to train a SLFN network. Such a network consists of D input (equal to the dimensionality of $\mathbf{x}_i$), L hidden and C output (equal to the number of classes involved in the classification problem) neurons. The number of hidden layer neurons is usually selected to be much greater than the number of classes [9, 16], i.e., $L \gg C$.

The network target vectors $\mathbf{t}_i = [t_{i1}, \dots, t_{iC}]^T$, each corresponding to a training vector $\mathbf{x}_i$, are set to $t_{ik} = 1$ for vectors belonging to class k, i.e., when $c_i = k$, and to $t_{ik} = -1$ when $c_i \neq k$. In ELMs, the network input weights $\mathbf{W}_{in} \in \mathbb{R}^{D \times L}$ and the hidden layer bias values $\mathbf{b} \in \mathbb{R}^L$ are randomly assigned, while the network output weights $\mathbf{W}_{out} \in \mathbb{R}^{L \times C}$ are analytically calculated. Let us denote by $\mathbf{v}_j$ the j-th column of $\mathbf{W}_{in}$, by $\mathbf{w}_k$ the k-th row of $\mathbf{W}_{out}$ and by w_{kj} the j-th element of $\mathbf{w}_k$. Given activation function for the network hidden layer $\Phi(\cdot)$ and by using a linear activation function for the network output layer, the output $\mathbf{o}_i = [o_1, \dots, o_C]^T$ of the network corresponding to $\mathbf{x}_i$ is calculated by:

$$o_{ik} = \sum_{j=1}^{L} w_{kj} \, \Phi(\mathbf{v}_j, b_j, \mathbf{x}_i), \; k = 1, \dots, C. \tag{1}$$

It has been shown [4, 9, 19] that, almost any nonlinear piecewise continuous activation function $\Phi(\cdot)$ can be used for the calculation of the network hidden layer outputs, like the sigmoid, sine, Gaussian, hard-limiting and Radial Basis Function (RBF), Fourier series, etc. The most widely adopted choice is the sigmoid function, defined by:

$$\Phi(\mathbf{v}_j, b_j, \mathbf{x}_i) = \frac{1}{1 + e^{-(\mathbf{v}_j^T \mathbf{x}_i + b_j)}}. \tag{2}$$

By storing the network hidden layer outputs corresponding to the training vectors $\mathbf{x}_i$, $i = 1, \dots, N$ in a matrix Φ:

$$\boldsymbol{\Phi} = \begin{bmatrix} \Phi(\mathbf{v}_1, b_1, \mathbf{x}_1) & \cdots & \Phi(\mathbf{v}_1, b_1, \mathbf{x}_N) \\ \cdots & \ddots & \cdots \\ \Phi(\mathbf{v}_L, b_L, \mathbf{x}_1) & \cdots & \Phi(\mathbf{v}_L, b_L, \mathbf{x}_N) \end{bmatrix}, \tag{3}$$

Equation (1) can be expressed in a matrix form as:

$$\mathbf{O} = \mathbf{W}_{out}^T \boldsymbol{\Phi}. \tag{4}$$

ELM [2] assumes that the predicted network outputs are equal to the network targets, i.e., $\mathbf{o}_i = \mathbf{t}_i$, $i = 1, \ldots, N$, $\mathbf{W}_{out}$ can be analytically calculated by solving the following set of equation:

$$\mathbf{W}_{out}^T \boldsymbol{\Phi} = \mathbf{T} \tag{5}$$

and are given by:

$$\mathbf{W}_{out} = \boldsymbol{\Phi}^\dagger \mathbf{T}^T, \tag{6}$$

where $\boldsymbol{\Phi}^\dagger = \left(\boldsymbol{\Phi}\boldsymbol{\Phi}^T\right)^{-1}\boldsymbol{\Phi}$ is the Moore-Penrose generalized pseudo-inverse of $\boldsymbol{\Phi}^T$ and $\mathbf{T} = [\mathbf{t}_1, \ldots, \mathbf{t}_N]$ is a matrix containing the network target vectors.

3 Regularized Extreme Learning Machine

The ELM algorithm assumes zero training error. In cases where the training data contain outliers, this assumption may reduce its potential in generalization. In addition, since the dimensionality of the ELM space is usually high, i.e., in some cases $L > N$, the matrix $\mathbf{B} = \boldsymbol{\Phi}\boldsymbol{\Phi}^T$ is singular and, thus, the adoption of (6) for the calculation of the network output weights is inappropriate. By allowing small training errors and trying to minimize the norm of the network output weights, $\mathbf{W}_{out}$ can be calculated by minimizing [9]:

$$\mathcal{J}_{RELM} = \frac{1}{2}\|\mathbf{W}_{out}\|_F^2 + \frac{c}{2}\sum_{i=1}^{N}\|\boldsymbol{\xi}_i\|_2^2, \tag{7}$$

$$\mathbf{W}_{out}^T \boldsymbol{\phi}_i = \mathbf{t}_i - \boldsymbol{\xi}_i, \quad i = 1, \ldots, N, \tag{8}$$

where $\boldsymbol{\xi}_i \in \mathbb{R}^C$ is the error vector corresponding to $\mathbf{x}_i$ and c is a parameter denoting the importance of the training error in the optimization problem. $\boldsymbol{\phi}_i$ is the i-th column of $\boldsymbol{\Phi}$, i.e., the hidden layer output corresponding $\mathbf{x}_i$. That is, $\boldsymbol{\phi}_i$ is the representation of $\mathbf{x}_i$ in $\mathbb{R}^L$. By substituting (8) in $\mathcal{J}_{RELM}$ (7) and determining the saddle point of $\mathcal{J}_{RELM}$, $\mathbf{W}_{out}$ is given by:

$$\mathbf{W}_{out} = \left(\boldsymbol{\Phi}\boldsymbol{\Phi}^T + \frac{1}{c}\mathbf{I}\right)^{-1}\boldsymbol{\Phi}\mathbf{T}^T \tag{9}$$

or

$$\mathbf{W}_{out} = \boldsymbol{\Phi}\left(\boldsymbol{\Phi}^T\boldsymbol{\Phi} + \frac{1}{c}\mathbf{I}\right)^{-1}\mathbf{T}^T \tag{10}$$

The adoption of (9) for $\mathbf{W}_{out}$ calculation, instead of (6), has the advantage that the matrices $\mathbf{B} = \left(\boldsymbol{\Phi}\boldsymbol{\Phi}^T + \frac{1}{c}\mathbf{I}\right)$ and $\tilde{\mathbf{B}} = \left(\boldsymbol{\Phi}^T\boldsymbol{\Phi} + \frac{1}{c}\mathbf{I}\right)$ are nonsingular, for $c > 0$.

4 Extreme Learning Machine Exploiting Dispersion Criteria

By allowing small training errors and trying to minimize both the norm of the network output weights and the within-class variance of the training vectors in the feature space determined by the network outputs, $\mathbf{W}_{out}$ can be calculated by minimizing [16]:

$$\mathcal{J}_{MCVELM} = \|\mathbf{S}_w^{\frac{1}{2}}\mathbf{W}_{out}\|_F^2 + \lambda \sum_{i=1}^{N} \|\boldsymbol{\xi}_i\|_2^2, \tag{11}$$

$$\mathbf{W}_{out}^T\boldsymbol{\phi}_i = \mathbf{t}_i - \boldsymbol{\xi}_i, \quad i = 1,\dots,N, \tag{12}$$

where $\mathbf{S}_w$ is the within-class scatter matrix used in Linear Discriminant Analysis (LDA) [20] describing the variance of the training classes in the ELM space and is defined by:

$$\mathbf{S}_w = \sum_{j=1}^{C} \sum_{i,c_i=j} \frac{1}{N_j}(\boldsymbol{\phi}_i - \boldsymbol{\mu}_j)(\boldsymbol{\phi}_i - \boldsymbol{\mu}_j)^T. \tag{13}$$

In (13), N_j is the number of training vectors belonging to class j and $\boldsymbol{\mu}_j = \frac{1}{N_j}\sum_{i,c_i=j}\boldsymbol{\phi}_i$ is the mean vector of class j. By calculating the within-class scatter matrix in the ELM space $\mathbb{R}^L$, rather than in the input space $\mathbb{R}^D$, nonlinear relationships between training vectors forming the various classes can be better described. By substituting (12) in $\mathcal{J}_{MCVELM}$ and determining the saddle point of $\mathcal{J}_{MCVELM}$, $\mathbf{W}_{out}$ is given by:

$$\mathbf{W}_{out} = \left(\boldsymbol{\phi}\boldsymbol{\phi}^T + \frac{1}{c}\mathbf{S}_w\right)^{-1}\boldsymbol{\phi}\mathbf{T}^T. \tag{14}$$

Since the matrix $\mathbf{B} = \left(\boldsymbol{\phi}\boldsymbol{\phi}^T + \frac{1}{c}\mathbf{S}_w\right)$ is not always nonsingular, an additional dimensionality reduction processing step performed by applying Principal Component Analysis [20] on $\boldsymbol{\phi}$ has been proposed in [16]. Another variants that exploiting the total scatter matrix of the entire training set and the within-class variance of multimodal classes have been proposed in [16, 17], respectively.

5 Extreme Learning Machine Exploiting Intrinsic Graph Structures

In this Section, we describe the an extension of the ELM algorithm exploiting local class dispersion criteria [18]. Similar to the ELM variance described in Sect. 2, the Local Class Variance ELM (LCVELM) algorithm exploits randomly assigned network input weights $\mathbf{W}_{in}$ and bias values $\mathbf{b}$, in order to perform a nonlinear mapping of the data in the (usually high-dimensional) ELM space $\mathbb{R}^L$. After the network hidden layer outputs calculation, we assume that the data representations in the ELM space $\boldsymbol{\phi}_i$, $i = 1, \ldots, N$ are embedded in a graph $\mathcal{G} = \{\mathcal{V}, \mathcal{E}, \mathbf{W}\}$, where $\mathcal{V}$ denotes the graph vertex set, i.e., $\mathcal{V} = \{\boldsymbol{\phi}_i\}_{i=1}^N$, $\mathcal{E}$ is the set of edges connecting $\boldsymbol{\phi}_i$, and $\mathbf{W} \in \mathbb{R}^{N \times N}$ is the matrix containing the weight values of the edge connections. Let us define a similarity measure $s(\cdot, \cdot)$ that will be used in order to measure the similarity between two vectors [21]. That is, $s_{ij} = s(\boldsymbol{\phi}_i, \boldsymbol{\phi}_j)$ is a value denoting the similarity between $\boldsymbol{\phi}_i$ and $\boldsymbol{\phi}_j$. $s(\cdot, \cdot)$ may be any similarity measure providing non-negative values (usually $0 \le s_{ij} \le 1$). The most widely adopted choice is the heat kernel (also known as diffusion kernel) [22], defined by:

$$s(\boldsymbol{\phi}_i, \boldsymbol{\phi}_j) = \exp\left(-\frac{\|\boldsymbol{\phi}_i - \boldsymbol{\phi}_j\|_2^2}{2\sigma^2}\right), \tag{15}$$

where $\|\cdot\|_2$ denotes the (squared) l_2 norm of a vector and σ is a parameter used in order to scale the Euclidean distance between $\boldsymbol{\phi}_i$ and $\boldsymbol{\phi}_j$.

In order to express the local intra-class relationships of the training data in the ELM space, we exploit the following two choices for the determination of the weight matrix $\mathbf{W}$:

$$W_{ij}^{(1)} = \begin{cases} 1 & \text{if } c_i = c_j \text{ and } j \in \mathcal{N}_i, \\ 0, & \text{otherwise,} \end{cases}$$

or

$$W_{ij}^{(2)} = \begin{cases} s_{ij} & \text{if } c_i = c_j \text{ and } j \in \mathcal{N}_i, \\ 0, & \text{otherwise.} \end{cases}$$

In the above, $\mathcal{N}_i$ denotes the neighborhood of $\boldsymbol{\phi}_i$ (we have employed 5-NN graphs in all our experiments). $\mathbf{W}^{(1)}$ has been successfully exploited for discriminant subspace learning in Marginal Discriminant Analysis (MDA) [21], while $\mathbf{W}^{(2)}$ can be considered to be modification of $\mathbf{W}^{(1)}$, exploiting geometric information of the class data. A similar weight matrix has also been exploited in Local Fisher Discriminant Analysis (LFDA) [23]. In both MDA and LFDA cases, it has been shown that by exploiting local class information enhanced class discrimination can be achieved, when compared to the standard LDA approach exploiting global class information, by using (13).

After the calculation of the graph weight matrix $\mathbf{W}$, the graph Laplacian matrix $\mathbf{L}^{N \times N}$ is given by [24]:

$$\mathbf{L} = \mathbf{D} - \mathbf{W}, \tag{16}$$

where $\mathbf{D}$ is a diagonal matrix with elements $D_{ii} = \sum_{j=1}^{N} W_{ij}$.

By exploiting $\mathbf{L}$, the network output weights $\mathbf{W}_{out}$ of the LCVELM network can be calculated by minimizing:

$$\mathcal{J}_{LCVELM} = \frac{1}{2}\|\mathbf{W}_{out}\|_F^2 + \frac{c}{2}\sum_{i=1}^{N}\|\xi_i\|_2^2 + \frac{\lambda}{2}tr\left(\mathbf{W}_{out}^T(\boldsymbol{\phi}\mathbf{L}\boldsymbol{\phi}^T)\mathbf{W}_{out}\right), \tag{17}$$

$$\mathbf{W}_{out}^T\boldsymbol{\phi}_i = \mathbf{t}_i - \xi_i, \quad i = 1,\ldots,N, \tag{18}$$

where $tr(\cdot)$ is the trace operator. By substituting the constraints (18) in $\mathcal{J}_{LCVELM}$ and determining the saddle point of $\mathcal{J}_{LCVELM}$, the network output weights $\mathbf{W}_{out}$ are given by:

$$\mathbf{W}_{out} = \left(\boldsymbol{\phi}\left(\mathbf{I} + \frac{\lambda}{c}\mathbf{L}\right)\boldsymbol{\phi}^T + \frac{1}{c}\mathbf{I}\right)^{-1}\boldsymbol{\phi}\mathbf{T}^T. \tag{19}$$

Similar to (9), the calculation of the network output weights by employing (19) has the advantage that the matrix $\mathbf{B} = \left(\boldsymbol{\phi}\left(\mathbf{I} + \frac{\lambda}{c}\mathbf{L}\right)\boldsymbol{\phi}^T + \frac{1}{c}\mathbf{I}\right)$ is nonsingular, for $c > 0$.

In addition, the calculation of the graph similarity values $s(\cdot,\cdot)$ in the ELM space $\mathbb{R}^L$, rather than the input space $\mathbb{R}^D$ has the advantage that nonlinear relationships between the training vectors forming the various classes can be better expressed.

6 Data Classification (Test Phase)

After the determination of the network output weights $\mathbf{W}_{out}$ by using (9), (10), (14) or (19), a test vector $\mathbf{x}_t \in \mathbb{R}^D$ can be introduced to the trained network and the corresponding network output is obtained:

$$\mathbf{o}_t = \mathbf{W}_{out}^T\boldsymbol{\phi}_t, \tag{20}$$

where $\boldsymbol{\phi}_t$ denotes the network hidden layer output for $\mathbf{x}_t$. $\mathbf{x}_t$ is finally classified to the class corresponding to the maximal network output:

$$c_t = \underset{k}{argmax}\ o_{tk}, \quad k = 1,\ldots,C. \tag{21}$$

7 Experimental Study

In this section, we present experiments conducted in order to evaluate the performance of the ELM algorithms described in Sects. 2, 4 and 5. We have employed six publicly available datasets to this end. These are: the ORL, AR and Extended YALE-B (face recognition) and the COHN-KANADE, BU and JAFFE (facial expression recognition). A brief description of the datasets is provided in the following sections. Experimental results are provided in Sect. 7.3.

In all the presented experiments we compare the performance of the LCVELM algorithm [18] with that of ELM [2], RELM [9] and MCVELM [16] algorithms. The number of hidden layer neurons has been set equal to $L = 1000$ for all the ELM variants, a value that has been shown to provide satisfactory performance in many classification problems [9, 16]. For fair comparison, in all the experiments, we make sure that the the same ELM space is used in all the ELM variants. That is, we first map the training data in the ELM space and, subsequently, calculate the network output weights according to each ELM algorithm. Regarding the optimal values of the regularization parameters c, λ used in the ELM-based classification schemes, they have been determined by following a grid search strategy. That is, for each classifier, multiple experiments have been performed by employing different parameter values ($c = 10^r$, $r = -3, \ldots, 3$ and $\lambda = 10^p$, $p = -3, \ldots, 3$) and the best performance is reported.

7.1 Face Recognition Datasets

7.1.1 The ORL Dataset

It consists of 400 facial images depicting 40 persons (10 images each) [25]. The images were captured at different times and with different conditions, in terms of lighting, facial expressions (smiling/not smiling) and facial details (open/closed eyes, with/without glasses). Facial images were taken in frontal position with a tolerance for face rotation and tilting up to $20°$. Example images of the dataset are illustrated in Fig. 1.

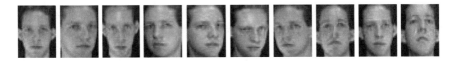

Fig. 1 Facial images depicting a person from the ORL dataset

Fig. 2 Facial images depicting a person from the AR dataset

Fig. 3 Facial images depicting a person from the Extended YALE-B dataset

7.1.2 The AR Dataset

It consists of over 4000 facial images depicting 70 male and 56 female faces [26]. In our experiments we have used the preprocessed (cropped) facial images provided by the database, depicting 100 persons (50 males and 50 females) having a frontal facial pose, performing several expressions (anger, smiling and screaming), in different illumination conditions (left and/or right light) and with some occlusions (sun glasses and scarf). Each person was recorded in two sessions, separated by two weeks. Example images of the dataset are illustrated in Fig. 2.

7.1.3 The Extended YALE-B Dataset

It consists of facial images depicting 38 persons in 9 poses, under 64 illumination conditions [27]. In our experiments we have used the frontal cropped images provided by the database. Example images of the dataset are illustrated in Fig. 3.

7.2 Facial Expression Recognition Datasets

7.2.1 The COHN-KANADE Dataset

It consists of facial images depicting 210 persons of age between 18 and 50 (69 % female, 31 % male, 81 % Euro-American, 13 % Afro-American and 6 % other groups) [28]. We have randomly selected 35 images for each facial expression, i.e., anger, disgust, fear, happiness, sadness, surprise and neutral. Example images of the dataset are illustrated in Fig. 4.

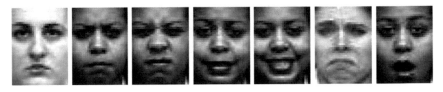

Fig. 4 Facial images from the COHN-KANADE dataset. From *left* to *right*: neutral, anger, disgust, fear, happy, sad and surprise

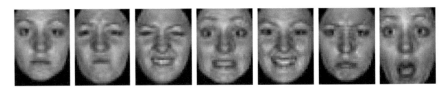

Fig. 5 Facial images depicting a person from the BU dataset. From *left* to *right*: neutral, anger, disgust, fear, happy, sad and surprise

7.2.2 The BU Dataset

It consists of facial images depicting over 100 persons (60 % female and 40 % male) with a variety of ethnic/racial background, including White, Black, East-Asian, Middle-east Asian, Hispanic Latino and others [29]. All expressions, except the neutral one, are expressed at four intensity levels. In our experiments, we have employed the images depicting the most expressive intensity of each facial expression. Example images of the dataset are illustrated in Fig. 5.

7.2.3 The JAFFE Dataset

It consists of 210 facial images depicting 10 Japanese female persons [30]. Each of the persons is depicted in 3 images for each expression. Example images of the dataset are illustrated in Fig. 6.

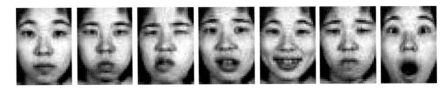

Fig. 6 Facial images depicting a person from the JAFFE dataset. From *left* to *right*: neutral, anger, disgust, fear, happy, sad and surprise

7.3 Results

In our first set of experiments, we have applied the competing algorithms on the face recognition datasets. Since there is not a widely adopted experimental protocol for these datasets, we randomly partition the datasets in training and test sets as follows: we randomly select a subset of the facial images depicting each of the persons in each dataset in order to form the training set and we keep the remaining facial images for evaluation. We create five such dataset partitions, each corresponding to a different training set cardinality. Experimental results obtained by applying the competing algorithms are illustrated in Tables 1, 2 and 3 for the ORL, AR and the Extended Yale-B datasets, respectively. As can be seen in these tables, the incorporation of local class information in the optimization problem used for the determination of the network output weights, generally increases the performance of the ELM network. In all the cases the best performance is achieved by one of the two LCVELM variants. By comparing the two LCVELM algorithms, it can be seen that the one exploiting the graph weight matrix used in MDA generally outperforms the remaining choice.

Table 1 Classification rates on the ORL dataset

	ELM (%)	RELM (%)	MCVELM (%)	LCVELM (1) (%)	LCVELM (2) (%)
10 %	30.78	40.65	41.01	**41.26**	**41.22**
20 %	20.67	39.76	**41.81**	**41.81**	**41.81**
30 %	38.17	52.11	55	**55.78**	**55.78**
40 %	38.31	53	57	**57.19**	**57.13**
50 %	47	77.62	75.54	**77.69**	**77.77**

Table 2 Classification rates on the AR dataset

	ELM (%)	RELM (%)	MCVELM (%)	LCVELM (1) (%)	LCVELM (2) (%)
10 %	66.47	67.79	68.87	**69.19**	**69.15**
20 %	70.49	80.24	80.91	80.86	**80.96**
30 %	65.26	82.98	81.81	**83.27**	83.1
40 %	75.33	91.9	92.94	**93.01**	**93.01**
50 %	80.33	94.16	94.65	**94.9**	**94.9**

Table 3 Classification rates on the YALE-B dataset

	ELM (%)	RELM (%)	MCVELM (%)	LCVELM (1) (%)	LCVELM (2) (%)
10 %	69.17	**72.22 %**	**72.22 %**	**72.22 %**	**72.22%**
20 %	83.44	84.38	84.38	**85**	84.38
30 %	82.86	85.36	85.36	**88.21**	85.36
40 %	90	92.08	92.08	**92.5**	92.08
50 %	91	93.5	**94.5**	**94.5**	**94.5**

Table 4 Classification rates on the facial expression recognition dataset

	ELM (%)	RELM (%)	MCVELM (%)	LCVELM (1) (%)	LCVELM (2) (%)
COHN-KANADE	49.8	79.59	80	**80.41**	80
BU	65	71, 57	71, 57	**72**	**72, 86**
JAFFE	47.62	58.57	59.05	**60**	**59.52**

In our second set of experiments, we have applied the competing algorithms on the facial expression recognition datasets. Since there is not a widely adopted experimental protocol for these datasets too, we apply the five-fold crossvalidation procedure [31] by employing the facial expression labels. That is, we randomly split the facial images depicting the same expression in five sets and we use five splits of all the expressions for training and the remaining splits for evaluation. This process is performed five times, one for each evaluation split. Experimental results obtained by applying the competing algorithms are illustrated in Table 4. As can be seen in this Table, the LCVELM algorithms outperform the remaining choices in all the cases.

Overall, enhanced facial image classification performance can be achieved by exploiting class data geometric information in the ELM optimization process.

8 Conclusion

In this chapter an overview of Extreme Learning Machine-based Single-hidden Layer Feedforward Neural networks training has been provided. Extended versions of the ELM algorithm that exploit (local) class data geometric information in the optimization process followed for the calculation of the network output weights have been also described. An experimental study comparing the two approaches on facial image classification problems has been finally presented, showing that the exploitation of class data geometric information in the ELM optimization process enhances the performance of the ELM network.

Acknowledgments The research leading to these results has received funding from the European Union Seventh Framework Programme (FP7/2007–2013) under grant agreement number 316564 (IMPART).

References

1. Scarselli, F., Tsoi, A.C.: Universal approximation using feedforward neural networks: a survey of some existing methods and some new results. Neural Netw. **11**, 15–37 (1998)
2. Huang, G. B., Zhu, Q. Y., Siew, C. K.: Extreme learning machine: a new learning scheme of feedforward neural networks. IEEE Int. Joint Conf. Neural Netw. (2004)

3. Bartlett, P.L.: The sample complexity of pattern classification with neural networks: the size of the weights is more important than the size of the network. IEEE Trans. Inf. Theory **44**(2), 525–536 (1998)
4. Huang, G.B., Chen, L., Siew, C.K.: Universal approximation using incremental constructive feedforward networks with random hidden nodes. IEEE Trans. Neural Netw. **17**(4), 879–892 (2006)
5. Zong, W., Huang, G.B.: Face recognition based on extreme learning machine. Neurocomputing **74**(16), 2541–2551 (2011)
6. Rong, H. J., Huang, G. B., Ong, S. Y.: Extreme learning machine for multi-categories classification applications. IEEE Int. Joint Conf. Neural Netw. (2008)
7. Lan, Y., Soh, Y. C., Huang, G. B.: Extreme learning machine based bacterial protein subcellular localization prediction. IEEE Int. Joint Conf. Neural Netw. (2008)
8. Helmy, T., Sashee, Z.: Multi-category bioinformatics dataset classification using extreme learning machine. IEEE Evol. Comput. (2009)
9. Huang, G.B., Zhou, H., DIng, X., Zhang, R.: Extreme learning machine for regression and multiclass classification. IEEE Trans. Syst. Man Cybern. Part B Cybern. **42**(2), 513–529 (2012)
10. Iosifidis, A., Tefas, A., Pitas, I.: Person identification from actions based on artificial neural networks. IEEE Symp. Ser. Comput. Intell. (2013)
11. Iosifidis, A., Tefas, A., Pitas, I.: Dynamic action recognition based on dynemes and extreme learning machine. Pattern Recogn. Lett. **34**, 1890–1989 (2013)
12. Iosifidis, A., Tefas, A., Pitas, I.: Active Classification for Human Action Recognition. IEEE Int. Conf. Image Process. (2013)
13. Iosifidis, A., Tefas, A., Pitas, I.: Human action recognition based on bag of features and multi-view neural networks. IEEE Int. Conf. Image Process. (2014)
14. Iosifidis, A., Tefas, A., Pitas, I.: Semi-supervised classification of human actions based on Neural Networks. IEEE Int. Conf. Pattern Recognit. (2014)
15. Wang, Y., Cao, F., Yuan, Y.: A study on effectiveness of extreme learning machine. Neurocomputing **74**(16), 2483–2490 (2011)
16. Iosifidis, A., Tefas, A., Pitas, I.: Minimum class variance extreme learning machine for human action recognition. IEEE Trans. Circuits Syst. Video Technol. **23**(11), 1968–1979 (2013)
17. Iosifidis, A., Tefas, A., Pitas, I.: Minimum variance extreme learning machine for human action recognition. IEEE Int. Conf. Acoust. Speech Signal Process. (2014)
18. Iosifidis, A., Tefas, A., Pitas, I.: Exploiting local class information in extreme learning machine. Int. Joint Conf. Comput. Intell. (2014)
19. Huang, G., Chen, L.: Convex incremental extreme learning machine. Neurocomputing **70**(16), 3056–3062 (2008)
20. Duda, R. O., Hart, P.E., Stork, D. G.: Pattern Classification, 2nd ed. Wiley-Interscience (2000)
21. Yan, S., Xu, D., Zhang, B., Zhang, H.J., Yang, Q., Lin, S.: Graph embedding and extensions: a general framework for dimensionality reduction. IEEE Trans. Pattern Anal. Mach. Intell. **29**(1), 40–50 (2007)
22. Kondor, R. I., Lafferty, J. D.: Diffusion kernels on graphs and other discrete input spaces. Int. Conf. Mach. Learn. (2002)
23. Sugiyama, M.: Dimensionality reduction of multimodal labeled data by local fisher discriminant analysis. J. Mach. Learn. Res. **8**, 1027–1061 (2007)
24. Belkin, M., Niyogi, P., Sindhwani, V.: Manifold regularization: a geometric framework for learning from labeled and unlabeled examples. J. Mach. Learn. Res. **7**, 2399–2434 (2007)
25. Samaria, F., Harter, A.: Parameterisation of a stochastic model for human face identification. IEEE Workshop Appl. Comput. Vis. (1994)
26. Martinez, A., Kak, A.: PCA versus LDA. IEEE Trans. Pattern Anal. Mach. Intell. **23**(2), 228–233 (2007)
27. Lee, K.C., Ho, J., Kriegman, D.: Acquiriing linear subspaces for face recognition under varialbe lighting. IEEE Trans. Pattern Anal. Mach. Intell. **27**(5), 684–698 (2007)
28. Kanade, T., Tian, Y., Cohn, J.: Comprehensive database for facial expression analysis. IEEE Int. Conf. Autom. Face Gesture Recognit. (2000)

29. Yin, L., Wei, X., Sun, Y., Wang, J., Rosato, M.: A 3D facial expression database for facial behavior research. IEEE Int. Conf. Autom. Face Gesture Recogn. (2006)
30. Lyons, M., Akamatsu, S., Kamachi, M., Gyoba, J.: Coding facial expressions with Gabor wavelets. IEEE Int. Conf. Autom. Face Gesture Recogn. (1998)
31. Devijver, R., Kittler, J.: Pattern Recognition: A Statistical Approach. Prentice-Hall (1982)

Mixtures of Product Components Versus Mixtures of Dependence Trees

Jiří Grim and Pavel Pudil

Abstract Mixtures of product components assume independence of variables given the index of the component. They can be efficiently estimated from data by means of EM algorithm and have some other useful properties. On the other hand, by considering mixtures of dependence trees, we can explicitly describe the statistical relationship between pairs of variables at the level of individual components and therefore approximation power of the resulting mixture may essentially increase. However, we have found in application to classification of numerals that both models perform comparably and the contribution of dependence-tree structures to the log-likelihood criterion decreases in the course of EM iterations. Thus the optimal estimate of dependence-tree mixture tends to reduce to a simple product mixture model.

Keywords Product mixtures · Mixtures of dependence trees · EM algorithm · NIST numerals

1 Introduction

In the last decades there is an increasing need of estimating multivariate and multimodal probability distributions from large data sets. Such databases are usually produced by information technologies in various areas like medicine, image processing, monitoring systems, communication networks and others. A typical feature of the arising "technical" data is a high dimensionality and a large number of measurements. The unknown underlying probability distributions or density functions are nearly always multimodal and cannot be assumed in a simple parametric form. For this reason, one of the most efficient possibilities is to approximate the unknown

J. Grim (✉)
Institute of Information Theory and Automation, Czech Academy of Sciences,
Prague, Czech Republic
e-mail: grim@utia.cas.cz

P. Pudil
Faculty of Management, Prague University of Economics Jindřichův Hradec,
Prague, Czech Republic

© Springer International Publishing Switzerland 2016
J.J. Merelo et al. (eds.), *Computational Intelligence*,
Studies in Computational Intelligence 620, DOI 10.1007/978-3-319-26393-9_22

multidimensional probability distributions by finite mixtures and, especially, by mixtures of components defined as products of univariate distributions [6, 8, 13, 17, 18, 23]. In case of discrete variables the product mixtures are universal approximators since any discrete distribution can be expressed as a product mixture [12]. Similarly, the Gaussian product mixtures approach the universality of non-parametric Parzen estimates with the increasing number of components. In addition, the mixtures of product components have some specific advantages, like easily available marginals and conditional distributions, a direct applicability to incomplete data and the possibility of structural optimization of multilayer probabilistic neural networks (PNN) [9, 10, 19, 21, 22].

Nevertheless, the simplicity of product components may become restrictive in some cases and therefore it could be advantageous to consider more complex mixture models. A natural choice is to use dependence-tree distributions [3] as components. By using the concept of dependence tree we can explicitly describe the statistical relationships between pairs of variables at the level of individual components and therefore the approximation "power" of the resulting mixture model should increase. We have shown [7] that mixtures of dependence-tree distributions can be optimized by EM algorithm in full generality. In the domain of probabilistic neural networks the mixtures of dependence trees could help to explain the role of dendritic branching in biological neurons [20].

In this paper we describe first the product mixture model (Sects. 2 and 3). In Sect. 4 we recall the concept of dependence-tree distribution in the framework of finite mixtures. In Sect. 5 we discuss different aspects of the two types of mixtures in a computational experiment—in application to recognition of numerals. The results are summarized in the conclusion.

1.1 Estimating Mixtures

Considering distribution mixtures, we approximate the unknown probability distributions by a linear combination of component distributions

$$P(x|w, \Theta) = \sum_{m \in \mathcal{M}} w_m F(x|\theta_m), \quad \mathcal{M} = \{1, \dots, M\}, \tag{1}$$

$$w = (w_1, w_2, \dots, w_M), \quad \theta_m = \{\theta_{m1}, \theta_{m2}, \dots, \theta_{mN}\},$$

where $x \in X$ are discrete or real data vectors, w is the vector of probabilistic weights, $\mathcal{M}$ is the component index set and $F(x|\theta_m)$ are the component distributions with the parameters θ_m.

Since the late 1960s the standard way to estimate mixtures is to use the EM algorithm [4–6, 24–26, 36, 38]. Formally, given a finite set S of independent observations of the underlying N-dimensional random vector

$$S = \{x^{(1)}, x^{(2)}, \dots \}, \quad x = (x_1, x_2, \dots, x_N) \in X, \tag{2}$$

we maximize the log-likelihood function

$$L(w, \Theta) = \frac{1}{|S|} \sum_{x \in S} \log \left[\sum_{m \in \mathcal{M}} w_m F(x|\theta_m) \right] \tag{3}$$

by means of the following EM iteration equations ($m \in \mathcal{M}, n \in \mathcal{N}, x \in S$):

$$q(m|x) = \frac{w_m F(x|\theta_m)}{\sum_{j \in \mathcal{M}} w_j F(x|\theta_j)}, \quad w'_m = \frac{1}{|S|} \sum_{x \in S} q(m|x), \tag{4}$$

$$Q_m(\theta_m) = \sum_{x \in S} \frac{q(m|x)}{\sum_{y \in S} q(m|y)} \log F(x|\theta_m), \quad \theta'_m = \arg\max_{\theta_m} \left\{ Q_m(\theta_m) \right\}. \tag{5}$$

Here the apostrophe denotes the new parameter values in each iteration. One can easily verify (cf. [6]) that the general iteration scheme (4) and (5) produces nondecreasing sequence of values of the maximized criterion (3). In view of the implicit relation (5) any new application of EM algorithm is reduced to the explicit solution of Eq. (5) for fixed conditional weights $q(m|x)$.

Considering product mixtures, we assume the product components

$$F(x|\theta_m) = \prod_{n \in \mathcal{N}} f_n(x_n|\theta_{mn}), \quad m \in \mathcal{M} \tag{6}$$

and therefore Eq. (5) can be specified for variables independently ($n \in \mathcal{N}$):

$$Q_{mn}(\theta_{mn}) = \sum_{x \in S} \frac{q(m|x)}{w'_m |S|} \log f_n(x_n|\theta_{mn}), \quad \theta'_{mn} = \arg\max_{\theta_{mn}} \left\{ Q_{mn}(\theta_{mn}) \right\}. \tag{7}$$

The mixtures of product components have some specific advantages as approximation tools. Recall that any marginal distribution of product mixtures is directly available by omitting superfluous terms in product components. Thus, in case of prediction, we can easily compute arbitrary conditional densities and for the same reason product mixtures can be estimated directly from incomplete data without estimating the missing values [18]. Product mixtures support a subspace modification for the sake of component-specific feature selection [10] and can be used for sequential pattern recognition by maximum conditional informativity [14]. Moreover, the product components simplify the EM iterations, support sequential version [11] and increase the numerical stability of EM algorithm.

2 Multivariate Bernoulli Mixtures

In case of binary data $x_n \in \{0,1\}$ the product mixture model (1), (6) is known as multivariate Bernoulli mixture based on the univariate distributions

$$f_n(x_n|\theta_{mn}) = (\theta_{mn})^{x_n}(1 - \theta_{mn})^{1-x_n}, \quad 0 \le \theta_{mn} \le 1, \tag{8}$$

$$F(x|\theta_m) = \prod_{n \in \mathcal{N}} (\theta_{mn})^{x_n}(1 - \theta_{mn})^{1-x_n}, \quad m \in \mathcal{M}. \tag{9}$$

The conditional expectation criterion $Q_{mn}(\theta_{mn})$ can be expressed in the form

$$Q_{mn}(\theta_{mn}) = \sum_{\xi \in \mathcal{X}_n} \left(\sum_{x \in S} \delta(\xi, x_n) \frac{q(m|x)}{w'_m |S|} \right) \log f_n(\xi|\theta_{mn}),$$

and therefore there is a simple solution maximizing the weighted likelihood (7):

$$f_n(\xi|\theta_{mn}) = \sum_{x \in S} \delta(\xi, x_n) \frac{q(m|x)}{w'_m |S|} \;\Rightarrow\; \theta'_{mn} = \sum_{x \in S} x_n \frac{q(m|x)}{w'_m |S|}. \tag{10}$$

We recall that the multivariate Bernoulli mixtures are not restrictive as an approximation tool since, for a large number of components, any distribution of a random binary vector can be expressed in the form (1), (9), (cf. [12]).

In case of multivariate Bernoulli mixtures we can easily derive the structural (subspace) modification [8, 10] by introducing binary structural parameters $\varphi_{mn} \in \{0,1\}$ in the product components

$$F(x|\theta_m) = \prod_{n \in \mathcal{N}} f_n(x_n|\theta_{mn})^{\varphi_{mn}} f_n(x_n|\theta_{0n})^{1-\varphi_{mn}}, \quad m \in \mathcal{M}. \tag{11}$$

It can be seen that by setting $\varphi_{mn} = 0$ in the formula (11), we can substitute any component-specific univariate distribution $f_n(x_n|\theta_{mn})$ by the respective common background distribution $f_n(x_n|\theta_{0n})$. The structural component (9) can be rewritten in the form

$$F(x|\theta_m) = F(x|\theta_0)G(x|\theta_m, \boldsymbol{\phi}_m), \quad m \in \mathcal{M}, \tag{12}$$

where $F(x|\theta_0)$ is a nonzero "background" probability distribution—usually defined as a fixed product of the unconditional univariate marginals

$$F(x|\theta_0) = \prod_{n \in \mathcal{N}} f_n(x_n|\theta_{0n}), \quad \theta_{0n} = \frac{1}{|S|} \sum_{x \in S} x_n, \; n \in \mathcal{N}.$$

In this way we obtain the subspace mixture model

$$P(x|w, \Theta, \Phi) = F(x|\theta_0) \sum_{m \in \mathcal{M}} w_m G(x|\theta_m, \phi_m), \tag{13}$$

where the component functions $G(x|\theta_m, \phi_m)$ include additional binary structural parameters $\varphi_{mn} \in \{0, 1\}$:

$$G(x|\theta_m, \phi_m) = \prod_{n \in \mathcal{N}} \left[\frac{f_n(x_n|\theta_{mn})}{f_n(x_n|\theta_{0n})} \right]^{\varphi_{mn}}, \quad \phi_m = (\varphi_{m1}, \ldots, \varphi_{mN}). \tag{14}$$

Consequently, the component functions $G(x|\theta_m, \phi_m)$ may be defined on different subspaces. In other words, for each component we can "choose" the optimal subset of informative features. The complexity and "structure" of the finite mixture (13) can be controlled by means of the binary parameters φ_{mn} since the number of parameters is reduced whenever $\varphi_{mn} = 0$. Thus we can estimate product mixtures of high dimensionality while keeping the number of estimated parameters reasonably small.

The structural parameters φ_{mn} can be optimized by means of the EM algorithm in full generality (cf. [8, 10, 21]) by maximizing the corresponding log-likelihood criterion:

$$L = \frac{1}{|S|} \sum_{x \in S} \log \left[\sum_{m \in \mathcal{M}} w_m F(x|\theta_0) G(x|\theta_m, \phi_m) \right].$$

In the following EM iteration equations the apostrophe denotes the new parameter values ($m \in \mathcal{M}, n \in \mathcal{N}$):

$$q(m|x) = \frac{w_m G(x|\theta_m, \phi_m)}{\sum_{j \in \mathcal{M}} w_j G(x|\theta_j, \phi_j)}, \quad w'_m = \frac{1}{|S|} \sum_{x \in S} q(m|x), \tag{15}$$

$$\theta'_{mn} = \sum_{x \in S} x_n \frac{q(m|x)}{w'_m |S|}, \quad \gamma'_{mn} = \frac{1}{|S|} \sum_{x \in S} q(m|x) \log \frac{f_n(x_n|\theta'_{mn})}{f_n(x_n|\theta_{0n})}. \tag{16}$$

Assuming a fixed number λ of component specific parameters we define the optimal subset of nonzero parameters φ'_{mn} by means of the λ highest values $\gamma'_{mn} > 0$. From the computational point of view it is more efficient to specify the structural parameters by simple thresholding

$$\varphi'_{mn} = \begin{cases} 1, \gamma'_{mn} > \tau \\ 0, \gamma'_{mn} \le \tau \end{cases}, \quad \left(\tau \approx \frac{\gamma_0}{MN} \sum_{m \in \mathcal{M}} \sum_{n \in \mathcal{N}} \gamma'_{mn} \right)$$

where the threshold τ is derived from the mean value of γ'_{mn} by a coefficient γ_0. The structural criterion γ'_{mn} can be rewritten in the form:

$$\gamma'_{mn} = w'_m \sum_{\xi=0}^{1} f_n(\xi|\theta'_{mn}) \log \frac{f_n(\xi|\theta'_{mn})}{f_n(\xi|\theta_{0n})} = w'_m I(f_n(\cdot|\theta'_{mn})\|f_n(\cdot|\theta_{0n})). \quad (17)$$

In other words, the structural criterion γ'_{mn} can be expressed in terms of Kullback-Leibler information divergence $I(f_n(\cdot|\theta'_{mn})\|f_n(\cdot|\theta_{0n}))$ [29] between the component-specific distribution $f_n(x_n|\theta'_{mn})$ and the corresponding univariate "background" distribution $f_n(x_n|\theta_{0n})$. Thus, only the most specific and informative distributions $f_n(x_n|\theta'_{mn})$ are included in the components.

It can be verified [10, 21] that, for a fixed λ, the iteration scheme (15)–(17) guarantees the monotonic property of the EM algorithm. Recently the subspace mixture model has been apparently independently proposed to control the Gaussian mixture model complexity [31] and to estimate Dirichlet mixtures [2].

The main motivation for the subspace mixture model (13) has been the statistically correct structural optimization of incompletely interconnected probabilistic neural networks [10, 13, 16, 21]. Note that the background probability distribution $F(x|\theta_0)$ can be reduced in the Bayes formula and therefore any decision-making may be confined to just the relevant variables. In particular, considering a finite set of classes $\omega \in \Omega$ with a priori probabilities $p(\omega)$ and denoting $\mathcal{M}_\omega$ the respective component index sets, we can express the corresponding class-conditional mixtures in the form:

$$P(x|\omega, w, \Theta, \Phi) = F(x|\theta_0) \sum_{m\in\mathcal{M}_\omega} w_m G(x|\theta_m, \phi_m), \quad \omega \in \Omega. \quad (18)$$

In this way, the Bayes decision rule is expressed in terms of a weighted sum of component functions $G(x|\theta_m, \phi_m)$ which can be defined on different subspaces:

$$\omega^* = d(x) = \arg\max_{\omega\in\Omega}\{p(\omega|x)\} = \arg\max_{\omega\in\Omega}\{p(\omega) \sum_{m\in\mathcal{M}_\omega} w_m G(x|\theta_m, \phi_m)\}. \quad (19)$$

3 Mixtures of Dependence Trees

As mentioned earlier, the simplicity of product components may appear to be limiting in some cases and a natural way to generalize product mixtures is to use dependence-tree distributions as components [7, 32–34]. Of course, marginal distributions of the dependence-tree mixtures are not trivially available anymore and we lose some of the excellent properties of product mixtures, especially the unique possibility of structural optimization of probabilistic neural networks. Nevertheless, in some cases such properties may be unnecessary, while the increased complexity of components could become essential.

The idea of the dependence-tree distribution refers to the well known paper of Chow and Liu [3] who proposed approximation of multivariate discrete probability distribution $P^*(x)$ by the product distribution

$$P(x|\pi, \beta) = f(x_{i_1}) \prod_{n=2}^{N} f(x_{i_n}|x_{j_n}), \quad j_n \in \{i_1, \dots, i_{n-1}\}. \tag{20}$$

Here $\pi = (i_1, i_2, \dots, i_N)$ is a suitable permutation of the index set $\mathcal{N}$ and β is the dependence structure

$$\beta = \{((i_2, j_2), \dots, (i_N, j_N))\}, \quad j_n \in \{i_1, \dots, i_{n-1}\}$$

which defines a spanning tree of the complete graph over the nodes $\{1, 2, \dots, N\}$ because the edges β do not contain any loop. In this paper we use a simplified notation of marginal distributions whenever tolerable, e.g.,

$$f(x_n) = f_n(x_n), \quad f(x_n|x_k) = f_{n|k}(x_n|x_k).$$

The above approximation model (20) can be equivalently rewritten in the form

$$P(x|\alpha, \theta) = \left[\prod_{n=1}^{N} f(x_n) \right] \left[\prod_{n=2}^{N} \frac{f(x_n, x_{k_n})}{f(x_n)f(x_{k_n})} \right], \tag{21}$$

because the first product is permutation-invariant and the second product can always be naturally ordered. Thus, in the last equation, the indices $(k_2, \dots, k_N)$ briefly describe the ordered edges (n, k_n) of the underlying spanning tree β and we can write

$$P(x|\alpha, \theta) = f(x_1) \prod_{n=2}^{N} f(x_n|x_{k_n}), \quad \alpha = (k_2, \dots, k_N), \quad \theta = \{f(x_n, x_{k_n})\}. \tag{22}$$

Here α describes the dependence structure and θ stands for the related set of two-dimensional marginals. Note that all univariate marginals uniquely follow from the bivariate ones.

The dependence-tree mixtures can be optimized by means of EM algorithm in full generality, as shown in the paper [7]. Later, the concept of dependence-tree mixtures has been reinvented in [32–34].

Considering binary variables $x_n \in \{0, 1\}$ we denote by $P(x|w, \alpha, \Theta)$ a mixture of dependence-tree distributions

$$P(x|w, \alpha, \Theta) = \sum_{m \in \mathcal{M}} w_m F(x|\alpha_m, \theta_m), \quad x \in X, \tag{23}$$

$$F(x|\alpha_m, \theta_m) = f(x_1|m) \prod_{n=2}^{N} f(x_n|x_{k_n}, m) \tag{24}$$

with the two-dimensional marginals $\theta_m = \{f(x_n, x_{k_n}|m), n = 2, \ldots, N\}$, the underlying dependence structures α_m and the weight vector w:

$$\Theta = \{\theta_1, \theta_2, \ldots, \theta_M\}, \quad \alpha = \{\alpha_1, \alpha_2, \ldots, \alpha_M\}, \quad w = (w_1, w_2, \ldots, w_M).$$

The related log-likelihood function can be expressed by the formula

$$L(w, \alpha, \Theta) = \frac{1}{|S|} \sum_{x \in S} \log [\sum_{m \in \mathcal{M}} w_m F(x|\alpha_m, \theta_m)]. \tag{25}$$

In view of Eq. (5), the EM algorithm reduces the optimization problem to the iterative maximization of the following weighted log-likelihood criteria $Q_m, m \in \mathcal{M}$ with respect to θ_m and α_m:

$$Q_m(\alpha_m, \theta_m) = \sum_{x \in S} \frac{q(m|x)}{w_m'|S|} \log F(x|\alpha_m, \theta_m) \tag{26}$$

$$= \sum_{x \in S} \frac{q(m|x)}{w_m'|S|} [\log f(x_1|m) + \sum_{n=2}^{N} \log f(x_n|x_{k_n}, m)].$$

By using usual δ-function notation we can write

$$Q_m(\alpha_m, \theta_m) = \sum_{x \in S} \frac{q(m|x)}{w_m'|S|} [\sum_{\xi_1=0}^{1} \delta(\xi_1, x_1) \log f(\xi_1|m)$$

$$+ \sum_{n=2}^{N} \sum_{\xi_n=0}^{1} \sum_{\xi_{k_n}=0}^{1} \delta(\xi_n, x_n)\delta(\xi_{k_n}, x_{k_n}) \log f(\xi_n|\xi_{k_n}, m)] \tag{27}$$

and further, using notation

$$\hat{f}(\xi_n|m) = \sum_{x \in S} \frac{q(m|x)}{w_m'|S|}\delta(\xi_n, x_n), \quad \hat{f}(\xi_n, \xi_{k_n}|m) = \sum_{x \in S} \frac{q(m|x)}{w_m'|S|}\delta(\xi_n, x_n)\delta(\xi_{k_n}, x_{k_n}),$$

we can write:

$$Q_m(\alpha_m, \theta_m) = \sum_{\xi_1=0}^{1} \hat{f}(\xi_1|m) \log f(\xi_1|m) \tag{28}$$

$$+ \sum_{n=2}^{N} \sum_{\xi_{k_n}=0}^{1} \hat{f}(\xi_{k_n}|m) \sum_{\xi_n=0}^{1} \frac{\hat{f}(\xi_n, \xi_{k_n}|m)}{\hat{f}(\xi_{k_n}|m)} \log f(\xi_n|\xi_{k_n}, m).$$

For any fixed dependence structure α_m, the last expression is maximized by the two-dimensional marginals $\theta'_m = \{f'(\xi_n, \xi_{k_n}|m), n = 2, \ldots, N\}$:

$$f'(\xi_n|m) = \hat{f}(\xi_n|m), \quad f'(\xi_n|\xi_{k_n}, m) = \frac{\hat{f}(\xi_n, \xi_{k_n}|m)}{\hat{f}(\xi_{k_n}|m)}. \tag{29}$$

Making substitutions (29) in (28) we can express the weighted log-likelihood criterion $Q_m(\alpha_m, \theta'_m)$ just as a function of the dependence structure α_m:

$$Q_m(\alpha_m, \theta'_m) = \sum_{n=1}^{N} \sum_{\xi_n=0}^{1} f'(\xi_n|m) \log f'(\xi_n|m)$$
$$+ \sum_{n=2}^{N} \sum_{\xi_n=0}^{1} \sum_{\xi_{k_n}=0}^{1} f'(\xi_n, \xi_{k_n}|m) \log \frac{f'(\xi_n, \xi_{k_n}|m)}{f'(\xi_n|m)f'(\xi_{k_n}|m)}.$$

Here the last expression is the Shannon formula for mutual statistical information between the variables x_n, x_{k_n} [37], i.e. we can write

$$\mathcal{I}(f'_{n|m}, f'_{k_n|m}) = \sum_{\xi_n=0}^{1} \sum_{\xi_{k_n}=0}^{1} f'(\xi_n, \xi_{k_n}|m) \log \frac{f'(\xi_n, \xi_{k_n}|m)}{f'(\xi_n|m)f'(\xi_{k_n}|m)} \tag{30}$$

$$Q_m(\alpha_m, \theta'_m) = \sum_{n=1}^{N} -H(f'_{n|m}) + \sum_{n=2}^{N} \mathcal{I}(f'_{n|m}, f'_{k_n|m}).$$

In the last equation, the sum of entropies $H(\cdot)$ is structure-independent and therefore the weighted log-likelihood criterion $Q_m(\alpha_m, \theta'_m)$ is maximized by means of the second sum, in terms of the dependence structure α_m.

The resulting EM iteration equations for mixtures of dependence-tree distributions can be summarized as follows (cf. [7], Eqs. (4.17)–(4.20)):

$$q(m|x) = \frac{w_m F(x|\alpha_m, \theta_m)}{\sum_{j\in\mathcal{M}} w_j F(x|\alpha_j, \theta_j)}, \quad w'_m = \frac{1}{|S|} \sum_{x\in S} q(m|x), \tag{31}$$

$$\alpha'_m = \arg\max_{\alpha} \left\{ \sum_{n=2}^{N} \mathcal{I}(f'_{n|m}, f'_{k_n|m}) \right\}, \quad f'(\xi_n|m) = \sum_{x\in S} \frac{q(m|x)}{w'_m|S|} \delta(\xi_n, x_n), \tag{32}$$

$$f'(\xi_n, \xi_{k_n}|m) = \sum_{x\in S} \frac{q(m|x)}{w'_m|S|} \delta(\xi_n, x_n)\delta(\xi_{k_n}, x_{k_n}), \quad n = 1, 2, \ldots, N. \tag{33}$$

Thus the optimal dependence structure α'_m can be found by constructing the maximum-weight spanning tree of the related complete graph with the edge weights $\mathcal{I}(f'_{n|m}, f'_{k|m})$ [3]. For this purpose we can use e.g. the algorithm of Kruskal [28] or Prim [35] (cf. Appendix for more details).

The concept of dependence-tree mixtures can be applied to continuous variables by using bivariate Gaussian densities (cf. [7, 15]).

4 Recognition of Numerals

In recent years we have repeatedly applied multivariate Bernoulli mixtures to recognition of hand-written numerals from the NIST benchmark database, with the aim to verify different decision-making aspects of probabilistic neural networks (cf. [13, 16]). In this paper we use the same data to compare performance of the product (Bernoulli) mixtures and mixtures of dependence trees. We assume that the underlying 45 binary (two class) subproblems may reveal even very subtle differences between the classifiers. Moreover, the relatively stable graphical structure of numerals should be advantageous from the point of view of dependence-tree mixtures.

The considered NIST Special Database 19 (SD19) contains about 400000 hand-written numerals in binary raster representation (about 40000 for each numeral). We normalized all digit patterns to a 32×32 binary raster to obtain 1024-dimensional binary data vectors. In order to guarantee the same statistical properties of the training- and test data sets, we have used the odd samples of each class for training and the even samples for testing. Also, to increase the variability of the binary patterns, we extended both the training- and test data sets four times by making three differently rotated variants of each pattern (by -4, -2 and $+2°$) with the resulting 80000 patterns for each class.

In order to make the classification test we estimated for all ten numerals the class-conditional distributions by using Bernoulli mixtures in the subspace modification (13) and also by using dependence-tree mixtures. Recall that we need 2048 parameters to define each component of the dependence-tree distribution (24). The marginal probabilities of dependence-tree components displayed in raster arrangement (cf. Fig. 1) correspond to the typical variants of the training numerals. Simultaneously, the figure shows the corresponding maximum-weight spanning tree α_m. Note that the superimposed optimal dependence structure naturally "reveals" how the numerals have been written because the "successive" raster points are strongly correlated.

For the sake of comparison we used the best solutions obtained in a series of experiments—both for the product mixtures and for the dependence-tree mixtures. The independent test patterns were classified by means of Bayes decision function (19). Each test numeral was classified by using mean Bayes probabilities obtained with the four differently rotated variants. Table 1 shows the classification error based on the product mixtures (18) as a function of model complexity. Number of para-

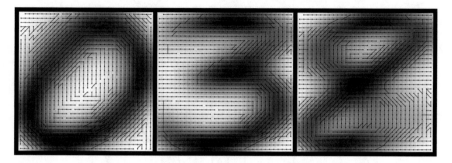

Fig. 1 Mixture of dependence trees for binary data—examples of marginal component probabilities in raster arrangement. Note that the superimposed optimal dependence structure (defined by maximum-weight spanning tree) reflects the way the respective numerals have been written

Table 1 Recognition of numerals from the NIST SD19 database by mixtures with different number of product components

Experiment no.	I	II	III	IV	V	VI	VII	VIII
Components	10	100	299	858	1288	1370	1459	1571
Parameters	10240	89973	290442	696537	1131246	1247156	1274099	1462373
Classif. error in %	11.93	4.28	2.93	2.40	1.95	1.91	1.86	1.84

In the third row the number of parameters denotes the total number of component specific parameters θ_{mn}

meters in the third row denotes the total number of component-specific parameters θ_{mn} (for which $\phi_{mn} = 1$). Similar to Table 1 we can see in Table 3 the classification error as a function of model complexity, now represented by different numbers of dependence-tree components.

The detailed classification results for the best solutions are described by the error matrix in Table 2 (ten class-conditional mixtures with the total number of M=1571 product components including 1462373 parameters) and Table 4 (ten mixtures with total number of M = 400 dependence tree components including 819200 parameters). As it can be seen the global recognition accuracy (right lower corner) is comparable in both cases. Note that in both tables the detailed frequencies of false negative and false positive decisions are also comparable.

Roughly speaking, the dependence-tree mixtures achieve only slightly better recognition accuracy with a comparable number of parameters, but the most complex model (M = 500) already seems to overfit. Expectedly, the dependence tree mixtures needed much less components for the best performance but they have stronger tendency to overfitting. The best recognition accuracy in Table 1 (cf. col. VIII) well illustrates the power of the subspace product mixtures.

The most surprising result of the numerical experiments is the decreasing importance of the component dependence structure during the EM estimation process. We have noticed that in each class the cumulative weight of all dependence trees expressed by the weighted sum

Table 2 Classification error matrix obtained by means of multivariate Bernoulli mixtures (the total number of components M = 1571, number of parameters: 1462373)

Class	0	1	2	3	4	5	6	7	8	9	False n. (%)
0	19950	8	43	19	39	32	36	0	38	17	1.1
1	2	22162	30	4	35	7	18	56	32	6	0.9
2	32	37	19742	43	30	9	8	29	90	16	1.5
3	20	17	62	20021	4	137	2	28	210	55	2.6
4	11	6	19	1	19170	11	31	51	30	247	2.1
5	25	11	9	154	4	17925	39	6	96	34	2.1
6	63	10	17	6	23	140	19652	1	54	3	1.6
7	7	12	73	10	73	4	0	20497	22	249	2.1
8	22	25	53	97	30	100	11	11	19369	72	2.1
9	15	13	25	62	114	22	3	146	93	19274	2.5
False p. (%)	0.9	0.7	2.7	2.0	1.7	2.3	0.7	1.6	3.3	3.5	**1.84**

The last column contains the percentage of false negative decisions. The last row contains the total frequencies of false positive rates in percent of the respective class test patterns with the global error rate in bold

$$\Sigma' = \sum_{m \in \mathcal{M}_\omega} w'_m \sum_{n=2}^{N} \mathcal{I}(f'_{n|m}, f'_{k_n|m}) \tag{34}$$

is decreasing in the course of EM iterations (cf. Fig. 2). In other words the optimal estimate of the dependence tree mixture tends to suppress the information contribution of the dependence structures in components, i.e. the component dependence trees tend to degenerate to simple products. Nevertheless, this observation is probably typical only for mixtures having a large number of components since a single product component is clearly more restrictive than a single dependence tree.

5 Conclusions

We compare the computational properties of mixtures of product components and mixtures of dependence trees in application to recognition of numerals from the NIST Special Database 19. For the sake of comparison we have used for each of the considered mixture models the best solution obtained in series of experiments. The detailed description of the classification performance (cf. Tables 2 and 4) shows that the recognition accuracy of both models is comparable. It appears that, in our case, the dependence structure of components does not improve the approximation power of the product mixture essentially and, moreover, the information contribution of the dependence structure decreases in the course of EM iterations as shown in Fig. 2. Thus, the optimal estimate of the dependence tree mixture tends to approach a simple product mixture model. However, this observation is probably related to large number of components only.

We assume that the dependence tree distribution is advantageous if we try to fit a small number of components to a complex data set. However, in case of a large number of multidimensional components the component functions are almost non-overlapping [17], the structural parameters tend to fit to small compact subsets of data and the structurally modified form of the components is less important. We can summarize the results of comparison as follows:

Table 3 Recognition of numerals from the NIST SD19 database by mixtures with different number of dependence trees

Experiment No.	I	II	III	IV	V	VI	VII	VIII
Components	10	40	100	150	200	300	400	500
Parameters	20480	81920	204800	307200	409600	614400	819200	1024000
Classif. error in %	6.69	4.13	2.64	2.53	2.22	2.13	1.97	2.01

For comparable number of parameters the dependence-tree mixtures achieve only slightly better recognition accuracy

Table 4 Classification error matrix obtained by means of dependence-tree mixtures (number of components M = 400, number of parameters: 819200)

Class	0	1	2	3	4	5	6	7	8	9	False n. (%)
0	19979	11	62	21	18	26	25	2	28	10	1.0
1	5	21981	78	13	74	1	20	155	21	4	1.7
2	22	15	19777	72	26	5	6	35	72	6	1.3
3	20	10	66	20169	1	120	1	20	122	27	1.9
4	12	16	13	4	19245	1	13	52	44	177	1.7
5	25	5	15	157	8	17874	45	9	129	36	2.3
6	100	19	38	25	43	90	19575	1	75	3	2.0
7	17	33	108	24	71	0	0	20367	28	299	2.8
8	18	30	47	167	27	55	22	17	19337	70	2.3
9	12	20	62	74	89	33	3	144	134	19196	2.9
False p. (%)	1.4	0.7	2.4	2.7	1.8	1.8	0.7	1.6	3.1	3.2	**1.97**

The last column contains percentage of false negative decisions. The last row contains false positive rates in percent of the respective class test patterns with the global error rate in bold

Fig. 2 The decreasing information contribution of the dependence structure to the estimated dependence-tree mixtures (the first eight iterations of the ten estimated class-conditional distributions). The EM algorithm tends to suppress the information contribution of the dependence structures to the optimal estimate

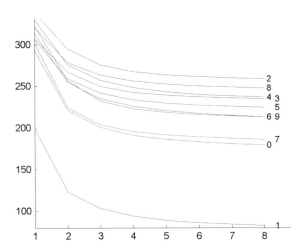

In Case of a Large Number of Components

- intuitively, the large number of components is the main source of the resulting approximation power
- dependence structure of components does not improve the approximation power of product mixtures essentially
- the total information contribution of the component dependence structures decreases in the course of EM iterations
- the optimal estimate of the dependence tree mixture tends to approach a simple product mixture model

In Case of a Small Number of Components

- a single dependence tree component is capable to describe the statistical relations between pairs variables
- consequently, the approximation power of a single dependence tree component is much higher than that of a product component
- information contribution of the dependence structure can increase in the course of EM iterations
- dependence structure of components can essentially improve the approximation quality

In this sense, the computational properties of dependence tree mixtures provide an additional argument to prefer the product mixture models in case of large multidimensional data sets. A large number of product components in the subspace modification (19) seems to outperform the advantage of the more complex dependence-tree distributions.

Acknowledgments This work was supported by the Czech Science Foundation Projects No. 14-02652S and P403/12/1557.

Appendix: Maximum-Weight Spanning Tree

The algorithm of Kruskal (cf. [3, 28]) assumes ordering of all $N(N - 1)/2$ edge weights in descending order. The maximum-weight spanning tree is then constructed sequentially, starting with the first two (heaviest) edges. The next edges are added sequentially in descending order if they do not form a cycle with the previously chosen edges. Multiple solutions are possible if several edge weights are equal, but they are ignored as having the same maximum weight. Obviously, in case of dependence-tree mixtures with many components, the application of the Kruskal algorithm may become prohibitive in high-dimensional spaces because of the repeated ordering of the edge-weights.

The algorithm of Prim [35] does not need any ordering of edge weights. Starting from any variable we choose the neighbor with the maximum edge weight. This first edge of the maximum-weight spanning tree is then sequentially extended by adding the maximum-weight neighbors of the currently chosen subtree. Again, any ties may be decided arbitrarily since we are not interested in multiple solutions.

Both Kruskal and Prim refer to an "obscure Czech paper" of Otakar Borůvka [1] from the year 1926 giving an alternative construction of the minimum-weight spanning tree and the corresponding proof of uniqueness. Moreover, the Prim's algorithm was developed in 1930 by Czech mathematician Vojtěch Jarník (cf. [27], in Czech). The algorithm of Prim can be summarized as follows (in C-code):

```
//   Maximum-weight spanning tree construction
//********************************************************************
//   spanning tree: {<2,A[2]>,...,<NN,A[NN]>}
//   NN........ number of nodes, N=1,2,...,NN
//   T[N]...... labels of the defined part of the spanning tree
//   E[N][K]... positive weight of the edge <N,K>
//   A[K]...... heaviest neighbor of K in the defined subtree
//   GE[K]..... greatest edge weight between K and defined subtree
//   K0........ the most heavy neighbor of the defined subtree
//   SUM....... total weight of the spanning tree
//********************************************************************
  for(N=1; N<=NN; N++) {GE[N]=-1;  T[N]=0;  A[N]=0;}
  N0=1; T[N0]=1; K0=0;        // initial values
  for(I=2; I<=NN; I++)   // spanning tree loop
  {  FMAX=-1E0;
     for(N=2; N<=NN; N++) if(T[N]<1)
     {  F=E[N0][N];
        if(F>GE[N]) {GE[N]=F; A[N]=N0;} else F=GE[N];
        if(F>FMAX)  {FMAX=F; K0=N;}
     }  //  end of N-loop
     N0=K0;  T[N0]=1;    SUM+=FMAX;
  }  // end of spanning tree construction
//********************************************************************
```

References

1. Borůvka, O.: On a minimal problem. Trans. Moravian Soc. Nat. Sci. (in Czech) No. 3 (1926)
2. Bouguila, N., Ziou, D., Vaillancourt, J.: Unsupervised learning of a finite mixture model based on the Dirichlet distribution and its application. IEEE Trans. Image Process. **13**(11), 1533–1543 (2004)
3. Chow, C., Liu, C.: Approximating discrete probability distributions with dependence trees. IEEE Trans. Info. Theory **IT-14**(3), 462–467 (1968)
4. Dempster, A.P., Laird, N.M., Rubin, D.B.: Maximum likelihood from incomplete data via the EM algorithm. J. Roy. Statist. Soc. B **39**, 1–38 (1977)
5. Day, N.E.: Estimating the components of a mixture of normal distributions. Biometrika **56**, 463–474 (1969)
6. Grim, J.: On numerical evaluation of maximum—likelihood estimates for finite mixtures of distributions. Kybernetika **18**(3), 173–190 (1982). http://dml.cz/dmlcz/124132
7. Grim, J.: On structural approximating multivariate discrete probability distributions. Kybernetika **20**(1), 1–17 (1984). http://dml.cz/dmlcz/125676
8. Grim, J.: Multivariate statistical pattern recognition with nonreduced dimensionality. Kybernetika **22**(2), 142–157 (1986). http://dml.cz/dmlcz/125022
9. Grim, J.: Design of multilayer neural networks by information preserving transforms. In: Pessa, E., Penna, M.P., Montesanto A. (eds.) Third European Congress on Systems Science. Edizioni Kappa, Roma, pp. 977–982 (1996)
10. Grim, J.: Information approach to structural optimization of probabilistic neural networks. In: Ferrer, L. et al. (eds.) Proceedings of the 4th System Science European Congress. Valencia, Soc. Espanola de Sistemas Generales, pp. 527–540 (1999)
11. Grim, J.: A sequential modification of EM algorithm. In: Gaul, W., Locarek-Junge, H. (eds.) Studies in Classification, Data Analysis and Knowledge Organization, pp. 163–170. Springer (1999)

12. Grim, J.: EM cluster analysis for categorical data. In: Yeung, D.Y., Kwok, J.T., Fred, A. (eds.) Structural, Syntactic and Statistical Pattern Recognition, LNCS 4109, pp. 640–648. Springer, Berlin (2006)
13. Grim, J.: Neuromorphic features of probabilistic neural networks. Kybernetika **43**(5), 697–712 (2007). http://dml.cz/dmlcz/135807
14. Grim, J.: Sequential pattern recognition by maximum conditional informativity. Pattern Recogn. Lett. **45C**, 39–45 (2014). doi:10.1016/j.patrec.2014.02.024
15. Grim, J.: Approximating probability densities by mixtures of gaussian dependence trees. In: Hobza, T. (ed.) Proceedings of the SPMS 2014 Stochastic and Physical Monitoring Systems, pp. 43–56. Czech Technical University Prague (2014)
16. Grim, J., Hora, J.: Iterative principles of recognition in probabilistic neural networks. Neural Netw. **21**(6), 838–846 (2008)
17. Grim, J., Hora, J.: Computational properties of probabilistic neural networks. In: Artificial Neural Networks—ICANN 2010 Part II, LNCS 5164, pp. 52–61. Springer, Berlin (2010)
18. Grim, J., Hora, J., Boček P., Somol, P., Pudil, P.: Statistical model of the 2001 Czech census for interactive presentation. J. Official Stat. **26**(4), 673–694 (2010). http://ro.utia.cas.cz/dem.html
19. Grim, J., Kittler, J., Pudil, P., Somol, P.: Multiple classifier fusion in probabilistic neural networks. Pattern Anal. Appl. **5**(7), 221–233 (2002)
20. Grim, J., Pudil, P.: Pattern recognition by probabilistic neural networks—mixtures of product components versus mixtures of dependence trees. In: Proceedings of the International Conference on Neural Computation Theory and Applications NCTA2014. Rome, SCITEPRESS, 2014, s. 65–75 (2014)
21. Grim, J., Pudil, P., Somol, P.: Recognition of handwritten numerals by structural probabilistic neural networks. In: Bothe, H., Rojas, R. (eds.) Proceedings of the Second ICSC Symposium on Neural Computation, pp. 528–534. Berlin ICSC, Wetaskiwin (2000)
22. Grim, J., Pudil, P., Somol, P.: Boosting in probabilistic neural networks. In: Kasturi, R., Laurendeau, D., Suen, C. (eds.) Proceedings of the 16th International Conference on Pattern Recognition, pp. 136–139. IEEE Computer Society, Los Alamitos (2002b)
23. Grim, J., Somol, P., Haindl, M., Daneš, J.: Computer-aided evaluation of screening mammograms based on local texture models. IEEE Trans. Image Process. **18**(4), 765–773 (2009)
24. Hasselblad, V.: Estimation of prameters for a mixture of normal distributions. Technometrics **8**, 431–444 (1966)
25. Hasselblad, V.: Estimation of finite mixtures of distributions from the exponential family. J. Am. Stat. Assoc. **58**, 1459–1471 (1969)
26. Hosmer Jr, D.W.: A comparison of iterative maximum likelihood estimates of the parameters of a mixture of two normal distributions under three different types of sample. Biometrics 761–770 (1973)
27. Jarník, V.: About a certain minimal problem. Trans. Moravian Soc. Nat. Sci. (in Czech) No. 6, 57–63 (1930)
28. Kruskal, J.B.: On the shortest spanning sub-tree of a graph. Proc. Am. Math. Soc. **7**, 48–50 (1956)
29. Kullback, S., Leibler, R.A.: On information and sufficiency. Ann. Math. Stat. **22**(1), 79–86 (1951)
30. Lowd, D., Domingos, P.: Naive Bayes models for probability estimation. In: Proceedings of the 22nd International Conference on Machine Learning, ACM 2005, pp. 529–536 (2005)
31. Markley, S.C., Miller, D.J.: Joint parsimonious modeling and model order selection for multivariate Gaussian mixtures. IEEE J. Sel. Top. Sign. Process. **4**(3), 548–559 (2010)
32. Meila, M., Jordan, M.I.: Estimating dependency structure as a hidden variable. In: Proceedings of the 1997 Conference on Avances in Neural Information Processing Systems, vol. 10, pp. 584–590 (1998)
33. Meila, M., Jaakkola T.: Tractable bayesian learning of tree belief networks. In: Proceedings of the 16th Conference on Uncertainty in Artificial Intelligence, pp. 380–388 (2000)

34. Meila, M., Jordan, M.I.: Learning with mixtures of trees. J. Mach. Learn. Res. **1**(9), 1–48 (2001)
35. Prim, R.C.: Shortest connection networks and some generalizations. Bell Syst. Tech. J. **36**, 1389–1401 (1957)
36. Schlesinger, M.I.: Relation between learning and self learning in pattern recognition (in Russian). Kibernetika (Kiev) No. 2, 81–88 (1968)
37. Vajda, I.: Theory of Statistical Inference and Information. Kluwer Academic Publishers, Dordrecht and Boston (1989)
38. Wolfe, J.H.: Pattern clustering by multivariate mixture analysis. Multivar. Behav. Res. **5**, 329–350 (1970)

Recurrent Neural Networks Training Using Derivative Free Nonlinear Bayesian Filters

Branimir Todorović, Miomir Stanković and Claudio Moraga

Abstract We have implemented the recurrent neural networks training algorithms as joint estimation of synaptic weights and neuron outputs using approximate nonlinear recursive Bayesian estimators. We have considered two nonlinear derivative free estimators: Divided Difference Filter and Unscented Kalman filter and compared there computational efficiency and performances to the Extended Kalman Filter as training algorithms for different recurrent neural network architectures. Algorithms and architectures were tested on problems of long term, chaotic time series prediction.

1 Introduction

In this paper we consider the training of Recurrent Neural Networks (RNNs) as derivative free approximate Bayesian estimation. RNNs form a wide class of neural networks with feedback connections among processing units (artificial neurons). Neural networks with feed forward connections implement static input-output mapping, while recurrent networks implement the mapping of both input and internal state (represented by outputs of recurrent neurons) into the future internal state.

In general, RNNs can be classified as locally recurrent, where feedback connections exist only from a processing unit to itself, and globally recurrent, where

B. Todorović (✉)
Faculty of Natural Sciences and Mathematics, University of Niš, Niš, Serbia
e-mail: branimirtodorovic@pmf.ni.ac.rs

M. Stanković
Faculty of Occupational Safety, University of Niš, Niš, Serbia

C. Moraga
European Centre for Soft Computing, 33600 Mieres, Spain

C. Moraga
Technical University of Dortmund, 44221, Dortmund, Germany

© Springer International Publishing Switzerland 2016
J.J. Merelo et al. (eds.), *Computational Intelligence*,
Studies in Computational Intelligence 620, DOI 10.1007/978-3-319-26393-9_23

feedback connections exist among distinct processing units. The modelling capabilities of globally recurrent neural networks are much richer than that of the simple locally recurrent networks.

There exist a group of algorithms for training synaptic weights of recurrent neural networks that are based on the exact or approximate computation of the gradient of an error measure in the weight space. Well known approaches that use methods for exact gradient computation are back-propagation through time (BPTT) and real time recurrent learning (RTRL) [10, 11]. Since BPTT and RTRL are using only first-order derivative information, they exhibit slow convergence. In order to improve the speed of the RNN training, a technique known as *teacher forcing* has been proposed [10]. The idea is to use the desired outputs of the neurons instead of the obtained to compute the future outputs. In this way the training algorithm is focused on the current time step, given that the performance is correct on all earlier time steps.

However, in its basic form teacher forcing is not always applicable. It clearly cannot be applied in networks where feedback connections exist only from hidden units, for which the target outputs are not explicitly given. The second important case is the training on noisy data, where the target outputs are corrupted by noise. Therefore, to apply teacher forcing in such cases, a true target outputs of neurons have to be estimated somehow.

The well-known extended Kalman filter [1], as a second order sequential training algorithm and state estimator offers the solution to the both stated problems. It improves the learning rate by exploiting second order information on criterion function and generalizes the teacher forcing technique by estimating the true outputs of the neurons.

The extended Kalman filter can be considered as the approximate solution of the recursive Bayesian state estimation problem. The problem of estimating the hidden state of a dynamic system using observations which arrive sequentially in time is very important in many fields of science, engineering and finance. The hidden state of some dynamic system is represented as a random vector variable, and its evolution in time $\{x_k, k = 1, 2, \ldots\}$ is described by a so called dynamic or process equation:

$$x_k = f_k(x_{k-1}, u_k, d_k), \tag{1}$$

where $f_k: R^{n_x \times n_d} \to R^{n_x}$ is nonlinear function, and $\{d_k, k = 1, 2, \ldots\}$ is an i.i.d. process noise sequence, while n_x and n_d are dimensions of the state and process noise vectors respectively. The hidden state is known only through the measurement (observation) equation:

$$y_k = h_k(x_k, v_k), \tag{2}$$

where $h_k: R^{n_x \times n_v} \to R^{n_y}$ is nonlinear function, and $\{v_k, k = 1, 2, \ldots\}$ is an i.i.d. measurement noise sequence, and n_y and n_v are dimensions of the measurement and measurement noise vectors, respectively.

In a sequential or recursive Bayesian estimation framework, the state filtering probability density function (pdf) $p(x_k/y_{0:k})$, (where $y_{0:k}$ denotes the set of all observations $y_{0:k} = \{y_0, y_1, \ldots, y_k\}$ up to the time step k), represents the complete solution. The optimal state estimate with respect to any criterion can be calculated based on this pdf.

The recursive Bayesian estimation algorithm consists of two steps: **prediction** and **update**. In the prediction step the previous posterior $p(x_{k-1}/y_{0:k-1})$ is projected forward in time, using the probabilistic process model:

$$p(x_k/y_{0:k-1}) = \int p(x_k/x_{k-1})p(x_{k-1}/y_{0:k-1})dx_k, \qquad (3)$$

where the state transition density function $p(x_k/x_{k-1})$ is completely specified by $f(\cdot)$ and the process noise distribution $p(d_k)$.

In the second step, the predictive density is updated by incorporating the latest noisy measurement y_k using the observation likelihood $p(y_k/x_k)$ to generate the posterior:

$$p(x_k/y_{0:k}) = \frac{p(y_k/x_k)p(x_k/y_{0:k-1})}{\int p(y_k/x_k)p(x_k/y_{0:k-1})dx_k}. \qquad (4)$$

This recursive estimation algorithm can be applied to RNN training after representing the time evolution of neurons outputs and connection weights, as well as their observations, in the form of the state space model. The hidden state of the recurrent neural network x_k is a stacked vector of recurrent neurons outputs s_k and connection weights w_k. Its evolution in time can be represented by the following dynamic equation.

$$\begin{bmatrix} s_k \\ w_k \end{bmatrix} = \begin{bmatrix} f(s_{k-1}, u_k, w_{k-1}, d_{s_k}) \\ w_{k-1} + d_{w_k} \end{bmatrix}, \qquad (5)$$

where d_{s_k} and d_{w_k} represent dynamic noise vectors.

The outputs of the neurons are obtained through the following observation equation:

$$y_k = h(s_k, w_k, v_k). \qquad (6)$$

The recurrence relations (3) and (4) are only conceptual solutions and the posterior density $p(x_k/y_{0:k})$ cannot be determined analytically in general. The restrictive set of cases includes the famous Kalman filter, which represents the optimal solution of (3) and (4) if the prior state density $p(x_0/y_0) = p(x_0)$, the process noise as well as the observation noise densities are Gaussians, and $f(\cdot)$ and $h(\cdot)$ are linear functions.

In case of RNN training, $f(\cdot)$ and $h(\cdot)$ are nonlinear in general and an analytic solution is not tractable, therefore some approximations and suboptimal solutions have to be considered. The well known suboptimal solution is the Extended Kalman Filter (EKF), which assumes the Gaussian property of noise and uses the Taylor expansion of $f(\cdot)$ and $h(\cdot)$ (usually up to the linear term) to obtain the recursive estimation for $p(x_k/y_{0:k})$. The EKF has been successfully applied in RNN training [7, 8, 9] due to important advantages compared to RTRL and BPTT: faster convergence and generalization of teacher forcing. Recently, families of new derivative free filters have been proposed as an alternative to EKF for estimation in nonlinear systems. Divided Difference Filters (DDF), derived in [6], are based on polynomial approximation of nonlinear transformations using a multidimensional extension of Stirling's interpolation formula. The Unscented Kalman Filter (UKF) [2, 3] uses the true nonlinear models and approximates the state distribution using deterministically chosen sample points. Surprisingly, both the DDF and the UKF result in similar equations which do not require calculation of neither dynamic nor observation equation Jacobians, and are therefore usually called derivative free filters [9].

The rest of the paper is organized as follows. In the second section recursive Bayesian estimator is approximated by linear minimum mean square error estimator (MMSE), which recursively updates only the first two moments of the relevant probability densities. The problem that remains to be solved is propagating these moments through the nonlinear mapping of the process equation and the observation equation. In the third section we describe three approaches to this problem: a linearization of the nonlinear mapping using a Taylor series expansion, a derivative free polynomial approximation using a multidimensional extension of the Stirling's interpolation formula and a derivative free unscented transformation. In the fourth section we give the state space models of three globally recurrent neural networks: fully connected, Elman and Non-linear AutoRegresssive with eXogenous inputs (NARX) recurrent neural networks. We trained them by applying approximate recursive Bayesian joint estimation of the recurrent neurons outputs and synaptic weights. The results of applying three different estimation algorithms in training three different architectures of recurrent neural networks are given in the last section.

2   Linear MMSE Estimation of the Nonlinear State Space Model

An analytically tractable solution of the problem of recursive Bayesian estimation framework can be obtained based on the assumption that the state estimator $\hat{x}_k$ can be represented as a linear function of the current observation y_k:

$$\hat{x}_k = A_k y_k + b_k, \tag{7}$$

where matrix A_k and vector b_k are derived by minimizing mean square estimation error criterion:

$$R_k = \iint (x_k - \hat{x}_k)^T (x_k - \hat{x}_k) \cdot p(x_k, y_k / y_{0:\,k-1}) dx_k dy_k. \tag{8}$$

Note that the condition $\partial R_k / \partial b_k = 0$ is equivalent to the requirement that the estimator is unbiased:

$$\iint (x_k - A_k y_k - b_k) \cdot p(x_k, y_k / y_{0:\,k-1}) dx_k dy_k = 0, \tag{9}$$

from which we obtain $b_k = \hat{x}_k^- - A_k \hat{y}_k^-$, and consequently:

$$\hat{x}_k = \hat{x}_k^- - A_k(y_k - \hat{y}_k^-), \tag{10}$$

where

$$\hat{x}_k^- = E[x_k / y_{0:\,k-1}] = \int x_k p(x_k / y_{0:\,k-1}) dx_k$$

and

$$\hat{y}_k^- = E[y_k / y_{0:\,k-1}] = \int y_k p(y_k / y_{0:\,k-1}) dy_k.$$

Both the condition $\partial R_k / \partial A_k = 0$ and the unbiasedness of the estimator result in the so called orthogonality principle, which states that the estimation error is orthogonal to the current observation:

$$\iint (x_k - \hat{x}_k^- - A_k(y_k - \hat{y}_k^-))(y_k - \hat{y}_k^-)^T p(x_k, y_k / y_{0:\,k-1}) dx_k dy_k = 0. \tag{11}$$

From (11) we obtain the matrix $A_k = P_{x_k y_k} P_{y_k}^{-1}$, where

$$P_{y_k} = E[(y_k - \hat{y}_k^-)(y_k - \hat{y}_k^-)^T / y_{0:\,k-1}] \tag{12a}$$

$$P_{x_k y_k} = E[(x_k - \hat{x}_k^-)(y_k - \hat{y}_k^-)^T / y_{0:\,k-1}] \tag{12b}$$

Note that P_{y_k} must be invertible, that is measurements y_k have to be linearly independent.

Finally, after replacing $A_k = P_{x_k y_k} P_{y_k}^{-1}$ we obtain the linear MMSE estimator:

$$\hat{x}_k = \hat{x}_k^- + P_{x_k y_k} P_{y_k}^{-1}(y_k - \hat{y}_k^-). \tag{13a}$$

The matrix Mean Square Error (MSE) corresponding to (13a):

$$E[(x_k - \hat{x}_k)(x_k - \hat{x}_k)^T] = P_{x_k}^- - P_{x_k y_k} P_{y_k}^{-1} P_{x_k y_k}^T. \tag{13b}$$

is used as the approximation of the estimator covariance $P_{x_k} \approx E[(x_k - \hat{x}_k)(x_k - \hat{x}_k)^T]$.

If the dynamic and the observation models are linear and process and observation noises are Gaussian, the linear MMSE estimator is the best MMSE estimator and is equal to the conditional mean $E[x_k / y_{0:k}]$, otherwise it is the best within the class of linear estimators.

The problem that remains to be solved is the estimation of the statics of a random variable propagated trough the nonlinear transformation.

$$\hat{x}_k^- = \iint f(x_{k-1}, u_k, d_k) p(x_{k-1}/y_{0:k-1}) p(d_k) dx_{k-1} dd_k \tag{14a}$$

$$P_{x_k}^- = \iint (f(x_{k-1}, u_k, d_k) - \hat{x}_k^-)(f(x_{k-1}, u_k, d_k) - \hat{x}_k^-)^T p(x_{k-1}/y_{0:k-1}) p(d_k) dx_{k-1} dv_k \tag{14b}$$

$$\hat{y}_k^- = \iint h(x_k, u_k, v_k) p(x_k/y_{0:k-1}) p(v_k) dx_k dv_k \tag{14c}$$

$$P_{y_k}^- = \iint (h(x_k, u_k, v_k) - \hat{y}_k^-)(h(x_k, u_k, v_k) - \hat{y}_k^-)^T p(x_k/y_{0:k-1}) p(v_k) dx_k dv_k \tag{14d}$$

$$P_{x_k y_k} = \iint (x_k - \hat{x}_k^-)(h(x_k, u_k, v_k) - \hat{y}_k^-)^T p(x_k/y_{0:k-1}) p(v_k) dx_k dv_k \tag{14e}$$

The problem can be considered in a general way. Suppose that x is a random variable with mean $\bar{x}$ and covariance P_x. A random variable y is related to x through the nonlinear function $y = f(x)$. We wish to calculate the mean $\bar{y}$ and the covariance P_y of y.

2.1 Extended Kalman Filter

The extended Kalman filter is based on the multidimensional Taylor series expansion of $f(x)$. We shall consider only the first order EKF, obtained by excluding the nonlinear terms of Taylor series expansion:

$$f(x) = f(\hat{x} + \Delta x) \approx f(\hat{x}) + f_x'(\hat{x}) \Delta x \tag{15}$$

where $f_x'(\hat{x}) = \partial f / \partial x|_{x=\hat{x}}$ is the Jacobian of the nonlinear function and Δx is a zero mean random variable with covariance P_x.

After linearization of the dynamic equation (1) we obtain the following approximation:

$$x_k = f_k(\hat{x}_{k-1}, u_k, \bar{d}_k) + F_k(x_{k-1} - \hat{x}_{k-1}) + G_k(d_k - \bar{d}_k), \qquad (16)$$

where

$$F_k = \partial f_k(x_{k-1}, u_k, d_k)/\partial x_{k-1}\Big|_{\substack{x_{k-1}=\hat{x}_{k-1} \\ d_k = \bar{d}_k}} \qquad (17)$$

and

$$G_k = \partial f_k(x_{k-1}, u_k, d_k)/\partial d_k\Big|_{\substack{x_{k-1}=\hat{x}_{k-1} \\ d_k = \bar{d}_k}}, \qquad (18)$$

$\hat{x}_{k-1} = E[x_{k-1}/y_{1:k-1}]$ represents the estimate of the state in time step k-1 and $\bar{d}_k = E[d_k]$ is process noise mean.

Prediction of the state $\hat{x}_{k-1} = E[x_{k-1}/y_{1:k-1}]$ and prediction covariance $P_{x_k}^- = E[(x_k - \hat{x}_k^-)(x_k - \hat{x}_k^-)^T/y_{0:k-1}]$ are obtained after applying (14a) and (14b) to linearized dynamic equation (16):

$$\hat{x}_k^- = f(\hat{x}_{k-1}, u_k, \bar{d}_k) \qquad (19a)$$

$$P_{x_k}^- = F_k P_{x_{k-1}} F_k^T + G_k Q_k G_k^T \qquad (19b)$$

where $Q_k = E[(d_k - d_k)(d_k - d_k)^T]$ represents the process noise covariance.

After the linearization of the observation Eq. (2) we obtain:

$$x_k = f_k(\hat{x}_k^-, \bar{v}_k) + H_k(x_k - \hat{x}_k^-) + L_k(v_k - \bar{v}_k), \qquad (20)$$

where

$$H_k = \partial h_k(x_k, v_k)/\partial x_k\Big|_{\substack{x_k=\hat{x}_k^- \\ v_k = \bar{v}_k}} \qquad (21)$$

and

$$L_k = \partial h_k(x_{k-1}, v_k)/\partial v_k\Big|_{\substack{x_k=\hat{x}_k^- \\ v_k = \bar{v}_k}}, \qquad (22)$$

and $\bar{v}_k = E[v_k]$ is the mean of the observation noise.

The prediction of the observation is given by:

$$\hat{x}_k^- = h(\hat{x}_k^-, \bar{v}_k) \qquad\qquad (23a)$$

and the prediction covariance is:

$$P_{y_k} = H_k P_{x_k}^- H_k^T + L_k R_k L_k^T \qquad\qquad (23b)$$

and cross covariance:

$$P_{x_k y_k} = P_{x_k}^- H_k^T \qquad\qquad (23b)$$

where $R_k = E[v_k v_k^T]$ is the observation noise covariance.

2.2 Divided Difference Filter (DDF)

In [6] Nørgaard et al. proposed a new set of estimators based on a derivative free polynomial approximation of nonlinear dynamic and observation equation using Stirling's interpolation formula which uses central divided differences. Here we will consider only second order polynomial approximation, which is for arbitrary nonlinear function $f(x)$ given by:

$$f(x) \approx f(\bar{x}) + \tilde{D}_{\Delta x} f + \frac{1}{2!} \tilde{D}_{\Delta x}^2 f \qquad\qquad (24)$$

where $\tilde{D}_{\Delta x} f$ and $\tilde{D}_{\Delta x}^2 f$ are the first and second order central divided difference operators acting on $f(x)$:

$$\tilde{D}_{\Delta x} f = (x - \bar{x}) \frac{f(\bar{x} + h) - f(h - h)}{2h}$$

$$\tilde{D}_{\Delta x}^2 f = (x - \bar{x}) \frac{f(\bar{x} + h) - f(h - h)}{h^2}$$

h is the central difference step size and $\bar{x}$, around which we expand $f(x)$, is the prior mean of random variable x.

Previous formulation can be extended to the multidimensional case by stochastic decoupling of random variable x:

$$z = R_x^{-T} x, \qquad\qquad (25)$$

where R_x represents the upper triangular Cholesky factor of the covariance matrix $P_x = E[(x - \bar{x})(x - \bar{x})^T] = R_x^T R_x$. After decoupling we have:

$$f(x) = f(R_x^T z) = \tilde{f}(z). \tag{26}$$

Individual components of random variable z are mutually uncorrelated, with unity variance $P_z = E[(z - \bar{z})(z - \bar{z})^T] = I$ and consequently we can apply the first and the second order central difference operators independently to the components of $\tilde{f}(z)$, in order to obtain the following multidimensional central difference operators:

$$\tilde{D}_{\Delta z}\tilde{f} = \frac{1}{h}\left(\sum_{i=1}^{n} \Delta_{z_i}\mu_i\delta_i\right)\tilde{f}(\bar{z}) \tag{27}$$

$$\tilde{D}_{\Delta z}^2\tilde{f} = \frac{1}{h^2}\left(\sum_{i=1}^{n} \Delta_{z_i}^2\delta_i^2 + \sum_{i=1}^{n_x}\sum_{\substack{j=1\\j\neq i}}^{n_x} \Delta_{z_i}\Delta_{z_j}(\mu_i\delta_i)(\mu_j\delta_j)\right)\tilde{f}(\bar{z}) \tag{28}$$

where Δ_{z_i} represents the ith component of $(z - \bar{z})$. Partial first and second order difference operators δ_i and δ_i^2, and the mean operator μ_i are defined as:

$$\delta_i\tilde{f}(\bar{z}) = \tilde{f}(\bar{z} + \frac{h}{2}e_i) - \tilde{f}(\bar{z} - \frac{h}{2}e_i) \tag{29a}$$

$$\delta_i^2\tilde{f}(\bar{z}) = \tilde{f}(\bar{z} + he_i) + \tilde{f}(\bar{z} - he_i) - 2\tilde{f}(\bar{z}) \tag{29b}$$

$$\mu_i\tilde{f}(z) = \frac{1}{2}(\tilde{f}(\bar{z} + \frac{h}{2}e_i) + \tilde{f}(\bar{z} - \frac{h}{2}e_i)) \tag{29c}$$

where e_i is the ith unit vector.

When propagating multidimensional random variable x with prior mean $\bar{x} = E[x]$ and covariance $P_x = E[(x - \bar{x})(x - \bar{x})]$ through nonlinear function $y = f(x)$, we can use the second order multidimensional Stirling expansion of () to approximate posterior mean $\bar{y} = E[y]$, covariance $P_y = E[(y - \bar{y})(y - \bar{y})^T]$ and cross covariance $P_y = E[(x - x)(y - \bar{y})^T]$.

Approximation of the mean is given by:

$$\begin{aligned}\bar{y} &\approx E[\tilde{f}(\bar{z}) + \tilde{D}_{\Delta x}\tilde{f} + \frac{1}{2!}\tilde{D}_{\Delta x}^2\tilde{f}] = f(\bar{z}) + \frac{1}{2}E[\tilde{D}_{\Delta z}^2\tilde{f}]\\ &= \tilde{f}(\bar{z}) + \frac{\sigma_2}{2h^2}(\sum_{i=1}^{n}\Delta_{z_i}^2\delta_i^2)\tilde{f}(\bar{z})\\ &= \frac{h^2 - n_x}{h^2}\tilde{f}(\bar{z}) + \frac{1}{2h^2}\sum_{i=1}^{n_x}(\tilde{f}(\bar{z} + he_i) + \tilde{f}(\bar{z} - he_i))\end{aligned} \tag{30}$$

The ith moment of an arbitrary element in Δ_{z_i} is denoted by σ_i. All elements are assumed to be identically distributed therefore their moments are equal. Moreover, we have $E[(z-\bar{z})(z-\bar{z})^T]=I$, that is $\sigma_2=1$, while higher moments depend on the distribution of Δ_{z_i}.

Using $\tilde{f}(\bar{z}+he_i)=f(R_x^T(\bar{z}\pm he_i))=f(R_x^T\bar{z}\pm hR_x^Te_i)=f(x\pm hR_{x,i}^T)$, where $R_{x,i}^T$ represents the ith column of the upper triangular Cholesky factor transpose, we can rewrite the posterior mean approximation in terms of the prior statistic of x:

$$\bar{y}\approx\frac{h^2-n_x}{h^2}f(\bar{x})+\frac{1}{2h^2}\sum_{i=1}^{n_x}(f(\bar{x}+hR_{x,i}^T)+f(\bar{x}-hR_{x,i}^T)) \tag{31}$$

After applying the identity $\bar{y}=f(\bar{x})+E[y-f(\bar{x})]$, we obtain the posterior covariance in the following form:

$$\begin{aligned}P_y&=E[(y-\bar{y})(y-\bar{y})^T]\\&=E[(y-f(\bar{x}))(y-f(\bar{x}))^T]-E[y-f(\bar{x})]E[y-f(\bar{x})]^T\\&=E[(y-\tilde{f}(\bar{z}))(y-\tilde{f}(\bar{z}))^T]-E[y-\tilde{f}(\bar{z})]E[y-\tilde{f}(\bar{z})]^T\end{aligned} \tag{32}$$

As a consequence, the approximation of the posterior covariance can be written:

$$\begin{aligned}P_y&=E[\tilde{D}_x\tilde{f}+\frac{1}{2!}\tilde{D}_x^2\tilde{f}].\\&=E\left[(\tilde{D}_x\tilde{f}+\frac{1}{2!}\tilde{D}_x^2\tilde{f})(\tilde{D}_x\tilde{f}+\frac{1}{2!}\tilde{D}_x^2\tilde{f})^T\right]-E\left[\tilde{D}_x\tilde{f}+\frac{1}{2!}\tilde{D}_x^2\tilde{f}\right]E\left[\tilde{D}_x\tilde{f}+\frac{1}{2!}\tilde{D}_x^2\tilde{f}\right]^T\\&=E\left[\tilde{D}_x\tilde{f}(\tilde{D}_x\tilde{f})^T\right]+\frac{1}{4}E\left[\tilde{D}_x^2\tilde{f}(\tilde{D}_x^2\tilde{f})^T\right]-\frac{1}{4}E\left[\tilde{D}_x^2\tilde{f}\right]E\left[\tilde{D}_x^2\tilde{f}\right]^T\end{aligned} \tag{33}$$

After cancelling the identical terms when subtracted, and discarding the terms containing cross-differences, as it was explained in [3], we obtain the following approximation:

$$\begin{aligned}P_y&\approx\sigma_2\sum_{i=1}^n(\mu_i\delta_i\tilde{f}(\bar{z}))(\mu_j\delta_j\tilde{f}(\bar{z}))^T+\frac{\sigma_4-\sigma_2^2}{4}\sum_{i=1}^n\delta_i^2\tilde{f}(\bar{z})(\delta_i^2\tilde{f}(\bar{z}))^T.\\&=\frac{\sigma_2}{4h^2}\sum_{p=1}^{n_x}(\tilde{f}(\bar{z}+he_i)-\tilde{f}(\bar{z}+he_i)(\tilde{f}(\bar{z}+he_i)-\tilde{f}(\bar{z}+he_i)^T\\&\quad+\frac{\sigma_4-\sigma_2^2}{4h^4}\sum_{p=1}^{n_x}((\tilde{f}(\bar{z}+he_i)+\tilde{f}(\bar{z}+he_i)-2\tilde{f}(\bar{z}))\cdot\\&\quad(\tilde{f}(\bar{z}+he_i)+\tilde{f}(\bar{z}+he_i)-2\tilde{f}(\bar{z}))^T\end{aligned} \tag{34}$$

Using $\sigma_2 = 1$ and setting $h^2 = \sigma_4$ [3], we have:

$$
\begin{aligned}
P_y = {} & \frac{1}{4h^2} \sum_{p=1}^{n_x} (f(\bar{x} + hR_{x,i}^T) - f(\bar{x} - hR_{x,i}^T)) \cdot \\
& (f(\bar{x} + hR_{x,i}^T) - f(\bar{x} - hR_{x,i}^T))^T \\
& + \frac{h^2 - 1}{4h^4} \sum_{p=1}^{n_x} (f(\bar{x} + hR_{x,i}^T) + f(\bar{x} - hR_{x,i}^T) - 2f(\bar{x})) \cdot \\
& (f(\bar{x} + hR_{x,i}^T) + f(\bar{x} - hR_{x,i}^T) - 2f(\bar{x}))^T
\end{aligned}
\tag{35}
$$

Since $\sigma_4 - \sigma_2^2 = E[(\Delta_z)^4] - E[(\Delta_z)^2]^2 = Var[(\Delta_z)^2] > 0$, for all distributions, we always select $h^2 > 1$, and consequently the covariance approximation will always be positive semi definite.

Nørgaard et al. have derived the alternative covariance estimate as well [6]:

$$
\begin{aligned}
P_y = {} & \frac{h^2 - n_x}{h^2} (f(\bar{x}) - \bar{y}) f(\bar{x}) - \bar{y})^T \\
& + \frac{1}{2h^2} \sum_{p=1}^{n_x} (f(\bar{x} + hR_{x,i}^T) - \bar{y}) f(\bar{x} + hR_{x,i}^T) - \bar{y})^T \\
& + \frac{1}{2h^2} \sum_{i=1}^{n_x} (f(\bar{x} - hR_{x,i}^T) - \bar{y}) f(\bar{x} - hR_{x,i}^T) - \bar{y})^T
\end{aligned}
\tag{36}
$$

This estimate is less accurate than (35). Moreover, for $h^2 < n$ the last term becomes negative semi-definite with a possible implication that the covariance estimate (36) becomes non-positive definite. The reason why this estimate is considered here is to provide a comparison with the covariance estimate obtained by the Unscented Transformation described in the next subsection.

The estimate of the cross-covariance matrix is:

$$
\begin{aligned}
P_{xy} &= E[(x - x)(\tilde{f}(\bar{z}) + \tilde{D}_{\Delta z} \tilde{f} - \tilde{f}(\bar{z}))^T] = E[R_x^T \Delta_z (\tilde{D}_{\Delta z} \tilde{f})^T] \\
&= E\left[\sum_{i=1}^{n_x} R_{x,i}^T \Delta_{z_i} \left(\sum_{i=1}^{n} \Delta_{z_i} \mu_i \delta_i \tilde{f}(\bar{z}) \right)^T \right] = E\left[\sum_{i=1}^{n_x} R_{x,i}^T \Delta_{z_i} \left(\sum_{i=1}^{n} \Delta_{z_i} \mu_i \delta_i \tilde{f}(\bar{z}) \right)^T \right] \\
&= \sigma_2 \left[\sum_{i=1}^{n_x} R_{x,i}^T \left(\mu_i \delta_i \tilde{f}(\bar{z}) \right)^T \right] = \sum_{p=1}^{n_x} (f(\bar{x} - hs_{x,p}) - \bar{y}) f(\bar{x} - hs_{x,p}) - \bar{y})^T
\end{aligned}
\tag{37}
$$

2.3 Unscented Kalman Filter (UKF)

The unscented transformation [2, 3, 4] is a method for calculating the statistics of a random variable which undergoes a nonlinear transformation. It is based on the intuition that is easier to approximate a probability distribution than arbitrary function.

Again we consider propagating the n_x-dimensional continuous random variable x with prior mean $\bar{x} = E[x]$ and covariance $P_x = E[(x - \bar{x})(x - \bar{x})^T]$ through nonlinear function $y = f(x)$. To calculate the first two moments of random variable y using unscented transformation we first select the set of $2n_x + 1$ samples $\mathcal{X}_i$ called sigma points with corresponding weights ω_i. The weights and sample locations are selected to accurately capture the prior mean and covariance of a random variable and to capture the posterior mean and covariance accurately up to and including second order terms in the Taylor series expansion of the true quantities [3]. Sigma points and their weights which satisfy previous constraints are given by:

$$
\begin{aligned}
\mathcal{X}_0 &= \bar{x}, & \omega_0 &= \kappa/(n_x + \kappa) & i &= 0 \\
\mathcal{X}_i &= \bar{x} + \sqrt{n_x + \lambda} \cdot s_{x,i} & \omega_i &= 0.5/(n_x + \kappa) & i &= 1, 2, \ldots, n_x \\
\mathcal{X}_{i+n_x} &= \bar{x} - \sqrt{n_x + \lambda} \cdot s_{x,i} & \omega_{i+n_x} &= 0.5/(n_x + \kappa) & i &= 1, 2, \ldots, n_x
\end{aligned}
\tag{38}
$$

where $\kappa \in \mathfrak{R}$ is the scaling parameter, $s_{x,i}$ is the ith row or column of the matrix square root of P_x. For weights associated with sigma points it holds $\sum_{i=0}^{2n_x} w_i = 1$. Note that using this idea is possible to capture the higher order moments of posterior random variable, but at the cost of a larger set of sigma points [3].

After propagating sigma points through the nonlinear function $\mathcal{Y}_i = f(\mathcal{X}_i)$, approximations of the posterior mean, covariance and cross covariance are:

$$
\bar{y} = \sum_{i=0}^{2n_x} \omega_i \mathcal{Y}_i
\tag{39a}
$$

$$
P_y = \sum_{i=0}^{2n_x} \omega_i (\mathcal{Y}_i - \bar{y})(\mathcal{Y}_i - \bar{y})^{\mathrm{T}}.
\tag{39b}
$$

$$
P_{xy} = \sum_{i=0}^{2n_x} \omega_i (\mathcal{X}_i - \bar{x})(\mathcal{Y}_i - \bar{y})^{\mathrm{T}}
\tag{39c}
$$

The approximations are accurate to the second order of the Taylor series expansion of $f(x)$ (third order for Gaussian prior). Errors in the third and higher moments can be scaled by appropriate choice of scaling parameter κ. When prior random variable is Gaussian a useful heuristic is to select $\kappa = 3 - n_x$ [3].

It can be easily verified that for $h = \sqrt{n + \lambda}$, the estimates of the mean (34) and the covariance (36) obtained by Stirling's interpolation formula are equivalent to the estimates (39a) and (39b) obtained by unscented transformation.

The spread of the selected sigma points depends on dimensionality of prior random variable and in case of significant nonlinearity of considered mapping $f(x)$, it can cause problems by possibly sampling non-local effects. The scaled unscented transformation was introduced in [5] as a possible solution of this problem. It has been shown that sigma point selection and scaling can be combined into a single step by setting $\lambda = \alpha^2(n_x + \kappa) - n_x$, and selecting the sigma points by:

$$
\begin{aligned}
&\chi_0 = \bar{x}, &&\omega_0^{(m)} = \lambda/(n_x + \lambda), &&\omega_0^{(c)} = \lambda/(n_x + \lambda) + (1 - \alpha^2 + \beta) \\
&\chi_i = \bar{x} + \sqrt{n_x + \lambda} \cdot s_{x,i} &&\omega_i = 0.5/(n_x + \lambda), &&i = 1, 2, \ldots, n_x \\
&\chi_{i+n_x} = \bar{x} + \sqrt{n_x + \lambda} \cdot s_{x,i} &&\omega_{i+n_x} = 0.5/(n_x + \lambda), &&i = 1, 2, \ldots, n_x
\end{aligned}
$$

By choosing $\kappa \geq 0$ we can guarantee positive semi definiteness of the covariance matrix. Parameter α controls the spread of sigma points around $\bar{x}$ and is a small number (usually $1.e - 4 \leq \alpha \leq 1$) to avoid sampling non-local effects in case of significant nonlinearity. β is a non negative weighting term which can be used to incorporate knowledge of the higher order moments of the distribution. Optimal choice for Gaussian prior is $\beta = 2$ [5].

The true posterior mean and the mean calculated using the unscented transformation or Stirling's interpolation agrees exactly to the third order. Errors are introduced in the forth and higher order terms [3, 6]. The extended Kalman filter linearization approach calculates the posterior mean which agrees with the true mean only up to the first order. UKF and DDF calculate the posterior covariance accurately in the first two terms of Taylor series expansion, with errors only introduced at the fourth and higher order moments. It was shown in [3] that absolute term-by-term errors of these higher order moments are consistently smaller for the nonlinear derivative free approach of UKF and DDF than for the linearized EKF case. In [6] Nørgaard et al. shows that DDF has slightly smaller absolute error compared to UKF in the fourth order terms and also guarantees positive semi definiteness of the posterior covariance.

3   Efficient Implementation of Training Algorithms

In order to apply recursive approximate Bayesian estimators as training algorithms of recurrent neural networks we need to represent dynamics of RNN in a form of state space model. In this section we define the state space models of three representative architectures of globally recurrent neural networks: Elman, fully connected, and NARX recurrent neural network.

3.1   Elman Network State Space Model

In Elman RNNs adaptive feedbacks are provided between every pair of hidden units. The network is illustrated in Fig. 1a, and the state space model of the Elman network is given by equations

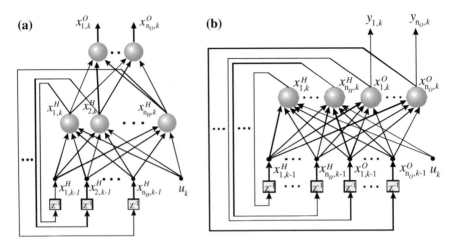

Fig. 1 Elman (**a**) and fully (**b**) connected RNN

$$\begin{bmatrix} x_k^H \\ w_k^O \\ w_k^H \end{bmatrix} = \begin{bmatrix} f(x_{k-1}^H, w_{k-1}^H, u_k) \\ w_{k-1}^O \\ w_{k-1}^H \end{bmatrix} + \begin{bmatrix} d_{x_k^H} \\ d_{w_k^O} \\ d_{w_k^H} \end{bmatrix} \tag{40a}$$

$$y_k = x_k^O + v_k, \quad x_k^O = h(x_k^H, w_k^O) \tag{40b}$$

where x_k^H represents the output of the hidden neurons in the kth time step, x_k^O is the output of the neurons in the last layer, w_{k-1}^O is the vector of synaptic weights between the hidden and the output layer and w_{k-1}^H is the vector of recurrent adaptive connection weights. Note that in the original formulation of Elman, these weights were fixed. Random variables $d_{x_k^H}$, $d_{w_k^O}$, $d_{w_k^H}$ represent the process noises.

It is assumed that the output of the network $x_k^O = h(x_k^H, w_k^O)$ is corrupted by the observation noise v_k.

3.2 Fully Connected Recurrent Network State Space Model

In fully connected RNNs adaptive feedbacks are provided between each pair of processing units (hidden and output). The state vector of a fully connected RNN consists of outputs (activities) of hidden x_k^H and output neurons x_k^O, and their synaptic weights w_k^H and w_k^O. The activation functions of the hidden and thee output neurons are $f^H(x_k^O, x_k^H, w_{k-1}^H, u_k)$ and $f^O(x_k^O, x_k^H, w_{k-1}^O, u_k)$, respectively. The network structure is illustrated in Fig. 1b.

The state space model of the network is given by:

$$
\begin{bmatrix} x_k^O \\ x_k^H \\ w_k^O \\ w_k^H \end{bmatrix} = \begin{bmatrix} f^O(x_k^O, x_k^H, w_{k-1}^O, u_k) \\ f^H(x_k^O, x_k^H, w_{k-1}^H, u_k) \\ w_{k-1}^O \\ w_{k-1}^H \end{bmatrix} + \begin{bmatrix} d_{x_k^O} \\ d_{x_k^H} \\ d_{w_k^O} \\ d_{w_k^H} \end{bmatrix} \tag{41a}
$$

$$
y_k = H \cdot \begin{bmatrix} x_k^O \\ x_k^H \\ w_k^O \\ w_k^H \end{bmatrix} + v_k, \quad H = [I_{n_O \times n_O} \ 0_{n_O \times (n_S - n_O)}] \tag{41b}
$$

The dynamic equation describes the evolution of neuron outputs and synaptic weights. In the observation equation, the matrix H selects the activities of output neurons as the only visible part of the state vector, where n_S is the number of hidden states which are estimated: $n_S = n_O + n_H + n_{W^O} + n_{W^H}$, n_O and n_H are the numbers of output and hidden neurons respectively, n_{W^O} is the number of adaptive weights of the output neurons, n_{W^H} is the number of adaptive weights of the hidden neurons.

3.3 NARX Recurrent Neural Network State Space Model

The non-linear AutoRegressive with eXogenous inputs (NARX) recurrent neural network usually outperforms the classical recurrent neural networks, like Elman or fully connected RNN, in tasks that involve long term dependencies for which the desired output depends on inputs presented at times far in the past (Fig. 2).

Here we define the state space model of a NARX RNN. Adaptive feedbacks are provided between the output and the hidden units. These feedback connection and possible input connections are implemented as FIR filters. The state vector consists

Fig. 2 NARX recurrent neural network

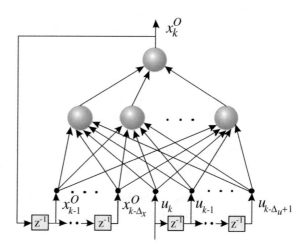

of outputs of the network in Δ_x time steps $x_k^O, x_{k-1}^O,\ldots,x_{k-\Delta_x+1}^O$, the output w_k^O, and hidden synaptic w_k^H weights.

$$
\begin{bmatrix} x_k^O \\ x_{k-1}^O \\ \vdots \\ x_{k-\Delta_x+1}^O \\ w_k^O \\ w_k^H \end{bmatrix} = \begin{bmatrix} f(x_{k-1}^O, .., x_{k-\Delta_x}^O, u_{k-1}, .., u_{k-\Delta_u}, w_{k-1}) \\ x_{k-1}^O \\ \vdots \\ x_{k-\Delta_x+1}^O \\ w_{k-1}^O \\ w_{k-1}^H \end{bmatrix} + \begin{bmatrix} d_{x_k^O} \\ 0 \\ \vdots \\ 0 \\ d_{w_k^O} \\ d_{w_k^H} \end{bmatrix} \quad (42a)
$$

The dynamic equation describes the evolution of network outputs and synaptic weights.

$$
y_k = H \cdot \begin{bmatrix} x_k^O \\ x_{k-1}^O \\ \vdots \\ x_{k-\Delta_x+1}^O \\ w_k^O \\ w_k^H \end{bmatrix} + v_k, \quad H = [I_{n_O \times n_O} \; 0_{n_O \times (n_S - n_O)}] \quad (42b)
$$

As in previous examples, n_O represents the number of output neurons. n_S is the number of hidden states of the NARX RNN: $n_S = n_O + n_{W^O} + n_{W^H}$, n_{W^O} is the number of adaptive weights of output neurons, n_{W^H} is number of adaptive weights of hidden neurons.

All considered models have nonlinear hidden neurons and linear output neurons. Two types of nonlinear activation functions have been used in the following tests: the sigmoidal and the Gaussian radial basis function.

3.4 Square Root Implementation of Recursive Bayesian Estimators as RNN Training Algorithms

Straightforward implementation of Unscented Kalman Filter and Divided Difference Filter requires calculation of the prior state covariance matrix, which has $O(n^3/6)$ computational complexity. However it is the full covariance matrix of the estimate which is recursively updated. The square root implementations of UKF and DDF recursively update the Cholesky factors of the covariance matrices. Although the general complexity of the algorithms is still $O(n^3)$, they will have better numerical properties, comparable to the standard square root implementation of Kalman filter.

The square root implementations of EKF, UKF and DDF are based on three linear algebra algorithms: matrix orthogonal-triangular decomposition (triangularization), rank one update of a Cholesky factor and efficient solution of the over-determined least square problem.

The orthogonal-triangular decomposition of $m \times n$ matrix A ($[Q, R] = qr(A)$) produces $m \times n$ upper triangular matrix R, which is Cholesky factor of $A^T A$, and $m \times n$ unitary matrix Q such that $A = Q \times R$.

Rank one update of a Cholesky factor R, returns upper triangular Cholesky factor $\tilde{R}$ for which holds:

$$\tilde{R}^T \tilde{R} = R^T R \pm xx^T, \tag{43}$$

where x is column vector of appropriate length.

For overdetermined least squares problem $AX = B$, if A is an upper triangular matrix, then X is simply computed by back substitution algorithm.

Using Cholesky decomposition of the prior covariance $P_{x_{k-1}} = R_{x_{k-1}}^T R_{x_{k-1}}$ we can represent the state and observation prediction covariance in EKF as:

$$P_{x_k}^- = F_k R_{x_{k-1}}^T R_{x_{k-1}} F_k^T + R_{d_k}^T R_{d_k} = \begin{bmatrix} F_k R_{x_{k-1}}^T & R_{d_k}^T \end{bmatrix} \cdot \begin{bmatrix} R_{x_{k-1}} F_k^T \\ R_{d_k} \end{bmatrix} = (R_{x_k}^-)^T R_{x_k}^- \tag{44}$$

$$P_{y_k}^- = H_k (R_{x_k}^-)^T R_{x_k}^- H_k^T + R_{v_k}^T R_{v_k} = \begin{bmatrix} H_k (R_{x_k}^-)^T & R_{v_k}^T \end{bmatrix} \cdot \begin{bmatrix} R_{x_k}^- H_k^T \\ R_{v_k} \end{bmatrix} = (R_{y_k}^-)^T R_{y_k}^-. \tag{45}$$

where R_{d_k} and R_{v_k} represent the Cholesky factors of process and observation noise respectively.

We obtain the recursive update of the estimation covariance Cholesky factor using the numerically stable Joseph form of the covariance:

$$\begin{aligned} P_{x_k} &= (I - K_k H_k)(R_{x_k}^-)^T R_{x_k}^- (I - K_k H_k)^T + K_k R_{v_k}^T R_{v_k} K_k^T. \\ &= \begin{bmatrix} (I - K_k H_k)(R_{x_k}^-)^T & K_k R_{v_k}^T \end{bmatrix} \cdot \begin{bmatrix} R_{x_k}^- (I - K_k H_k)^T \\ R_{v_k} K_k^T \end{bmatrix} = R_{x_k}^T R_{x_k} \end{aligned} \tag{46}$$

Algorithm: Square Root Extended Kalman Filter

% Initialization

$$\hat{x}_0 = E[x_0], R_{x_0} = Chol(E[(x_0 - \hat{x}_0)(x - \hat{x}_0)^T])$$

for k = 1, 2,...

% State prediction

% Estimate the mean and covariance Cholesky factor of the process noise:
$\bar{d}_k, R_{d_k}$
% Calculate Jacobians of a dynamic equation

$$F_k = \frac{\partial f(x_{k-1}, u_k, d_k)}{\partial x_{k-1}}\Bigg|_{\substack{x_{k-1}=\hat{x}_{k-1} \\ d_k=\bar{d}_k}}, G_k = \frac{\partial f(x_{k-1}, u_k, d_k)}{\partial d_k}\Bigg|_{\substack{x_{k-1}=\hat{x}_{k-1} \\ d_k=\bar{d}_k}}$$

% State prediction and Cholesky factor of state prediction covariance

$$\hat{x}_k^- = f(\hat{x}_{k-1}, u_k, \bar{d}_k)$$

$$R_{x_k}^- = QR\left(\begin{bmatrix} R_{x_{k-1}} F_k^T \\ R_{d_k} \end{bmatrix}\right)$$

% Observation prediction

% Estimate the mean and covariance Cholesky factor of the observation noise:
$\bar{v}_k, R_{v_k}$
% Calculate Jacobians of a dynamic equation

$$H_k = \frac{\partial h(x_k, v_k)}{\partial x_k}\Bigg|_{\substack{x_k=\hat{x}_k^- \\ v_k=\bar{v}_k}}, L_k = \frac{\partial h(x_k, v_k)}{\partial v_k}\Bigg|_{\substack{x_k=\hat{x}_k^- \\ v_k=\bar{v}_k}}$$

% Observation prediction, prediction covariance Cholesky factor and cross covariance

$$\hat{y}_k^- = h(\hat{x}_k^-, \bar{v}_k)$$

$$R_{y_k}^- = QR\left(\begin{bmatrix} R_{x_k}^- H_k^T \\ R_{v_k} \end{bmatrix}\right)$$

$$P_{x_k y_k}^- = P_{x_k}^- H_k^T$$

% Update

% Kalman gain
$$K_k = (P^-_{x_k y_k}/R^-_{y_k})/(R^-_{y_k})^T)$$
% State estimation

$$\hat{x}_k = \hat{x}^-_k + K_k(y_k - \hat{y}^-_k)$$

% Estimation covariance Cholesky factor

$$R_{x_k} = QR\left(\begin{bmatrix} R^-_{x_k}(I - K_k H_k)^T \\ R_{v_k} K^T_k \end{bmatrix}\right)$$

end for
end algorithm

Algorithm: Square Root Unscented Kalman Filter

% Initialization

$$\hat{x}_0 = E[x_0], R_{x_0} = Chol(E[(x_0 - \hat{x}_0)(x - \hat{x}_0)^T])$$

for k = 1,2,...

% State prediction

% Estimate the mean and covariance matrix Cholesky factor of process noise:
$\bar{d}_k, R_{d_k}$
% Calculate positions and weights of sigma points

$$\lambda = \alpha^2\sqrt{n_x + \kappa} - n_x, \gamma = \sqrt{n_x + \lambda}$$

$$\mathcal{X}_{x_{k-1}} = \left[\hat{x}_{k-1}, \hat{x}_{k-1} + \gamma R^T_{x_{k-1}}, \hat{x}_{k-1} - \gamma R^T_{x_{k-1}}\right]$$

$$w^{(m)}_0 = \frac{\lambda}{n_x + \lambda}, w^{(c)}_0 = w^{(m)}_0 + 1 - \alpha^2 + \beta, w^{(m)}_1 = w^{(c)}_1 = \frac{1}{2(n_x + \lambda)}$$

% State prediction and prediction covariance Cholesky factor

$$\mathcal{X}_k^- = f(\mathcal{X}_{k-1}, u_k)$$

$$\hat{x}_k^- = w_i^{(m)} \mathcal{X}_{k,0}^- + w_1^{(m)} \sum_{i=0}^{2n_x} \mathcal{X}_{k,i}^- + \bar{d}_k$$

$$R_{x_k}^- = Cholupdate\left(QR\left(\left[\begin{matrix} \sqrt{w_1^{(c)}}(\mathcal{X}_{k,1:2n_x}^- - \hat{x}_k^-)^T \\ R_{d_k} \end{matrix} \right] \right), \sqrt{|w_0^{(c)}|}(\mathcal{X}_{k,0}^- - \hat{x}_k^-), sign(w_0^{(c)}) \right)$$

% Observation prediction

% Estimate the mean and covariance Cholesky factor of the observation noise: $\bar{v}_k, R_{v_k}$

% Calculate positions and weights of sigma points

$$\lambda = \alpha^2 \sqrt{n_x + \kappa} - n_x, \gamma = \sqrt{n_x + \lambda}$$

$$\mathcal{X}_k^- = \left[\hat{x}_k^-, \hat{x}_k^- + \gamma (R_{x_k}^-)^T, \hat{x}_k^- - \gamma (R_{x_k}^-)^T \right]$$

$$w_0^{(m)} = \frac{\lambda}{n_x + \lambda}, w_0^{(c)} = w_0^{(m)} + 1 - \alpha^2 + \beta, w_1^{(m)} = w_1^{(c)} = \frac{1}{2(n_x + \lambda)}$$

% Observation prediction, prediction covariance Cholesky factor and cross covariance

$$\mathcal{Y}_k^- = h(\mathcal{X}_k^-)$$

$$\hat{y}_k^- = w_0^{(m)} \mathcal{Y}_{k,0}^- + w_1^{(m)} \sum_{i=1}^{2n_x} \mathcal{Y}_{k,i}^- + \bar{v}_k$$

$$R_{y_k}^- = Cholupdate\left(QR\left(\left[\begin{matrix} \sqrt{w_1^{(c)}}(\mathcal{Y}_{k,1:2n_x}^- - \hat{y}_k^-)^T \\ R_{v_k} \end{matrix} \right] \right), \sqrt{|w_0^{(c)}|}(\mathcal{Y}_{k,0}^- - \hat{y}_k^-), sign(w_0^{(c)}) \right)$$

$$P_{x_k y_k}^- = w_0^{(c)}(\mathcal{X}_{k,0}^- - \hat{x}_k^-)(\mathcal{Y}_{k,0}^- - \hat{y}_k^-)^T + w_1^{(c)} \sum_{i=1}^{2n_x} (\mathcal{X}_{k,i}^- - \hat{x}_k^-)(\mathcal{Y}_{k,i}^- - \hat{y}_k^-)^T$$

% Update

% Kalman gain

$$K_k = (P^-_{x_k y_k}/R^-_{y_k})/(R^-_{y_k})^T$$

% State estimation

$$\hat{x}_k = \hat{x}_k^- + K_k(y_k - \hat{y}_k^-)$$

% Estimation covariance Cholesky factor

$$R_{x_k} = Cholupdate(R^-_{x_k}, (K_k(R^-_{y_k})^T)_i, '-') \qquad (4.6.67)$$

end for
end algorithm

Algorithm: Square Root Divided Difference Filter

% Initialization

$$\hat{x}_0 = E[x_0], R_{x_0} = Chol(E[(x_0 - \hat{x}_0)(x - \hat{x}_0)^T])$$

for k = 1,2,...

% State prediction

% Estimate the mean and covariance matrix Cholesky factor of process noise: $\bar{d}_k, R_{d_k}$
% Calculate positions and weights of sigma points

$$\mathcal{X}_{k-1} = \left[\hat{x}_{k-1}, \hat{x}_{k-1} + hR^T_{x_{k-1}}, \hat{x}_{k-1} - hR^T_{x_{k-1}}\right]$$

$$w_1^{(m)} = \frac{h^2 - n_x}{h^2}, w_1^{(c)} = \frac{1}{4h^2}, w_2^{(m)} = \frac{1}{2h^2}, w_2^{(c)} = \frac{h^2 - 1}{4h^2}$$

% State prediction, prediction covariance Cholesky factor and cross covariance

$$\mathcal{X}_k^- = f(\mathcal{X}_{k-1}, u_k)$$

$$\hat{x}_k^- = w_1^{(m)} \mathcal{X}_{k,0}^- + w_2^{(m)} \sum_{i=0}^{2n_x} \mathcal{X}_{k,i}^- + \bar{d}_k$$

$$R_{x_k}^- = QR \left(\begin{bmatrix} \sqrt{w_1^{(c)}} (\mathcal{X}_{k,1:n_x}^- - \mathcal{X}_{k,n_x+1:2n_x}^-)^T \\ \sqrt{w_2^{(c)}} (\mathcal{X}_{k,1:n_x}^- + \mathcal{X}_{k,n_x+1:2n_x}^- - 2\mathcal{X}_{k,0}^-)^T \\ R_{d_k} \end{bmatrix} \right)$$

% Observation prediction

% Estimate the mean and covariance Cholesky factor of the observation noise: $\bar{v}_k, R_{v_k}$
% Calculate positions and weights of sigma points

$$\mathcal{X}_k^- = \left[\hat{x}_k^-, \hat{x}_k^- + h(R_{x_k}^-)^T, \hat{x}_k^- - h(R_{x_k}^-)^T \right]$$

$$w_1^{(m)} = \frac{h^2 - n_x}{h^2}, w_1^{(c)} = \frac{1}{4h^2}, w_2^{(m)} = \frac{1}{2h^2}, w_1^{(c)} = \frac{h^2 - 1}{4h^2}$$

% Observation prediction, prediction covariance Cholesky factor and cross covariance

$$\mathcal{Y}_k^- = h(\mathcal{X}_k^-) \tag{4.6.102}$$

$$\hat{y}_k^- = \sum_{i=0}^{2n_y} w_i^{(m)} \mathcal{Y}_{k,i}^- + \bar{v}_k \tag{4.6.105}$$

$$R_{y_k}^- = QR \left(\begin{bmatrix} \sqrt{w_1^{(c)}} (\mathcal{Y}_{k,1:n_x}^- - \mathcal{Y}_{k,n_x+1:2n_x}^-)^T \\ \sqrt{w_2^{(c)}} (\mathcal{Y}_{k,1:n_x}^- + \mathcal{Y}_{k,n_x+1:2n_x}^- - 2\mathcal{Y}_{k,0}^-)^T \\ R_{v_k} \end{bmatrix} \right) \tag{4.6.106}$$

$$P_{x_k y_k}^- = \sqrt{w_1^{(c)}} ((\mathcal{Y}_{k,1:n_x}^- - \mathcal{Y}_{k,n_x+1:2n_x}^-) R_{x_k}^-)^T \tag{4.6.107}$$

% Update

% Kalman gain

$$K_k = (P_{x_k y_k}^- / R_{y_k}^-) / (R_{y_k}^-)^T$$

% State estimation

$$\hat{x}_k = \hat{x}_k^- + K_k(y_k - \hat{y}_k^-)$$

% Estimation covariance Cholesky factor

$$R_{x_k} = Cholupdate\{R_{x_k}^-, K_k(R_{y_k}^-)^T, -1\}$$

end for
end algorithm

4 Examples

In this section we compare derived algorithms for sequential training of RNN. We have evaluated the performance of algorithms in training three different architectures of globally recurrent neural networks: fully connected RNN, Elman RNN with adaptive recurrent connections and NARX recurrent neural network. The problem at hand was the long term prediction of chaotic time series. Implementation of Divided Difference Filter and Unscented Kalman filter did not required linearization of the RNN state space models. However, in order to apply Extended Kalman Filter we had to linearize the RNNs state space models that are to calculate Jacobian of the RNN outputs with respect to the inputs and synaptic weights. Note that we did not apply back propagation through time but standard back propagation algorithm to calculate the Jacobian. This was possible because of the joint estimation of RNN outputs and synaptic weights.

In the process of the evaluation, recurrent neural networks were trained sequentially on the certain number of samples. After that they were iterated for a number of samples, by feeding back just the predicted outputs as the new inputs of the recurrent neurons. Time series of iterated predictions were compared with the test parts of the original time series by calculating the Normalized Root Mean Squared Error (NRMSE):

$$NRMSE = \sqrt{\frac{1}{\sigma^2 N} \sum_{k=1}^{N} (y_k - \hat{y}_k^-)^2} \tag{30}$$

where σ is the standard deviation of chaotic time series, y_k is the true value of sample at time step k, and $\hat{y}_k^-$ is the RNN prediction.

Mean and variance of the NRMSE obtained on 30 independent runs, average time needed for training and number of hidden neurons and adaptive synaptic weights are given in tables for comparison.

The variance of the process noise $d_{s,k}$ and $d_{w,k}$ were exponentially decayed form 1.e − 1 and 1.e − 3 to 1.e − 10, and the variance of the observation noise v_k was also exponentially decayed from 1.e − 1 to 1.e − 10 during the sequential training.

4.1 Mackey Glass Chaotic Time Series Prediction

In our first example we have considered the long term iterated prediction of the Mackey Glass time series. We have applied Divided Difference Filter, Unscented Kalman Filter and Extended Kalman Filter for joint estimation of synaptic weights and neuron outputs of three different RNN architectures: Elman, fully connected and non-linear AutoRegressive with eXogenous inputs (NARX) recurrent neural network.

After sequential adaptation on 2000 samples, a long term iterated prediction of the next $N = 100$ samples is used to calculate the NRMSE.

Table 1 contains mean and variance of NRMSE obtained after 30 independent trials of each estimator applied on each architecture. We also give the number of hidden units, the number of adaptable parameters and time needed for training on 2000 samples. Given these results we can conclude that the NARX network is superior in both NRMSE of long term prediction and time needed for training,

Table 1 NRMSE of the long term iterated prediction for various RNN architectures and training algorithms

	Mean	Var	n_H	n_W	T[s]
DDF_ELMAN_SIG	0.316	8.77e − 3	10	121	20.88
UKF_ELMAN_SIG	0.419	3.43e − 2	10	121	20.91
EKF_ELMAN_SIG	0.429	5.89e − 2	10	121	14.98
DDF_FC_SIG	0.269	7.15e − 3	10	131	23.43
UKF_FC_SIG	0.465	8.51e − 2	10	131	23.78
EKF_FC_SIG	0.359	8.64e − 2	10	131	17.38
DDF_NARX_SIG	0.0874	2.91e − 4	5	41	5.64
UKF_NARX_SIG	0.119	1.89e − 3	5	41	5.68
EKF_NARX_SIG	0.153	3.37e − 3	5	41	4.76

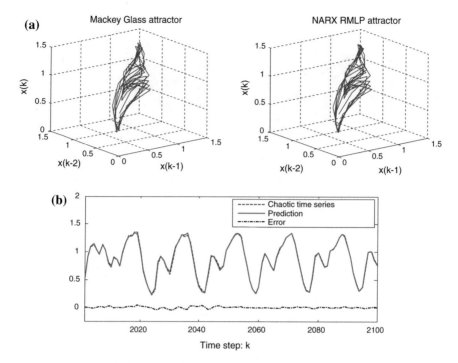

Fig. 3 Mackey Glass chaotic time series prediction. Phase plot of x_k versus x_{k-1} and x_{k-2} for the Mackey Glass time series and the NARX RMLP iterated prediction (**a**). Comparison of the original chaotic time series and the NARX RMLP iterated prediction (**b**)

compared to other two architectures. As for the approximate Bayesian estimators, although slightly slower in our implementation, derivative free filters (DDF and UKF) are consistently better than EKF, that is they produced RNN's with significantly lower NRMSE.

Sample results of long term prediction using NARX network with sigmoidal neurons, trained using DDF are shown in Fig. 3.

4.2 Hénon Chaotic Time Series Prediction

In our first example, we consider the prediction of the long-term behaviour of the chaotic Hénon dynamics:

$$x_k = 1 - 1.4x_{k-1}^2 + 0.3x_{k-2} \tag{31}$$

RNNs with sigmoidal and Gaussian hidden neurons (we call this network Recurrent Radial Basis Function network—RBF network) were trained sequentially

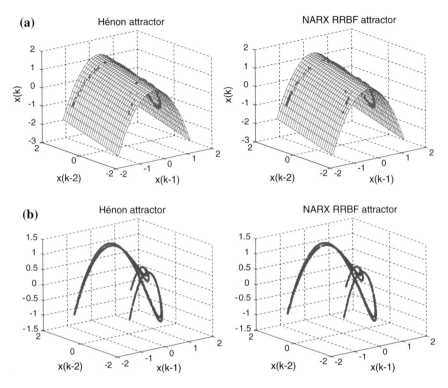

Fig. 4 Hénon chaotic time series prediction: Surface and phase plot. Surface plot of the Hénon and NARX_RRBF map; Dots—chaotic attractor and NARX_RRBF attractor (**a**). Phase plot of x_k versus x_{k-1} and x_{k-2} for the Hénon series and NARX RRBF iterated prediction (**b**)

on 3000 samples. After training networks were iterated for 20 samples by feeding back the current outputs of the neurons as the new inputs. Figures 4 and 5 show results of prediction using NARX_RBF network trained by DDF on a Hénon chaotic time series.

It can be seen from Fig. 4a and b, that, although the network was trained only using sample data chaotic attractor, which occupies small part of the surface defined by Eq. (31), the recurrent neural network was able to reconstruct that surface closely to the original mapping (Fig. 4a), as well as to reconstruct the original attractor (Fig. 4b).

Results presented in Table 2 show that both DDF and UKF produce more accurate RNNs than EKF with comparable training time.

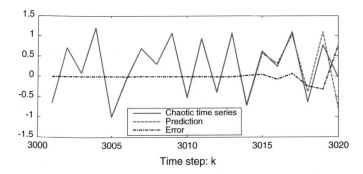

Fig. 5 Comparison of the original chaotic time series and the NARX RRBF iterated prediction

Table 2 Results of long term predictions of the Hénon chaotic time series

	Mean	Var	n_H	n_W	T[s]
DDF_ELMAN_SIG	1.73e − 2	6.19e − 5	4	25	8.34
DDF_ELMAN_RBF	6.02e − 2	3.89e − 4	3	22	7.76
UKF_ELMAN_SIG	7.29e − 2	9.46e − 3	4	25	8.53
UKF_ELMAN_RBF	7.24e − 2	1.79e − 3	3	22	7.91
EKF_ELMAN_SIG	1.69e − 1	5.17e − 2	4	25	7.69
EKF_ELMAN_RBF	1.01e − 1	7.50e − 3	3	22	7.96
DDF_NARX_SIG	7.46e − 3	3.39e − 6	4	17	6.21
DDF_NARX_RBF	4.36e − 3	4.15e − 6	3	16	5.85
UKF_NARX_SIG	1.28e − 2	2.68e − 5	4	17	6.37
UKF_NARX_RBF	5.72e − 3	7.14e − 6	3	16	6.00
EKF_NARX_SIG	1.57e − 2	1.65e − 5	4	17	5.76
EKF_NARX_RBF	7.07e − 3	1.35e − 6	3	16	6.17

5   Conclusions

We considered the problem of recurrent neural network training as an approximate recursive Bayesian state estimation. Results in chaotic time series long term prediction show that derivative free estimators Divided Difference Filter and Unscented Kalman Filter considerably outperform Extended Kalman Filter as RNN learning algorithms with respect to the accuracy of the obtained network, while retaining comparable training times.

Experiments also show that of tree considered architectures: Elman, fully connected and non-linear AutoRegressive with eXogenous inputs (NARX) recurrent neural network, NARX is by far superior in both training time and accuracy of trained networks in long term prediction.

References

1. Anderson, B., Moore, J.: Optimal Filtering. Englewood Cliffs, NJ, Prentice-Hall (1979)
2. Julier, S., Uhlmann, J., Durrant-Whyte, H.: A new approach for filtering nonlinear systems. In: Proceedings of the American Control Conference, pp. 1628–1632 (1995)
3. Julier, S.J., Uhlmann, J.K.: A general method for approximating nonlinear transformations of probability distributions. Technical report, RRG, Department of Engineering Science, University of Oxford (1996)
4. Julier, S.J.: A skewed approach to filtering. In: SPIE Conference on Signal and Data Processing of Small Targets, vol. 3373, pp. 271–282. SPIE, Orlando, Florida (1998)
5. Julier, S.J.: The scaled unscented transformation. In: Proceedings of the American Control Conference, vol. 6, pp. 4555–4559 (2002)
6. Nørgaard, M., Poulsen, N.K., Ravn, O.: Advances in derivative free state estimation for nonlinear systems, Technical Report, IMM-REP-1998-15. Department of Mathematical Modelling, DTU (2000)
7. Todorović, B., Stanković, M., Moraga, C.: On-line learning in recurrent neural networks using nonlinear Kalman Filters. In: Proceedings of the ISSPIT 2003, Darmstadt, Germany (2003)
8. Todorović, B., Stanković, M., Moraga, C.: Nonlinear Bayesian estimation of recurrent neural networks. In: Proceedings of the IEEE 4th International Conference on Intelligent Systems Design and Applications ISDA, Budapest, Hungary, pp. 855–860, 26–28 Aug 2004
9. Van der Merwe, R., Wan, E.A.: Efficient derivative-free Kalman Filters for online learning. In: Proceedings of the ESSAN, Bruges, Belgium (2001)
10. Williams, R.J., Zipser, D.: A learning algorithm for continually running fully recurrent neural networks. Neural Comput. **1**, 270–280 (1989)
11. Williams, R.J., Zipser, D.: Gradient-based learning algorithms for recurrent connectionist networks. TR NU_CCS_90-9. Northeastern University, Boston (1990)

Author Index

© Springer International Publishing Switzerland 2016
J.J. Merelo et al. (eds.), *Computational Intelligence*,
Studies in Computational Intelligence 620, DOI 10.1007/978-3-319-26393-9